T0122844

BIOLOGY AND CONSERVATION OF NORTH AMERICAN TORTOISES

Biology and Conservation of North American Tortoises

EDITED BY DAVID C. ROSTAL,

EARL D. MCCOY,

AND HENRY R. MUSHINSKY

JOHNS HOPKINS UNIVERSITY PRESS | BALTIMORE

Printed in the United States of America on acid-free paper
9 8 7 6 5 4 3 2 1

Johns Hopkins University Press
2715 North Charles Street
Baltimore, Maryland 21218-4363
www.press.jhu.edu

Library of Congress Cataloging-in-Publication Data

Biology and conservation of North American tortoises / edited by
David C. Rostal, Earl D. McCoy, and Henry R. Mushinsky.
pages cm
Includes bibliographical references and index.
ISBN-13: 978-1-4214-1377-8 (hardcover : alk. paper)
ISBN-13: 978-1-4214-1378-5 (electronic)
ISBN-10: 1-4214-1377-9 (hardcover : alk. paper)
ISBN-10: 1-4214-1378-7 (electronic)
1. Gopherus—North America. I. Rostal, David C., 1957– II. McCoy,
Earl D. III. Mushinsky, Henry R.
QL666.C584B56 2014
597.92'4097—dc23 2013036036

A catalog record for this book is available from the British Library.

Johns Hopkins University Press uses environmentally friendly book materials, including recycled text paper that is composed of at least 30 percent post-consumer waste, whenever possible.

CONTENTS

CONTRIBUTORS

Gustavo Aguirre L.
Instituto de Ecología, A. C.
Xalapa, Veracruz, Mexico

Linda J. Allison
Desert Tortoise Monitoring Coordinator
US Fish and Wildlife Service
Reno, Nevada

Matthew J. Aresco
Director, Nokuse Plantation
Bruce, Florida

Roy C. Averill-Murray
Desert Tortoise Recovery Coordinator
US Fish and Wildlife Service
Reno, Nevada

Joan E. Berish
Florida Fish and Wildlife Conservation Commission
Wildlife Research Laboratory
Gainesville, Florida

Kristin H. Berry
US Geological Survey
Biological Resources Discipline
Western Ecological Research Center
Riverside, California

Dennis M. Bramble
Department of Biology
The University of Utah
Salt Lake City, Utah

K. Kristina Drake
US Geological Survey
Biological Resources Discipline
Henderson, Nevada

Taylor Edwards
Tucson, Arizona

Todd C. Esque
Research Ecologist
US Geological Survey
Henderson, Nevada

Richard Franz
P.O. Box 210
Putnam Hall, Florida

Craig Guyer
Department of Biological Sciences
Auburn University
Auburn, Alabama

J. Scott Harrison
Department of Biology
Georgia Southern University
Statesboro, Georgia

Sharon M. Hermann
Department of Biological Sciences
Auburn University
Auburn, Alabama

J. Howard Hutchison
University of California Museum of Paleontology
Berkeley, California

Elliott R. Jacobson
College of Veterinary Medicine
University of Florida
Gainesville, Florida

Valerie M. Johnson
Department of Biological Sciences
Auburn University
Auburn, Alabama

Richard T. Kazmaier
Department of Life, Earth and Environmental Sciences
West Texas A&M University
Canyon, Texas

Earl D. McCoy
Department of Integrative Biology
University of South Florida
Tampa, Florida

Philip A. Medica
US Geological Survey
Biological Resources Discipline
Henderson, Nevada

Robert W. Murphy
Centre for Biodiversity and Conservation Biology
Royal Ontario Museum
Toronto, Ontario

Henry R. Mushinsky
Department of Integrative Biology
University of South Florida
Tampa, Florida

Kenneth E. Nussear
US Geological Survey
Biological Resources Discipline
Henderson, Nevada

Michael P. O'Connor
Department of Biology
Drexel University
Philadelphia, Pennsylvania

Thomas A. Radzio
Department of Biology
Drexel University
Philadelphia, Pennsylvania

David C. Rostal
Department of Biology
Georgia Southern University
Statesboro, Georgia

Lora L. Smith
Joseph W. Jones Ecological Research Center
Newton, Georgia

James R. Spotila
Department of Biology
Drexel University
Philadelphia, Pennsylvania

Craig B. Stanford
Department of Ichthyology and Herpetology
Natural History Museum of Los Angeles County
Los Angeles, California
South Pasadena, California

C. Richard Tracy
Department of Biology
University of Nevada, Reno
Reno, Nevada

Tracey D. Tuberville
University of Georgia
Savannah River Ecology Lab
Aiken, South Carolina

Michael Tuma
SWCA Environmental Consultants
South Pasadena, California

Thane Wibbels
Department of Biology
University of Alabama at Birmingham
Birmingham, Alabama

PREFACE

WHY A BOOK ABOUT TORTOISES?

Tortoises hold a special place in the minds of most persons. They are symbols of many things, including longevity—much to their detriment, sometimes—wisdom, and cunning. We have been fascinated by their slow, plodding ability to outwit mammals with hyperactivity disorders for centuries. Unfortunately, tortoises also represent a free lunch, or a roadside souvenir, or an annoying impediment to development for some persons. For these reasons and others, tortoises, which have proliferated over the past 50 million years or so, are now under serious threat of extinction. Although saving tortoises, other turtles, or any group of organisms ultimately will require the social will to do so, scientists are the persons who can provide the means. It is important, therefore, for scientists to take stock from time to time of what is known and what is not known about a group of organisms; and that is what this book does for the North American tortoises. We have brought together a distinguished group of tortoise researchers to take stock, and these researchers have responded impressively to the call.

The book was originally envisioned as a compendium and comparison of knowledge about two particular species of North American tortoise, *Gopherus (Xerobates) agassizii* (desert tortoise) and *G. polyphemus* (gopher tortoise), for which a wealth of information is available. Readers will no doubt see the difference between what we know about these two species of North American tortoises and the other species. Nevertheless, readers will find plenty of information, analysis, and interpretation concerning all of the species within the 18 chapters. The book provides one of the largest literature databases on North American tortoises currently available. We think that the book will serve well as a textbook; a field manual; a source of information for government agencies, nonprofits, and other organizations working with North American tortoises; a valuable resource for turtle biologists and hobbyists; and a good read for anyone interested in studies of the natural world. Clearly, authors were selected based upon experience and reputation in studying North American tortoises. Where feasible, authors who have studied different species were paired to address common topics. Among the authors is an outstanding group of young researchers, on which the future of North American tortoise research and management will depend. We hope the book will inspire other young scientists to study this fascinating group of organisms.

WHAT'S IN A NAME?

We would have preferred to use common names throughout, for the sake of clarity for those readers not used to dealing with binomial nomenclature. Although a recognized species has only one Latin binomial at any particular time—even Latin binomials may change (chapter 1)—not so for common names. To introduce an element of consistency to the common names of amphibians and reptiles, several organizations, such as the Center for North American Herpetology and the major herpetological societies (American Society of Ichthyology and Herpetology, Herpetologists' League, Society for the Study of Amphibians and Reptiles) have created lists of "official" names. When the North American tortoises included four recognized species, little controversy existed concerning the "official" names. These names were Desert Tortoise (for *Gopherus agassizii*), Texas Tortoise (for *G. berlandieri*), Bolsón Tortoise (for *G. flavomarginatus*), and Gopher Tortoise (for *G. polyphemus*). Perhaps the largest dissention was from those individuals who maintained a preference for Berlandier's Tortoise, rather than Texas Tortoise. According to some herpetologists, the North American tortoises now include five recognized species, although not all herpetologists agree. The splitting of the former "Desert Tortoise" into two species (chapter 3) has emphasized the problem with multiple common names. According to the joint committee of the ASIH/HL/SSAR, the two new species should have the common names Mohave Desert Tortoise (for *G. agassizii*) and Sonoran Desert Tortoise (for *G. morafkai*). Several herpetologists have objected to the use of these names. One position among these objectors is a preference for the use of Agassizi's Desert Tortoise and Morafka's Desert Tortoise, respectively. A second position is a spelling preference for "Mojave," as apposed to "Mohave." Although common names for several other species use "Mohave" (e.g., Mohave Rattlesnake, Mohave Ground Squirrel), the major objection is that this name is not linked to major geographical entities as they are currently recognized. The use of "Mohave" apparently is intended to recognize the Native Americans who inhabited the area, but the name seems to have little historical validity (e.g., Sherer 1967). The authors of the various chapters could not agree on common names to be used throughout the book, so we have opted to use Latin binomials.

BIOLOGY AND CONSERVATION OF NORTH AMERICAN TORTOISES

1

Morphology, Taxonomy, and Distribution of North American Tortoises

An Evolutionary Perspective

Dennis M. Bramble
J. Howard Hutchison

Modern gopher tortoises are the remains of a once abundant and diverse North American tortoise fauna (Williams 1950a, Auffenberg 1974, Bramble 1971). The group is endemic to North America and has an especially rich fossil record. As the name suggests, gopher tortoises are notable for their digging abilities. Although digging is reported for some Old World tortoises (e.g., *Centrochelys sulcata* Miller 1779), the gopher tortoises are unique in representing the only known example of true fossorial specialization among terrestrial turtles. Fossorial vertebrates are those specialized for digging and spending appreciable time underground (Hildebrand 1985). They also exhibit associated modifications—i.e., morphological, physiological, and behavioral (Gans 1974). As this and other chapters of this volume make clear, and despite very significant advances in recent years, important questions remain as to the evolution, biogeographic and life histories, and appropriate taxonomic treatment of gopher tortoises. Unfortunately, unless efforts to stem the severe threats facing the remaining wild populations of these tortoises succeed (see chapters 17, 18), we may never fully secure the answers.

This chapter consists of several parts. We provide an overview of the taxonomy, distribution, and important morphological distinctions among the five living species of gopher tortoise. We then highlight key morphological distinctions that characterize the two primary lineages of extant gopher tortoises and discuss how they reflect differences in the behavior and ecology of the two groups. We next consider controversies regarding the major evolutionary patterns within gopher tortoises and how to best classify these tortoises in light of those patterns. In that context, we note important evolutionary patterns (parallelism, convergence) and mechanisms (paedomorphosis) that have been largely overlooked in these debates. Lastly, we list some unresolved questions concerning the form, function, behavior, and evolution of these unique tortoises and ones that we hope will be the targets of future research.

Comparative skeletal morphology has long played a central role in deciphering the evolutionary relationships of land tortoises (e.g., Williams 1950a, Loveridge and Williams 1957), including the North American gopher tortoises (Auffenberg 1974, 1976; Bramble 1971, 1982; Crumly 1994). Thus, skeletal morphology is emphasized throughout this chapter because it reflects some of the most profound evolutionary patterns within gopher tortoises, contributes directly to the fossil record (chapter 2), and has been the focus of phylogenetic analyses (e.g., Bramble 1982, Crumly 1994) central to the debate over how best to classify these tortoises. Some skeletal features, including those used below to define various taxonomic groups, will be unfamiliar to the general reader. Chapter 2 offers an illustrated introduction to some of the most important features related to shell morphology. Informative illustrations and discussions of other skeletal features can be found in selected publications (skulls—Gaffney 1979, Bramble 1982; limbs and vertebrae—Bramble 1982, Crumly 1994).

SYSTEMATICS AND TAXONOMY

Murphy et al. (2011; chapter 3) summarize the complex and sometimes confused systematic and nomenclatorial histories of the North American gopher tortoises. For the purposes of this chapter we have elected to recognize the two clear species groups within gopher tortoises as the genera *Xerobates* and *Gopherus*. The same classification is adopted by Franz (chapter 2), but other contributors to this volume employ *Xerobates* only as an informal subgroup within the genus *Gopherus* or simply assign all species (living and extinct) to *Gopherus*. We have also used the taxon *Oligopherus* Hutchison 1996 as a "placeholder" for the most widely accepted ancestral gopher tortoises, whose proper taxonomic assignment is uncertain at this moment. Evidence in support of our classification of gopher tortoises is discussed in parts two and three of this chapter.

Subfamily *Xerobatinae* Agassizi 1857

North American testudinids containing the gopher tortoises and perhaps the extinct genus *Hesperotestudo* Williams 1950a, but, importantly, not *Stylemys* Leidy 1851 or primitive Old World tortoises in the genus *Manouria* Gray 1852 (Meylan and Sterrer 2000). Late Eocene to Recent.

Definition: North American testudinids specialized for burrowing. Skull shape ranging from dolicocephalic (*Oligopherus*) through mesocephalic (*Xerobates*) to brachycephalic (*Gopherus*). Trochlear process formed chiefly or entirely by the quadrate; external adductor muscle tendon containing a unique sesamoid bone (= *Os transiliens*). Vomerine foramen and prefrontal pits present, except when secondarily lost; a well-formed median premaxillary ridge. Inner ear chambers varying from normal (*Oligopherus*) to reduced (*Xerobates*) or greatly enlarged (*Gopherus*). Manus with reduced phalangeal formula (2-2-2-2-1); posture ranging from subplantigrade to unguligrade. Carpus widened with 2–4 subradial bones. Pisiform typically present, but reduced or lost in *Gopherus*. Humeral shaft curved or straight, distal end variably expanded, entepicondylar foramen present. Shell depressed; cervical scute (= scale) typically wider than long; carapace thin, costal wedging pronounced, anterior peripheral plates short, not flared. Gular region of plastron often strongly projecting; gular scutes usually contacting entoplastron; pectoral scale wide, not contacting entoplastron; inguinal scute small, not contacting femoral scute. Osteoderms absent or sparsely developed on forearms and feet. Type I mental glands present on chin.

GENUS *OLIGOPHERUS* HUTCHISON 1996

Genotype: *Testudo laticunea* Cope 1873. Included species: *O. laticunea;* plus additional nominal species (chapter 2). Late Eocene to Middle Oligocene.

Definition: *Skull:* dolicocephalic; middle ear relatively large; inner ear (prootic and opisthotic) bones well exposed dorsally; tomial (labial) ridge low, no postmaxillary process; postorbital bar wide; posterior pterygoid well ossified; no vomer-basisphenoid contact. *Vertebrae:* zygopophyses of cervical vertebrae small and vertically oriented; first dorsal with small zygopopyses and neural arch restricted to first neural plate. *Manus:* moderately broadened, semiplantigrade and with 2 subradial bones (= carpal 1 + medial centrale); pisiform well developed. Proximal phalanges relatively long, with well-formed retroarticular processes. Mesocarpal joint fully functional. *Shell:* vertebral scutes relatively narrow with arcuate borders; vertebral 1 wider than 2. Gular region projecting in males. Inguinal scute single, not contacting femoral.

GENUS *XEROBATES* AGASSIZ 1857

Synonyms: *Scaptochelys* Bramble 1982. Genotype: *Xerobates agassizii* Cooper 1861, (Brown 1908). Included species: *X. agassizii, X. berlandieri* Agassiz 1857, *X. morafkai* Murphy et al. 2011, and fossil forms dating to at least the Early Miocene. Early Miocene to Recent.

Definition: *Skull:* Mesocephalic. Middle ear enlarged, but inner ear reduced and containing a small saccular otolith; prootic and opisthotic bones narrowly exposed on surface of skull; tympanic cavity large. Tomial ridge of upper jaw high; well-developed postmaxillary process; postorbital bar narrowed; posterior pterygoid well ossified; no vomer-basisphenoid contact. *Vertebrae:* Essentially as in *Oligopherus. Manus:* moderately broadened, digitigrade, with 3 subradial bones (carpals 1, 2 + medial centrale); pisiform typically present; proximal phalanges disc-like, lacking retroarticular processes; ungual phalanges of moderate size, round in section. Mesocarpal joint well developed. *Shell:* Vertebral scutes as in *Oligopherus* to wider than long, borders straight; vertebral 1 narrower than 2. Gular projection usually pronounced (esp. in older males) and often divided at tip. Frequently having multiple inguinal scutes.

GENUS *GOPHERUS* RAFINESQUE 1832

Synonyms: *Bysmachelys* Johnston 1937. Genotype: *Testudo polyphemus* Daudin 1802. Included species: *Gopherus polyphemus; G. flavomarginatus* Legler 1959, and fossil forms dating to at least the Late Oligocene. Late Oligocene to Recent.

Definition: Skull brachycephalic. Middle ear reduced, but inner ear inflated and housing a relatively massive saccular otolith; prootic and opisthotic bones widely exposed; tympanic cavity reduced; posterior pterygoid often incompletely ossified; tomial ridge low to high; well-developed postmaxillary process or not; postorbital bar wide to narrow. *Vertebrae:* neck short with robust vertebrae having enlarged, horizontally oriented zygopophyses; last cervical with elongated postzygopophyses that engage enlarged zygapophyses on the first dorsal vertebra; first dorsal attached to nuchal plate in addition to first neural. *Manus:* distinctly broadened, unguligrade, with 4 subradial bones (metacarpal 1, carpals 1, 2 + medial centrale); pisiform reduced or lost, allowing carpal 5 to contact ulna; proximal phalanges disc-like, lacking retroarticular processes; ungual phalanges enlarged, flattened, spatulate-shaped. Mesocarpal joint weakly developed or eliminated. *Shell:* Vertebral scutes as in *Oligopherus* to wider than long, with straight borders; vertebral 1 narrower than 2. Inguinal scute single, not contacting femoral. Gular projection modest or lacking, blunt, and usually undivided or weakly divided at tip.

THE LIVING SPECIES: CHARACTERISTICS AND DISTRIBUTION

Photographs of the five species are presented in figure 1, and distributional maps in figure 2. For additional detail on genetic evidence for geographic variation and distribution according to habitat type, see chapters 3 and 9, respectively.

Xerobates agassizii, Cooper 1861 (desert tortoise, Mohave desert tortoise, Mojave desert tortoise, Agassizi's desert dortoise).

Characteristics: A large tortoise with carapace lengths to 37 cm; males larger than females. Carapace oblong and flattened; scutes black to tan with yellow or orange centers

Fig. 1.1. Gopherus flavomarginatus (top left; photo courtesy of Gustavo Aguirre), *Xerobates (Gopherus) berlandieri (top right;* photo courtesy of Richard Kazmaier), *Xerobates (Gopherus) agassizii (center;* photo courtesy of David C. Rostal), *Gopherus polyphemus (bottom left;* photo courtesy of David C. Rostal), and *Xerobates (Gopherus) morafkai (bottom right;* photo courtesy of Roy Averill-Murray).

(= aerolae) in juveniles. Carapace flares strongly over hind limbs. Gular region of plastron strongly projecting, especially in older males, and often slightly notched at tip. Head rounded, eyes with greenish yellow iris, tympanic membrane large and naked. Forefeet shovel-like, with thick claws that are round in section; palm convex and covered with large scales. Fore- and hind feet about same size.

Distribution: Mojave Desert north and west of the Colorado River in southern California, southern Nevada, and extreme southwestern Utah as well as tortoises in the Colorado Desert of southern California. Preferred habitats are valleys and alluvial fans in desert scrublands and yucca tree woodlands. Tortoises of the Sonoran Desert, formerly included in *X. agassizii,* have recently been assigned to *X. morafkai,* thereby reducing the range of *X. agassizii* to about one-third its former size (Murphy et. al. 2011).

Xerobates lepidocephalus (Ottley and Velázques Solis 1989) has been identified as a species, based on tortoises from the Cape Region of Baja California Sur, Mexico. Genotyping shows the type specimens belong to a matriline from the Mojave Desert (*X. agassizii*), not the adjacent *X. morafkai* (Murphy et al. 2011). This and other factors reinforce the opinion that *X. lepidocephalus* is not a valid species, but instead a junior synonym of *X. agassizii* (Crumly and Grismer 1994).

Xerobates morafkai, Murphy et al. 2011 (desert tortoise, Sonoran desert tortoise, Morafka's desert tortoise)

Characteristics: Similar in size and most external features to *X. agassizii. X. morafkai* differs from *X. agassizii* in having a narrower, pear-shaped carapace and a plastron in which the projection of the gular and anal scutes is less pronounced. Coloration varies considerably with location, size, and tortoise age, but tends toward darker hues of brown, grey, and black. In juveniles the areolae are orange to reddish brown while lamellar areas are reddish brown to dark brown.

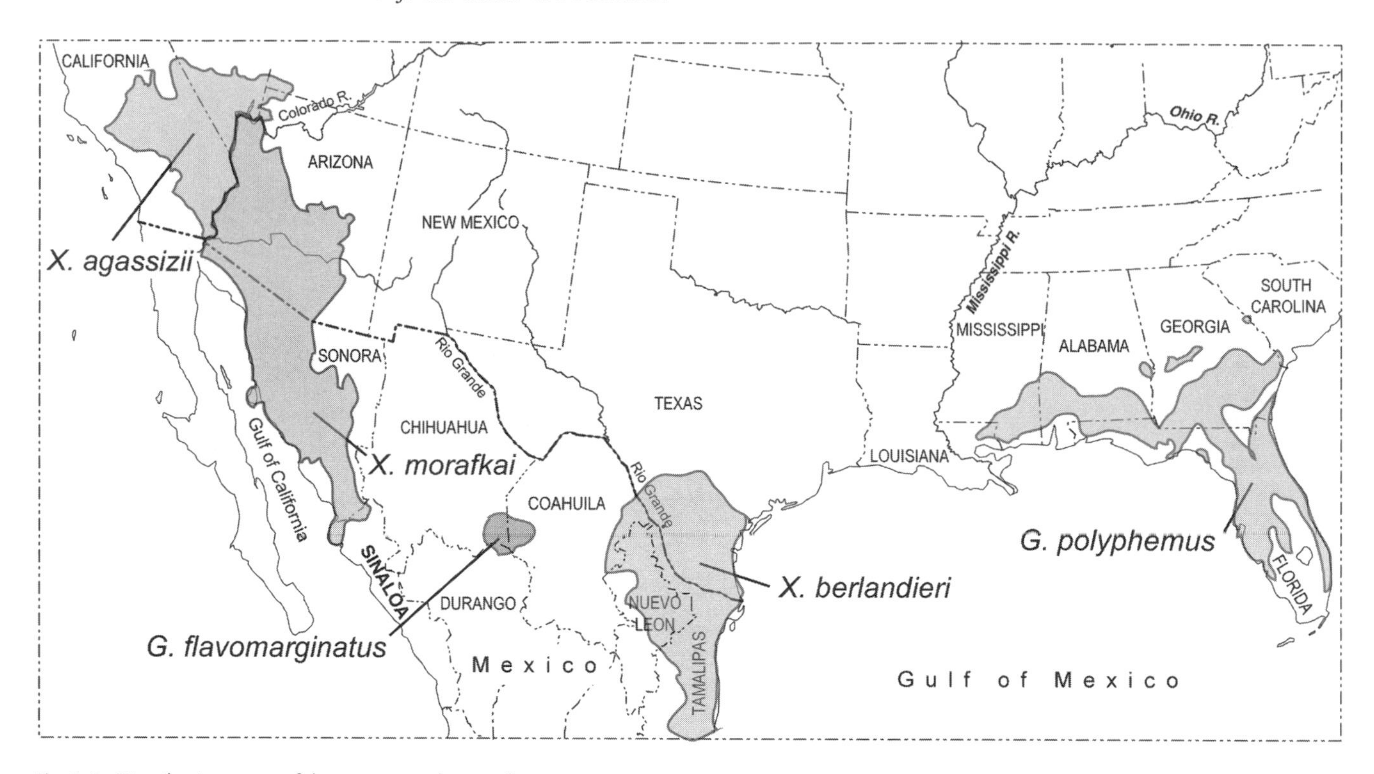

Fig. 1.2. Distribution map of the extant gopher turtles. Ranges after Auffenberg and Franz (1978), Iverson (1992b), and Murphy et al. (2011).

Distribution: Found in thorn-scrub and grassland habitats on slopes and hillsides of the Sonoran Desert south and east of the Colorado River in California and Arizona, extending southward into the Mexican states of Sonora (including Tiburon Island) and Sinaloa along the west flank of the Sierra Madre Occidental.

Xerobates berlandieri, Agassiz 1857 (Texas tortoise, Berlandier's tortoise)

Characteristics: The smallest of the living gopher tortoises, with maximum carapace length near 22.8 cm; males larger than females. The carapace is higher domed than in *X. agassizii* and *X. morafkai*, and often roughened by distinct growth lines on the scutes, unless worn smooth in old individuals. Coloration of the shell varies from black through reddish brown to a dull tan depending on location and specimen age. The gular projection is well developed, often bending upward (esp. in older males) and typically forked at the end. The anterior rim of the carapace is commonly incised in larger males and the cervical scute either reduced or absent. In adult males the posterior carapace is strongly recurved, causing the pygal scute to nearly or actually contact the anal scutes. The head is wedge-shaped with a pointed snout; the eyes have a greenish-yellow iris and the tympanic membrane is naked. Morphology and relative sizes of the fore- and hind feet are about the same as in *X. agassizii* and *X. morafkai*.

Distribution: Southeastern Texas and southward into Mexico, including northern Coahuila, northern and eastern Nuevo Leon, northeastern San Louis Potosi and Tamaulipas (possibly as far south as Tampico). Habitats range from near desert (Mexico) to scrub forests (Texas). In coastal areas, local populations tend to be associated with "lomas"—that is, semi-isolated clay and sand dunes (Auffenberg and Weaver 1969).

Gopherus polyphemus, Daudin 1802 (gopher tortoise)

Characteristics: A large tortoise with maximum carapace length near 38 cm. Males and females of similar size. Shell oblong and widest just in front of the bridge. Shell coloration a uniform dark brown to grayish black; scutes of juveniles have light centers; growth ridges are prominent on shell of younger specimens, but are often worn smooth on old individuals. Gular projection is well developed, blunt, and frequently turned upward to produce a "plow-like" structure. Head large and wide; the iris is a dark brown; the tympanic membrane is relatively small and overlain with small scales. The manus is large and spade-like and armed with large, closely spaced claws that are spatulate in shape. The palm is flat and invested by small scales. The forefeet are distinctly larger than the hind, the latter being reduced in size, elephantine in shape and lacking conspicuous heel spikes (contra *Xerobates*).

Distribution: Southeastern U.S. with the center of distribution in Florida and ranging westward along the Gulf Coast to Louisiana and northward to South Carolina (fig. 1.2). Preferred habitats include sandy, well-drained soils with good understory of grass and other herbaceous plants, often within fire-adapted woodlands (e.g., sand pine-scrub oak, longleaf pine-oak, beach scrub).

Gopherus flavomarginatus, Legler 1959 (Bolsón tortoise)

Characteristics: The largest of the living gopher tortoises, with carapace lengths reaching at least 40 cm and still larger sizes possible. Females are larger than males. Shell is flat, wide, and less oblong than in *G. polyphemus.* Posterior marginal scutes flare strongly in adults. Scutes of carapace range from orange beige to brown with darker markings at the centers; juveniles have distinctly yellowish borders along the bridge area, but these fade with age. Supernumery scutes (= extra vertebrals and costals) are common in this species. Plastron a dull yellow; gular projection moderately developed and blunt in adult males, but weak to absent in females. Head robust and wide; iris color dark brown; tympanic membrane somewhat reduced but naked. Keratinaceous coverings of the jaws bear prominent "tooth-like" structures in adults. Forefoot large and spade-like with very stout, blunt tipped claws; larger than hind foot.

Distribution: A relict distribution in the Mapimian subprovince of the Chihuahuan Desert, north-central Mexico near the intersection of the states of Chihuahua, Coahuila, and Durango (fig. 1.2). This tortoise occurs in near desert and brush grassland habitats, especially those in which tobosagrass (*Hilaria mutica*) is a major component (Legler and Webb 1961, Morafka 1988).

GOPHERUS VS. *XEROBATES:* MORPHOLOGICAL, BEHAVIORAL, AND ECOLOGICAL CONTRASTS

Members of *Gopherus* and *Xerobates* are easily distinguished in the construction of the skull, and especially the inner ear component (Bramble 1971, 1982). Head shape and size in *Xerobates* is similar to other testudinids (fig. 1.3A). In contrast, the head of *Gopherus* is relatively large and distinctly broader owing to inflation of the inner ear region, as reflected in much broader dorsal exposure of the prootic and opisthotic bones (fig. 1.3B). X-rays show that the inner ear is grossly enlarged and contains a massive otolith residing within the saccular chamber (fig. 1.4B). Small otolithic masses are normal components of the inner ear of many tetrapod vertebrates, including turtles (Wever 1979). Still, the otolith of *Gopherus* exceeds both the relative (to head size) and absolute size of that found in any other modern terrestrial vertebrate. By contrast, the inner ear chamber of the living species of *Xerobates,* including the size of the otolithic mass, is tiny by comparison and similar to that of other tortoises (fig. 1.4A). The external morphology of the head also separates *Gopherus* from *Xerobates* (Bramble 1982). The external ear of *Gopherus* is smaller than that of *Xerobates* and is overgrown with small scales in *G. polyphemus.* The cranium of advanced *Gopherus* is further characterized by failure of the posterior end of the pterygoid bone to fully ossify, leaving the inner ear chamber exposed from below (fig. 1.3B).

The cervical region of *Gopherus* also departs from *Xerobates* and all other known testudinids (Bramble 1982; figs. 3, 4). The neck vertebrae of *Xerobates* are generally similar to those of other tortoises in having zygopophyses of modest size and largely vertical orientation. The neck region of *Gopherus* is significantly shorter and the individual vertebrae possess much wider, more robust zygopophyses whose orientation is more nearly horizontal. Additionally, the last cervical vertebra (8th) carries enlarged, elongate posterior zygopophyses engaging equally enlarged zygopophyses on the first dorsal vertebra. The strongly interlocking union permits extensive, anteroposterior rotation of the base of the neck on the shell but also prohibits any significant side-to-side movement. The neural spine of the first dorsal extends forward to join a robust strut on the underside of the nuchal plate (Bramble 1982; fig. 5). The specialized construction of the cervico-dorsal complex of *Gopherus* is absolutely unique among tortoises, and provides what is likely the strongest neck-shell connection of any turtle. The same articulation of *Xerobates* closely mimics that of other tortoises.

The forelimb is the primary digging tool in all gopher tortoises, but the manus (hand) is distinctly more specialized in *Gopherus* than *Xerobates* (Bramble 1982). In *Xerobates,* it is club-shaped with a rounded palm and stout claws. Except for the enlarged claws, the shape is much like that of nonburrowing tortoises. In *Gopherus,* the manus is relatively larger, wider, and much flatter. It is also armed with very robust, distinctly flattened (spatulate) claws. The evolution of the stiffened, shovel-like hand of *Gopherus* involves suppression of the mesocarpal joint, the normal plane of movement within the carpus found in all other tortoises, including *Xerobates.*

The robust cervical column of *Gopherus,* together with the broad skull, implies a neck subjected to unusual mechanical loading. Behavioral observations indicate that *G. polyphemus* sometimes uses the head to buttress the body while pushing soil backward with the forelimbs. This involves placing the head against the back wall of the burrow with the neck partly retracted (Bramble 1982; figs. 2, 6). Progressive robusticity of the cervical column from the first to the last vertebrae and especially the remarkable reinforcement of the neck-shell connection are consistent with stresses arising from such a mechanical function. Use of the head is widespread among burrowing tetrapods, although among reptiles the head is most often used as a digging tool in animals that are also limbless (Gans 1974, Wake 1993). No evidence exists that *Gopherus* uses the head to actively dislodge or move soil.

The spectacular otolithic ear of modern *Gopherus* certainly begs an explanation, although at this time only informed speculation is possible. The specialization does not suggest any unusual ability related to conventional (aerial) hearing because the saccular region is instead concerned with recording linear acceleration. The detailed histological structure of the otolith ear of *Gopherus* (Bramble, unpublished) strongly suggests that its primary role is the detection of substrate vibrations reaching the head (either directly or via the shell, or shell and neck). Increased sensitivity to substrate vibration is

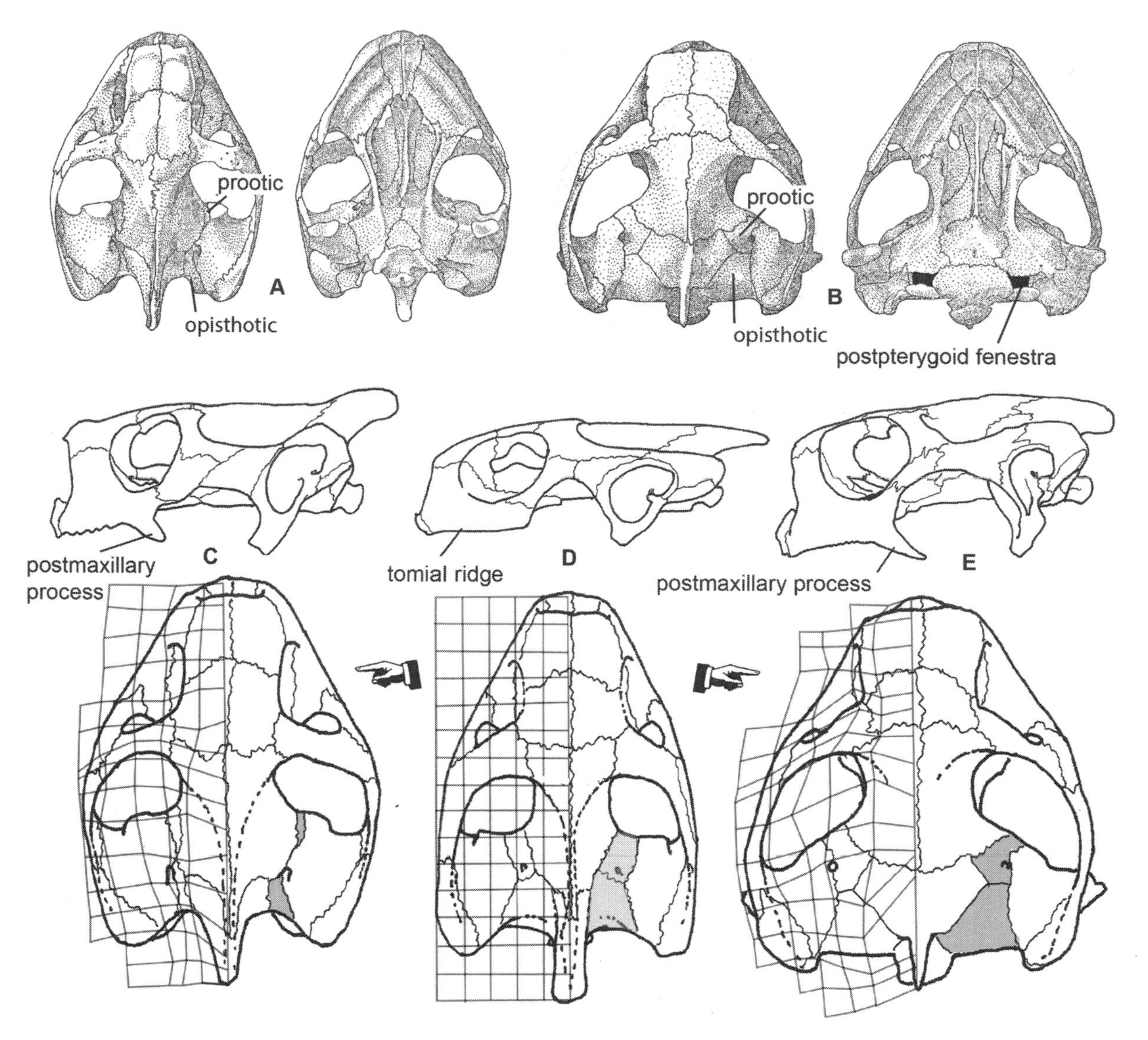

Fig. 1.3. Skulls. *A, Xerobates agassizii,* dorsal and ventral views. *B, Gopherus polyphemus,* dorsal and ventral views. *C–E,* left lateral and dorsal views with prootic and opisthotic highlighted and Cartesian grids modified from *Oligopherus laticuneus* and indicating the proposed evolutionary polarity. *C, X. agassizii. D, O. laticuneus. E, G. polyphemus.* Figures modified from Auffenberg (1976), Bramble (1971), Gaffney (1979), Gilmore (1946), and Hutchison (1996).

common among fossorial vertebrates, although this usually involves the auditory rather than the vestibular region of the inner ear (Lombard and Hetherington 1993). On first principles (i.e., vibration mechanics), such large otolithic masses in *Gopherus* would imply extraordinary sensitivity to low frequency "seismic" vibrations. What sorts of vibration might be of interest to a tortoise is uncertain, but a reasonable guess would be events that occur on the surface while the tortoise is safely within its burrow, and therefore unable to utilize other sensory clues (e.g., aerial sound; vision) to assess the situation. An ability to detect potentially threatening conditions before committing to leaving the safety of the burrow could be of considerable value to an animal that is both slow and without active defense mechanisms. Another (but less likely) possibility is that of facilitating intraspecific communication while underground. Any plausible communication signaling system would probably involve scraping the walls of the burrow while digging or in some other activity. Observations on *G. polyphemus,* in particular, suggest a surprising level of social organization built around spatially restricted local populations or "cliques" (chapter 12). The acknowledged sensory cues for social interaction are, thus far, those restricted to surface activity (olfaction, vision). The possibility of subsurface social interaction is largely unexplored.

The diminished size and scale covered tympanic region of *G. polyphemus* is consistent with protection of the delicate tympanic membrane when employing the head in digging. Reduction or loss of the external ear is a convergent trait in many unrelated fossorial vertebrates (Wake 1993). This, plus the reduced size of the underlying tympanic cavity, suggests a diminished reliance on aerial hearing as well as enhanced mechanical protection for the external ear. The larger, naked

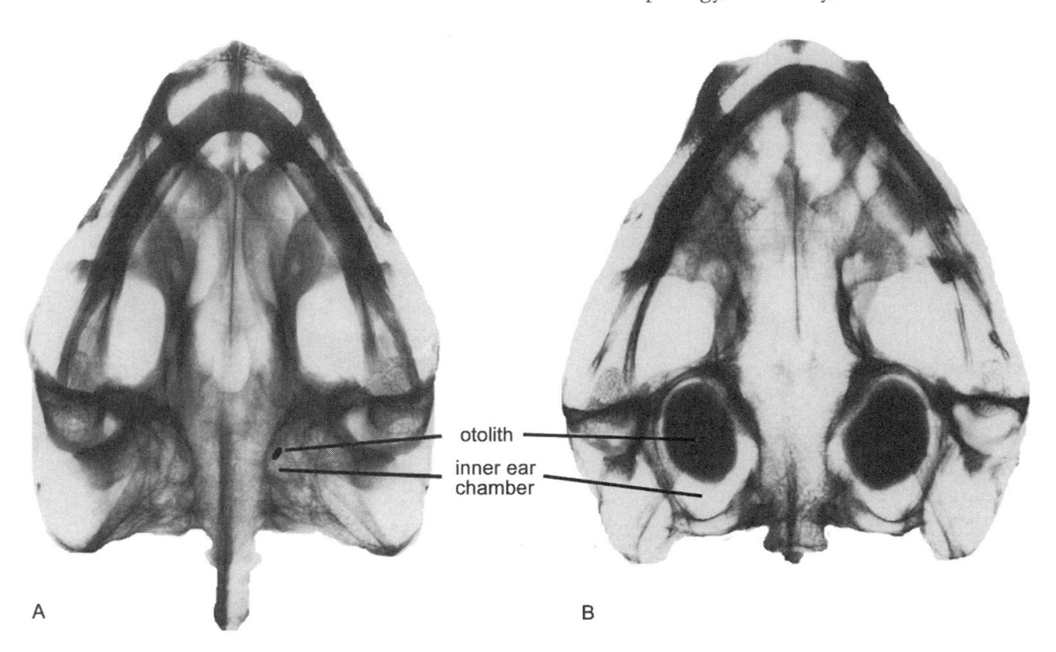

Fig. 1.4. X-rays (in ventral view) of the skulls of *A, Xerobates agassizii,* and *B, Gopherus polyphemus.* Figures adjusted to same length. Modified from Bramble (1971).

tympanic membrane of *Xerobates* reflects a lesser commitment to fossorial life and a mechanism of digging that does not directly engage the head.

Contrasts in the structure of the forelimbs of *Gopherus* and *Xerobates* reflect the differing physical characteristics of the environments with which they are commonly associated. In overall form, the forefoot of *Xerobates* is much closer to that of the club-shaped forefeet of nonburrowing testudinids. This signals a compromise between a foot modified for digging in resistant soils and one also suited to efficient longer-distance travel. Unlike *Xerobates,* the manus of *Gopherus* is best suited for shovel-like action in friable, sandy soils, a substrate type with which the genus has been closely associated since the Miocene (Bramble 1982).

The arid environments inhabited by *Xerobates* (esp. *X. agassizii*) commonly feature not only hard, indurated soils, but also food resources that tend to be patchy and widely dispersed. For this reason *Xerobates,* on average, would be confronted with longer foraging distances than *Gopherus,* which tends to occupy more mesic habitats, in which food resources are typically more abundant, reliable, and proximate (chapter 11). The rounded, digitigrade manus of *Xerobates* facilitates a slow and fairly smooth gait. The flattened, unguligrade forefoot of *Gopherus* is associated with a somewhat faster and choppier gait. Such kinematic differences in walking suggest that the metabolic cost of transport might be greater for *Gopherus.* The potentially less efficient gait of *Gopherus* might well be offset, however, by the fact that these tortoises may obtain adequate food resources over shorter foraging distances. In fact, when conditions permit, *G. polyphemus* may confine its foraging to well-defined feeding trails in close proximity to its burrow(s) (Auffenberg and Iverson 1979, Ashton and Ashton 2008). Surprisingly, while there are data on the metabolic consequences of body temperature, food availability, and reproductive effort in gopher tortoises (chapter 4), the energetic consequences of locomotion, including the behavior for which these tortoises are best known (i.e., burrowing), remain unknown.

RELATIONSHIPS, CONTROVERSIES, AND TAXONOMIC CONSEQUENCES

Population Genetics and Phylogeography

Whereas the past two decades have witnessed little progress in resolving the larger patterns of evolution within gopher tortoises, important advances have been made at finer scales. Chapter 3 summarizes new studies that detail the genetic relationships and historical patterns of distribution (i.e., "phylogeography" *sensu* Avise et al. 1987) at the population level. The most dramatic outcome of this work has been the recent naming of a fifth living species of gopher tortoise, *Xerobates morafkai* (Murphy et al. 2011).

The *Xerobates* Question

Substantial disagreement exists between the phylogenetic conclusions reached by studies employing molecular techniques and those using morphological features (see also chapters 2, 3). The former support the existence of a natural (monophyletic) lineage, *Xerobates,* including *X. berlandieri, X. agassizii,* and *X. morafkai.* Most morphological analyses, combining modern and fossil species, have instead concluded that *Xerobates* is an artificial taxon based on shared primitive traits (symplesiomorphies) rather than shared derived traits (synapomorphies) as required by modern rules of systematics and taxonomy. As discussed below, there are good reasons to question the methodologies of the morphology-based analyses and as well as their conclusions.

Recognition of two distinct species complexes within the modern gopher tortoises was first proposed on morphological grounds (Auffenberg 1966a, 1976; Bramble 1971, 1982) and subsequently supported by biochemical and molecular evidence (Lamb et al. 1989; Lamb and Lydeard 1994, Morafka et. al. 1994, Thomson and Shaffer 2010). Nonetheless, there is significant disagreement about how to best acknowledge these two lineages within a formal taxonomic framework. Bramble (1982) suggested generic separation and proposed *Scaptochelys* for the more conservative lineage (including *X. agassizii, X. berlandieri*), while restricting *Gopherus* to the more derived species complex (including *G. polyphemus, G. flavomarginatus*). Bour and Dubois (1984), however, noted that *Scaptochelys* was a junior synonym of *Xerobates*. Crumly (1994) claimed that *Xerobates* was a paraphyletic grouping and, therefore, that *Xerobates* should not be recognized and that all known gopher tortoises should be retained in the genus *Gopherus*. Other analyses, combining both morphological traits and the temporal patterning of fossil species (i.e., stratocladistic; McCord 1997, 2002), more clearly identified *Xerobates* as a legitimate clade, distinct from *Gopherus*. Still, the tendency among recent workers (including most contributors to this volume) is not to recognize the two groups as distinct at the generic level, and often to decline recognition of *Xerobates* at any level. Such a taxonomic treatment of gopher tortoises not only blurs evolutionary relationships within the group but also diminishes the unquestioned uniqueness of *Gopherus* (*sensu stricto*).

This taxonomic conundrum underscores our failure to decipher the true history of these turtles, despite striking levels of morphological differentiation (Bramble 1971, 1982) and what is arguably the richest fossil record for any group of land tortoises. It is also an example of the sometimes discordant evolutionary interpretations stemming from two different types of data—i.e., morphological and molecular. In the present case, both types offer robust support for a natural, monophyletic clade *Gopherus* (*sensu stricto*) (cf., Reynoso and Montellano-Ballesteros 2004), but render conflicting conclusions regarding *X. agassizii* (including *X. morafkai*), *X. berlandieri*, and putative fossil relatives. The molecular studies tend to support a natural grouping, *Xerobates* (Lamb et al. 1989, Lamb and Lydeard 1994, Morafka et al. 1994, Thomson and Shaffer 2010). On the other hand, morphologically based cladistic analyses, beginning with Crumly's (1994) influential study, have consistently concluded that tortoises of this group constitute an unnatural paraphyletic assemblage based on shared primitive traits or "transitional traits" (sympleisomorphies) or a polytomy whose relationships cannot be resolved on present evidence. The result is a standoff between two discordant views of gopher tortoise evolution, and at least one must be wrong.

In such disputes, molecular studies typically are favored, because they are viewed as more objective and better quantified. Although most such conflicts are decided in favor of the molecular studies, even state-of-the-art molecular investigations can sometimes yield very misleading conclusions (Losos et. al. 2012). In the case of gopher tortoise evolution, most workers have sided with the morphological analyses. It is important to acknowledge, however, that the molecular conclusions stem from a series of independent studies utilizing different approaches (e.g., allozyme, mtDNA, nuDNA), whereas all the morphological studies rely heavily on Crumly's (1994) original data matrix. Because subsequent analyses continued to be heavily weighted by Crumly's characters and scorings (i.e., polarity), they are not truly independent evaluations and have, as expected, largely replicated Crumly's results. They also perpetuate any shortcomings of the original study.

Given the striking morphological distinctions between *Xerobates* and *Gopherus* (figs. 1.3, 1.4), it is fair to ask why *Xerobates* cannot be resolved as a legitimate taxon distinct from *Gopherus*. In our view, the answer rests with at least four factors that complicate morphological analyses and whose proper consideration will be essential to any satisfactory resolution of the evolutionary patterns and systematic treatment of gopher tortoises: unequal rates of evolution, homoplasy, outgroup choice, and the influences of ontogeny and allometry.

Unequal Rates of Evolution

Xerobates has been a relatively conservative lineage since its separation from *Gopherus* (Bramble 1982). Although structural adaptation to digging has occurred, its consequences are far fewer and less obvious than those of the *Gopherus* lineage. The two lineages have experienced very unequal rates of morphological evolution since sharing a last common ancestor, and *Xerobates* is necessarily defined by fewer and subtler traits.

Homoplasy

Failure to detect homoplasy (i.e., parallel or convergent evolution) often causes discordance between phylogenetic interpretations based on molecular vs. morphological evidence (Wake et al. 2011). Tortoises seem particularly prone to parallelism (Bramble 1971, Auffenberg 1974, Crumly 1994, Pritchard 1994, Meylan and Sterrer 2000), and modern *Xerobates* and *Gopherus* may have independently acquired many of their morphological similarities. If so, characters used to define *Xerobates* (i.e., *Scaptochelys*, Bramble 1982) and later discounted as symplesiomorphies (e.g., Crumly 1994, McCord 1997, Reynoso and Montellano-Ballesteros 2004) may actually constitute convergent morphologies and thus, homoplasies. No recent morphological analysis seems to have considered this possibility.

The skull of gopher tortoises presents clear evidence of parallel evolution. *Oligopherus latecunea* is the most primitive gopher tortoise for which adequate cranial materials exist, and it is appropriately used as the ancestral condition (fig. 1.3D). Compared to this Oligocene form, both modern *Xerobates* (fig. 1.3C) and *Gopherus* (fig. 1.3E) have evolved significantly wider crania—i.e., mesocepahlic and brachycephalic, respectively. Crumly (1994) interprets the condition in *Xerobates* as "intermediate" between the ancestral shape and the very wide skull of *Gopherus*. As figure 3 illustrates, however, *Xerobates* and *Gopherus* have almost certainly evolved wider skulls independently. In *Xerobates*, this involves expansion of the middle ear but reduction of the inner ear; exactly the opposite process has produced the even wider skull of *Gopherus*. Thus, the developmental and evolutionary pathway to cranial widening

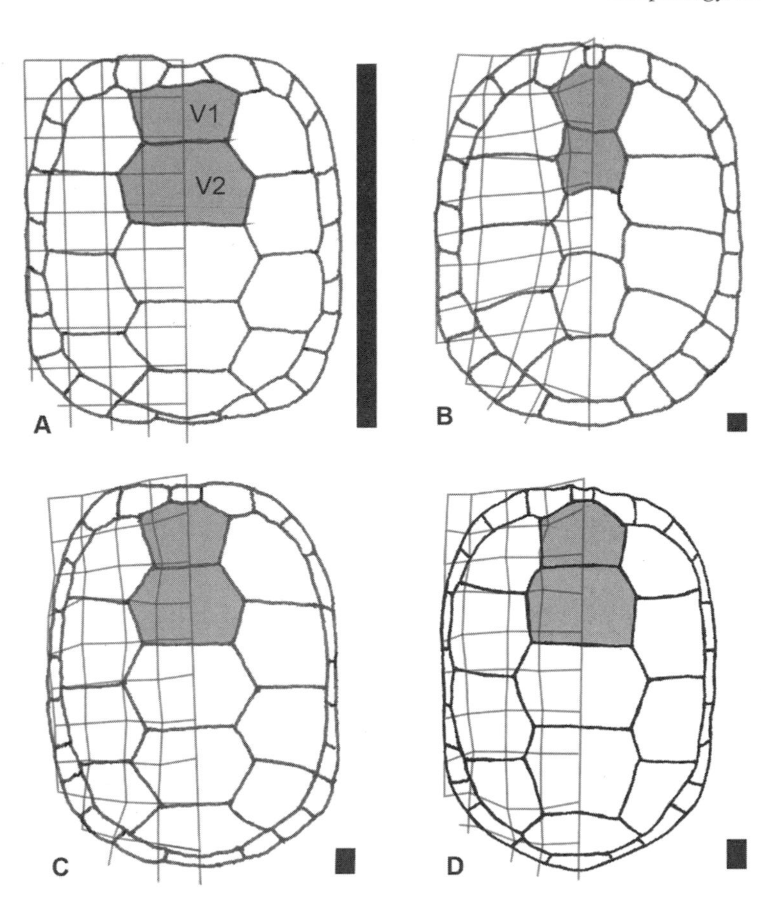

Fig. 1.5. Dorsal views of carapace with superimposed Cartesian grids and first two vertebral scutes identified (V1, V2). *A*, *Xerobates agassizii*, post-hatchling-sized juvenile, base grid. *B*, *Oligopherus laticuneus*, adult. *C*, *X. agassizii*, adult. *D*, *Gopherus polyphemus*, adult. Shells adjusted to same length, with the scale bars indicating the relative size of *A* to *B–D*. Figures modified from Bramble (1971) and Auffenberg (1976).

in *Xerobates* does not lead to *Gopherus*. In this (complex) trait *Xerobates* does not represent an intermediate or transitional morphology: it represents homoplasy.

Other similarities in the cranial morphology have likewise been achieved independently. The earliest, most basal and widely accepted member of *Gopherus* is *G. brevisternus* (Loomis 1909), from Early Miocene deposits in Wyoming. Primitive features of its skull include a low tomial ridge lacking a defined posterior maxillary process (as *in O. laticunea;* fig. 1.3D). High tomial ridges with pronounced posterior maxillary processes characterize modern *Gopherus* and *Xerobates* and have unambiguously been acquired in parallel (figs. 1.3C, 1.3E). These traits are part of a larger functional complex that reflects changes in feeding mechanics and shifts from softer to coarser food types (Bramble 1971, Bramble 1974, Bramble and Wake 1985). A more extreme version of this same feeding complex also emerged quite independently in yet another line of North American tortoises, *Hesperotestudo* (Bramble 1971, Meylan and Sterrer 2000).

If the modern representatives of *Xerobates* and *Gopherus* share convergent cranial morphologies, do other structural similarities have a similar explanation? The shell offers a quick confirmation (fig. 1.5). The primitive form of the vertebral scutes (fig. 1.5B) was also retained in basal *Gopherus* (i.e., *G. brevisternus*), but a derived pattern is expressed by members of both *Gopherus* (fig. 1.5D) and *Xerobates* (fig. 1.5C), beginning in the Late Miocene. Like the cranial features just cited, there can be little doubt that the similar shell morphologies have separate origins.

Outgroup Choice

Outgroup models root phylogenetic trees during cladistic analyses and strongly influence the outcomes. Appropriate outgroup selection is therefore an essential component of such evaluations. Molecular studies of gopher tortoises have tended to use the Asiatic tortoise *Manouria* Gray 1852 as the relevant outgroup (Lamb and Lydeard 1994, Spinks et al. 2004, Le et al. 2006, Thomson and Schaffer 2010). *Manouria* and gopher tortoises are sister clades near the base of the Testudinidae (Spinks et al. 2004, Thomson and Shaffer 2010), although the relationship may not be especially close (Le et al. 2006). Auffenberg (1974) proposed that the earliest recognized fossil tortoises in North America, traditionally assigned to the Eocene genus *Hadrianus* Cope 1872, should be submerged into the subgenus *Manouria* within the genus *Geochelone*. This decision was based primarily on primitive aspects of their morphologies. There is, however, compelling evidence that North American *"Geochelone"* is an independent lineage, *Hesperotestudo* Williams 1950a (Bramble 1971, Meylan and Sterrer 2000). Although the morphology of extant *Manouria* meets expectations for a primitive testudinid in some respects (skull traits), other features are highly derived (e.g., shell features; Hutchison and Bramble 1981; fig. 1.6). Other studies have indicated that New World gopher tortoises plus *Hesperotestudo* represent a monophyletic clade, Xerobatinae, from which *Manouria* is excluded (Meylan and Sterrer 2000).

North American *Hadrianus,* in its known morphology, is quite distinct from *Manouria.* Fossil evidence suggests that all the major post-Eocene clades of testudinids in North America could be derived from various species of *Hadrianus,* and that gopher tortoises are among its earliest recognizable derivatives (Bramble 1971, McCord 2002, Meylan and Sterrer 2000; see chapter 2). Phylogenetic trees rooted in Eocene *Hadrianus,* rather than living *Manouria,* might well foster greater resolution of gopher tortoise evolution. Crumly's (1994) cladistic analysis employs modern *Manouria* as an outgroup and completely ignores *Hadrianus.* Importantly, the substitution of *Hadrianus* for *Manouria* in that exercise would force reevaluation of a number of traits, including polarity reversals (Bramble and Hutchison, in preparation). Even more appropriate would be the inclusion of primitive Eocene gopher tortoises as an additional, more proximate outgroup. Although such fossils exist (Gilmore 1915; Bramble 1971, 1982), they are still incompletely studied and have never been included in a modern cladistic analyses.

Ontogenetic and Allometric Influences

There has also been little attempt to weigh the impact of ontogenetic and allometric influences on gopher tortoise phylogeny, although it is widely acknowledged that both can contribute importantly to evolutionary patterns (Alberch et al. 1979). It is quite evident, for example, that both lineages of gopher tortoise have been strongly influenced by one common form of evolutionary developmental morphology, paedomorphosis (Bramble 1971). This is an evolutionary mechanism by which novel morphologies arise through the selective transfer of ancestral juvenile traits to the adult stages of the descendent (Gould 1977). Thus, the "modernized" configuration of the carapacial scutes observed in post-Miocene gopher tortoises is that of hatchling and juvenile tortoises (fig. 1.5A) retained into adulthood (figs. 1.5C, D). The process draws on ancestral ontogenetic programming and selectively retards the growth trajectory of targeted traits in descendent species. The extremely similar, but independently derived pattern of carapacial scutes of modern *Xerobates* and *Gopherus* is not, therefore, particularly surprising, since both arise from modifications of developmental programs inherited from a last common ancestor, probably in the Oligocene. Under common selective influences (e.g., fossorial adaptation) very similar morphological responses may be expected with little genetic innovation.

Paedomorphism is more apparent in the evolution of *Gopherus* than *Xerobates* and explains such traits in *G. polyphemus* as the failure of the pterygoids to fully ossify, the enlarged size of the basisphenoid, and the abbreviated supraoccipital spine in the adult skull (fig. 1.3B). All are normal features of the juvenile gopher tortoise cranium, including that of *Xerobates.* Even the extraordinary inflation of the *Gopherus* inner ear likely owes its origins, in part, to paedomorphosis. Disproportionally large otic capsules are a universal feature of the later embryonic and juvenile stages of head development in all amniotes, including turtles. Failure of adjacent carpal bones to fuse, even in the adults of giant species (e.g., *G. canyonensis* [Johnston 1937]; Bramble 1982, fig. 8), is a distinguishing feature of advanced *Gopherus,* and again the result of developmental retardation. Future phylogenetic studies of gopher tortoises would benefit from a more acute awareness of the interplay between size, shape, development, and morphological transformation.

CONCLUSIONS

The extant tortoises of North America are but a remnant of their former diversity and distribution. As we have illustrated (fig. 1.6), this group is represented by two very distinct lineages whose morphology, behavior, and ecological characteristics attest to substantial independent histories.

Indeed, the fossil record (examined in more detail in chapter 2) documents the two groups as separate entities by at least the late Oligocene, considerably earlier than current estimates based on "molecular clocks." No other group of land tortoises possesses such a rich fossil record, in part because the gopher tortoises' burrowing habits increase the chances of preservation. Beginning in the middle Eocene and surviving into the present, the geologic span of gopher tortoises in North America very nearly equals that of the entire known history of the tortoise family (Testudinidae). Increasingly, the fossils are associated with very reliable geochronologic ages. These facts, coupled with increasingly sophisticated molecular tools for exploring the genetic relatedness and phylogeography within and between the two lineages (chapter 3), cast gopher tortoises as one of the most promising reptilian models with which to explore deeper patterns of morphological, biogeographical, and paleontological history. Below we list a few of the more important directions for future studies.

New Phylogenetic Analyses

We fully expect that new phylogenetic analyses using more appropriate outgroups, as well as character assessments informed by early (primitive) North American tortoises (e.g., *Hadrianus*) and the recognition of significant homoplasy, will unquestionably resolve *Xerobates* as a natural, monophyletic assemblage.

Calibration of Molecular Clocks

It is apparent that molecular estimates of the time of divergence of *Gopherus* and *Xerobates* are too recent, based on direct evidence from the fossil record (fig. 1.6). Recalibration of the existing molecular clock, based chiefly on data from distantly related turtle groups (Avise et al. 1992), would be a worthwhile exercise if it instead capitalized on well-dated fossil gopher tortoises to estimate evolutionary rates.

Historical Biogeography

The curious distribution of modern gopher tortoises has long been a matter of speculation (Bramble 1982, Morfka 1988,

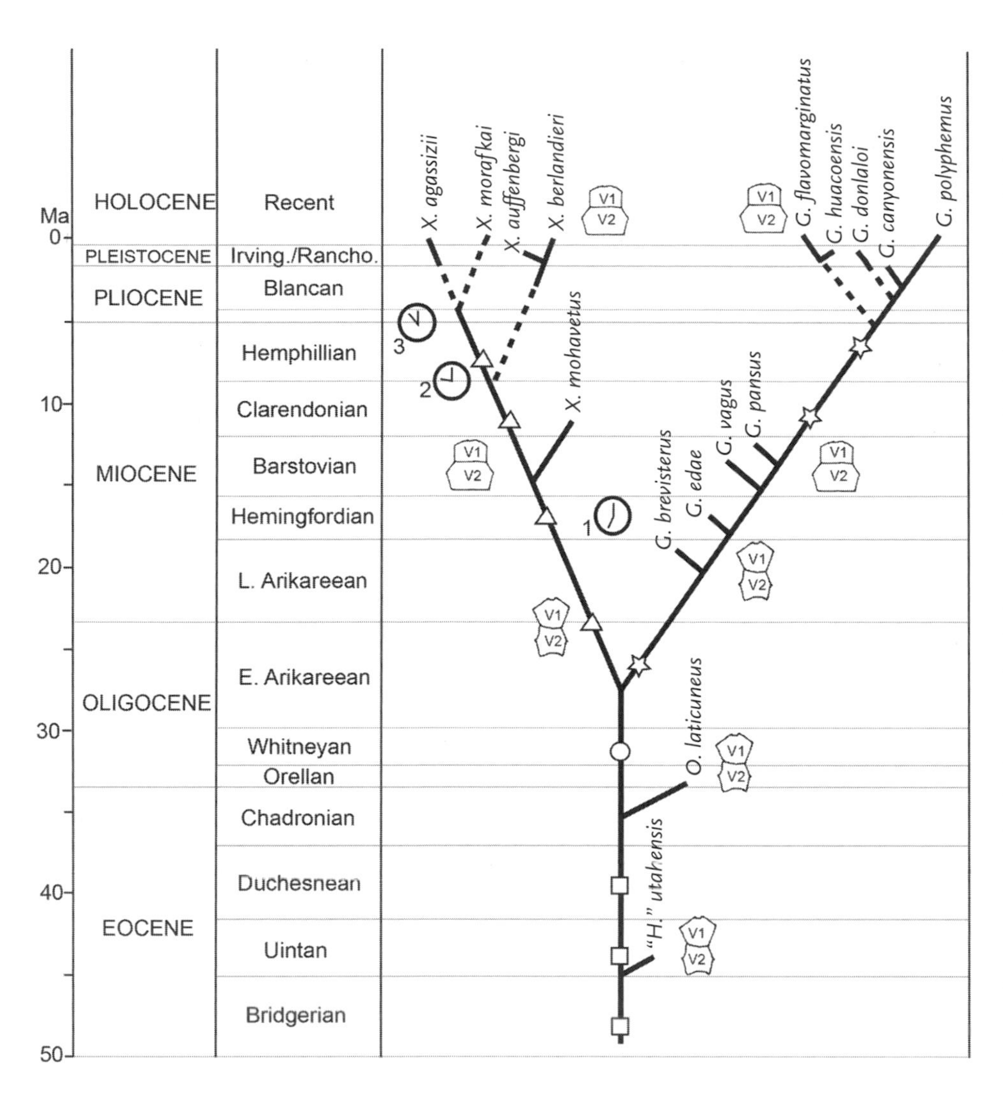

Fig. 1.6. Chronologic distribution and proposed relationships of gopher tortoises. Clock faces indicate estimated molecular divergence times: *1,* Divergence of *Gopherus* and *Xerobates* lineages at 17 Ma from Lamb and McLuckie (2002); presence of *Gopherus* group at 27 Ma (see chap. 2) indicates molecular date is too young. *2,* Suggested divergence of Sinaloan population of *X. morafkai* from other *morafkai* and *X. agassizii,* after Lamb and McLuckie (2002); divergence of *X. berlandieri* must precede this event. *3,* Divergence of *X. agassizii* and *X. morafkai* at 5 Ma from Lamb and Lydeard (1994). Symbols on the lines indicate presence of undescribed records of gopher tortoises: rectangle = Eocene taxa, circle = Oligocene taxa, star = *Gopherus,* triangle = *Xerobates* and Pacific slope records. V1, V2 depict the first two vertebral scutes (as in fig. 1.5). For discussion of the meaning of land mammal ages (e.g., Brigerian, Uintan, etc.), see chapter 2.

Reynoso and Montellano-Ballesteros 2004). A synergistic coupling of phylogeographic and paleontological approaches will move us closer to a resolution. The origin of *X. berlandieri* is among the most problematic issues, since its distribution is interposed between sister taxa of *Gopherus* and far from its closest living relatives, *X. agassizii* and *X. morfkai* (figs. 1.1, 1.2, 1.3). Various dispersal scenarios have attempted to derive *X. berlandieri* from an *X. agassizii*-like ancestor spreading eastward through the southwestern U.S. (Bramble 1982, Reynoso and Montellano-Ballesteros 2004). Although Late Pleistocene *X. agassizii* occurs in New Mexico and Texas (McCord 2002), its presence is much too late to be associated with the origin of *X. berlandieri.* An *X. berlandieri*-like *Xerobates* in Pleistocene deposits from costal Sonora hints that the distant ancestry of *X. berlandieri* might be in the Pacific Southwest and that its present range involved dispersal through southern, not northern, Mexico (Hutchison and Bramble, in preparation). Undescribed fossils from the southeastern U.S. suggest that even the biogeographic history of *Gopherus* may have heretofore unappreciated dimensions north of Mexico (chapter 2).

A poorly documented fossil record hampers the longer-term history of *Xerobates.* Late Miocene *X. mohavetus* (Merriam 1919) from southern California is the only pre-Pleistocene species generally recognized, but some analyses (e.g., McCord 1997, 2002; Reynoso and Montellano-Ballesteros 2004) have implied that it may instead be a primitive sister group to all gopher tortoises or not a gopher tortoise at all. Unfortunately, these assessments of *X. mohavetus* are compromised by the inclusion of hypothetical character states (cranial, carpal, pedal) for which no actual fossil materials seem to exist (Crumly 1994, personal communication). Additional new species of *Xerobates* from southern California and adjacent Mexico remain to be described.

Otolithic Ears

By far the most striking morphological feature of *Gopherus* is its unique inner ear, containing the largest saccular otolith of any living tetrapod. Combined neuro-physiological and behavioral studies will be required to tease out the adaptive significance of these bizarre sensory structures. One obvious question is how such tortoises avoid over-stimulation of the associated hair cells when voluntarily subjecting the head to high linear accelerations during such routine behaviors as feeding, digging, and courtship gestures. The physiological responses of gopher tortoises to a fossorial lifestyle have been studied with respect to thermoregulation, water balance, and respiratory gas exchange (chapter 4), but sensory mechanisms have so far escaped serious investigation.

Acknowledgements

We extend our sincere thanks to the following individuals for providing information that was essential to the completion of this article: Charles R. Crumly (Taylor and Francis Group, Publishers), Richard Franz (Florida State Museum), John Legler (University of Utah), and Kate Wellspring (Beneski Museum of Natural History, Amherst College).

2 The Fossil Record for North American Tortoises

RICHARD FRANZ

Gopher tortoises are terrestrially adapted turtles in the family Testudinidae. The modern members of these specialized tortoises are grouped into two lineages, *Gopherus* Rafinesque 1832 and *Xerobates* Agassiz 1857. They are represented by the following living species, one in the Southeast (*G. polyphemus*) and four in the Southwest and Mexico (*G. flavomarginatus, Xerobates agassizii, X. morafkai*, and *X. berlandieri*). *Xerobates morafkai* recognizes the distinctive populations of *X. agassizii* east and south of the Colorado River; other populations from Mexico may eventually be recognized as distinct (see Murphy et al. 2011). The genus *Scaptochelys*, erected for *X. agassizii* and its relatives by Bramble (1982), was placed in the synonymy of *Xerobates* by Bour and Dubois (1984). I follow the three genera arrangement—*Oligopherus* (extinct genus), *Gopherus*, and *Xerobates*—for gopher tortoises, as presented by Bramble and Hutchison (chapter 1).

Gopher tortoises evolved in North America, and their fossil record is restricted to this continent. Bramble (1971) suggested the middle Eocene tortoise *Hadrianus utahensis* Gilmore 1916 as a possible ancestor. Gopher tortoises are well-represented in the fossil record, with the number of recognized fossil taxa changing over time, depending on interpretations. Auffenberg (1974), for example, listed 24 taxa in his interpretation of *Gopherus*, with fossils ranging in ages from the late Eocene through the Pleistocene. Another nine taxa have been added to this list since then. Recent phylogenetic studies have reduced the number to 10–12 taxa (Crumly 1994; McCord 1997, 2002; Reynoso and Montellano-Ballesteros 2004). Each new accounting presents slightly different lists and taxonomic alignments. I recognize 16 taxa in this paper (shown with asterisk in the annotated list), more in line with Bramble's assessments (1971, 1982). Bramble's list and those acceptable taxa listed here are based on similar typological approaches, with a view toward functional and ecological morphologies. I project that the numbers of taxa will increase as the fossil gopher fauna becomes known more completely. This prediction, in part, is based on continual reinterpretations of named fossils and new discoveries from fieldwork and museum collections.

ORIGINS OF NORTH AMERICAN TORTOISES

The *Hadrianus* line was derived from Old World pond turtles (family Geoemydidae, formerly Bataguridae) in Asia, possibly as early as the earliest Eocene, 50–55 million years ago (Ma) (Gaffney and Meylan 1988, Shaffer et al. 1997, and others). Auffenberg (1974) considered North American *Hadrianus*-like tortoises close to living Asian brown tortoises in the genus *Geochelone* (subgenus *Manouria*). This assignment, however, is controversial (see chapter 1 for a discussion of their interpretations).

The lineage first appeared in North America during the early Eocene, with *Hadrianus majusculus* Hay 1904 representing the most ancient named tortoise reported from this continent (Auffenberg 1974). Exactly how *Hadrianus* arrived here from the Old World is open to speculation, but their sudden appearance in North America suggests that they came from elsewhere. Discoveries of fossil *Hadrianus* in early Eocene beds on Ellesmere Island in the High Arctic, near Greenland, suggest entry from Eurasia via northern latitude routes (Estes and Hutchison 1980). The presence of hundreds of stumps and roots of large buttressed trees in early Eocene coal beds on Ellesmere Island indicates the passage occurred when this polar region experienced dense riverside forests, high rainfall, and warm temperate climates (Francis 1988). These northern portals apparently closed when climates deteriorated near the Eocene-Oligocene boundary (Zachos and Kump 2005, Lear et al. 2008, Lui et al. 2009).

Climates changing toward drier conditions and the expansion of the prairie environment in the late Eocene of North America (see Retallack 1997) may have been the impetus

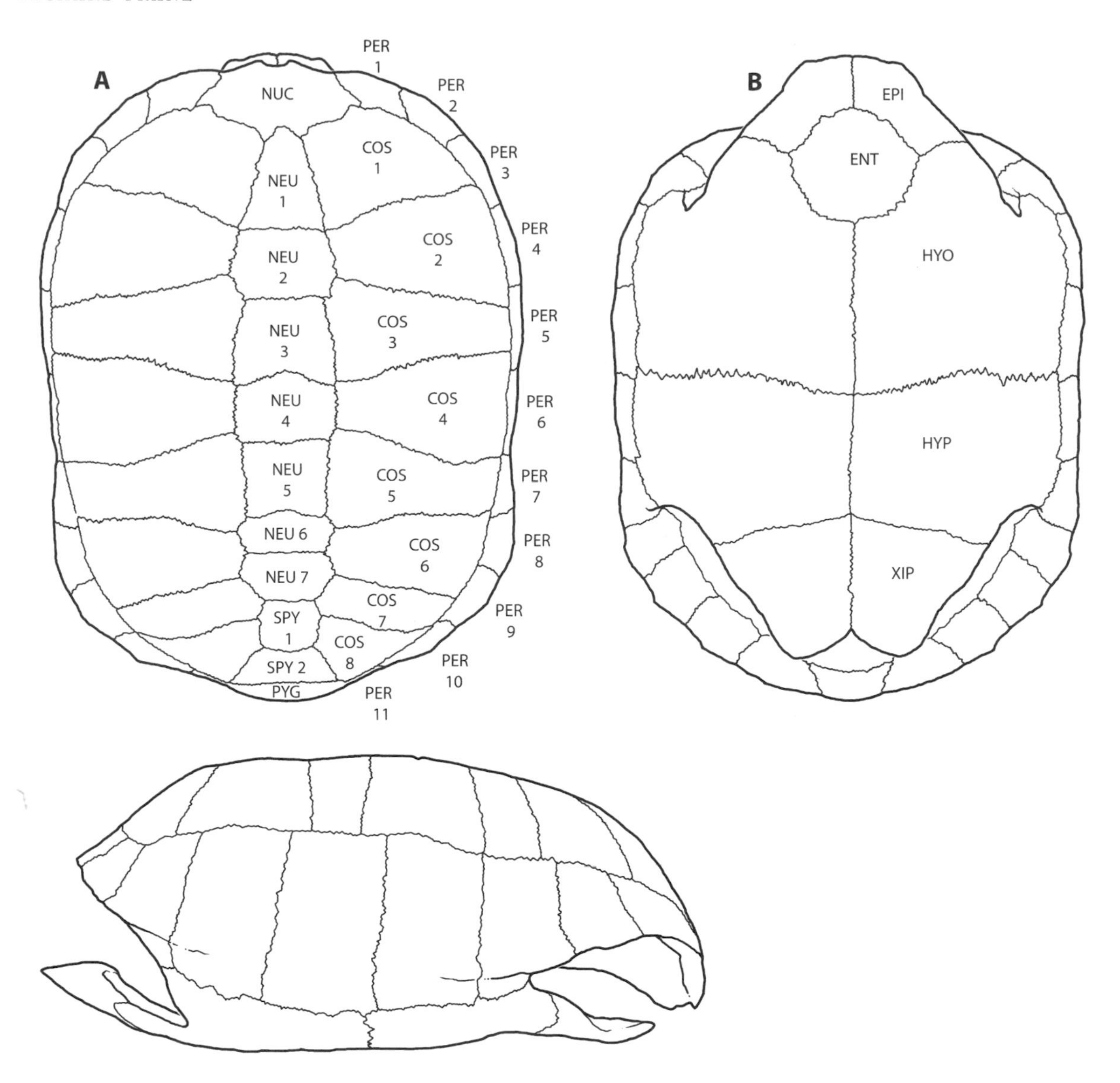

Fig. 2.1. (*above and opposite*) Shell morphology. *A,* carapace bones: NUC = nuchal plate, NEU = neurals, COS = costals, PER = peripherals, SPY = suprapygals 1 and 2, PYG = pygal (not visible). *B,* plastron bones: EPI = epiplastron, ENT = entoplastron, HYO = hyoplastron, HYP = hypoplastron, XIP = xiphiplastron. *C,* carapace scutes: CER = cervical (also underlap scute), VER= vertebrals, PLE = pleurals, MAR = marginals (12th is the posterior terminal marginal). *D,* plastron scutes: GUL = gulars, HUM = humerals, PEC = pectorals, ABD = abdominals, FEM = femorals, ANA = anals, AXI = axillaries, ING = inguinals.

for shifts from archaic forest tortoises toward more modern lineages, such as *Gopherus* (and their relatives), *Stylemys,* and *Hesperotestudo.* These new tortoises were probably better equipped to withstand the new aridity that was changing the North American landscape.

TURTLES AS FOSSILS

Identification of fossil turtles is based largely on shell and scute morphology, skull features, and to a lesser extent other skeletal parts. Complete shells or portions of them are among the most common fossils encountered in North American Cenozoic sites. The size, robustness, and sheer numbers (59) of elements in a turtle shell account, in part, for its persistence.

Shell

Turtle shells are composed of the carapace (top shell) and plastron (bottom shell). The two parts are connected to one another by specialized peripheral bones on either side of the shell, collectively known as the bridge. The carapace and plastron consist of separate bones (50 and 9, respectively) that are joined together along bone sutures. Each bone has a distinctive shape, often with growth rings (alternating ridges) or other surface textures, and is easy to identify. Shell bones have specific names that permeate the fossil literature (fig. 2.1A, B). Be aware, though, that bone anomalies are common in the shells of gopher tortoises (see discussion in Auffenberg 1976).

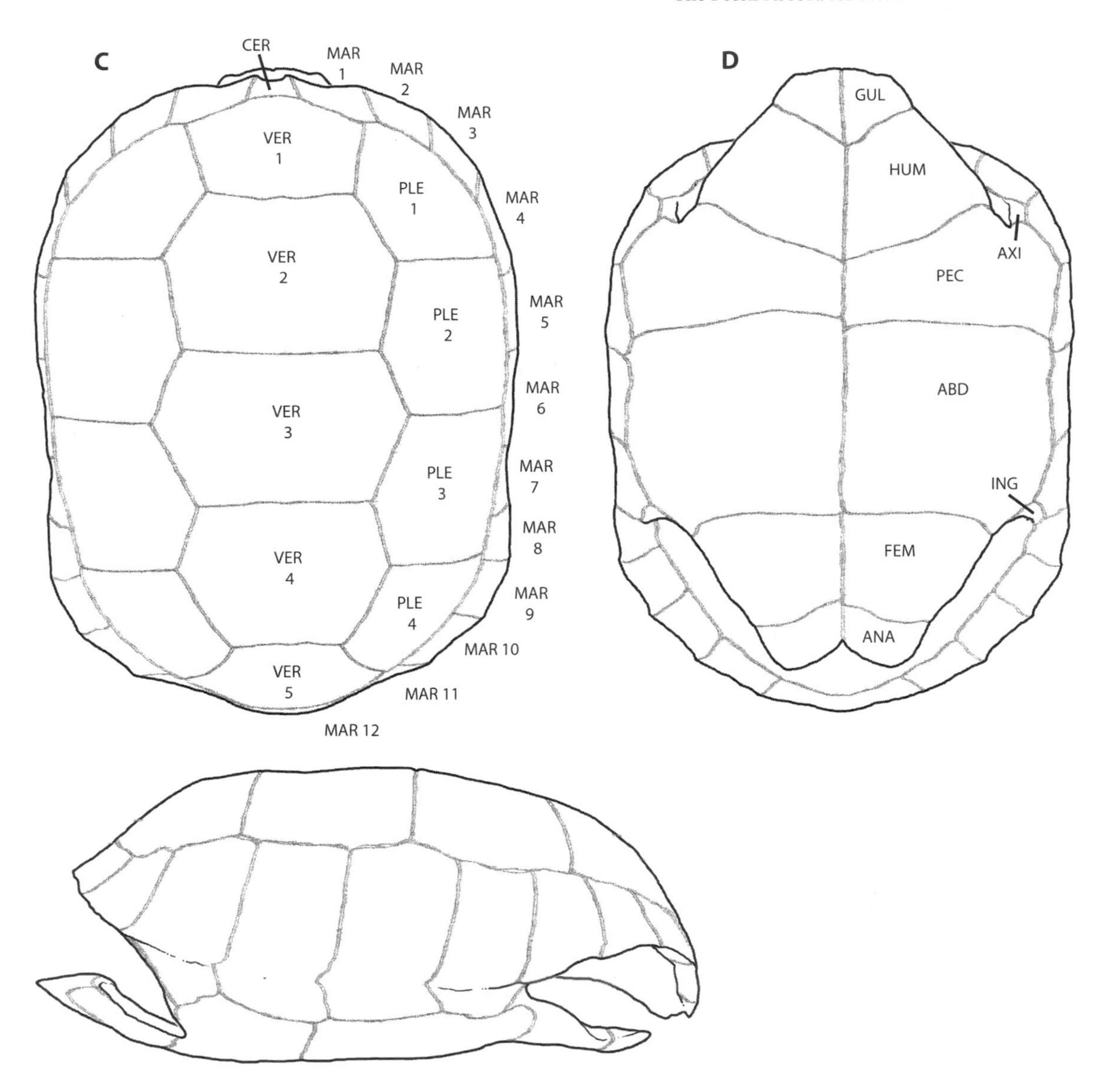

Scute Pattern

The exterior surfaces of the shell in living turtles are covered with horny scutes (or scales). They mostly overlap the bone sutures, which tend to strengthen suture connections where bones meet. Scutes can protect the shell from damage and provide crypsis as a deterrent against predators. The scute margins are embedded in channels on the external surfaces of the shell bones. These channels are prominently displayed after the animal has died and the scutes have fallen away. The channels, termed sulcus (sing.) and sulci (pl.), delineate the former margins of the scutes even on fossil specimens. Like the bones, scutes have specific identifying names (fig. 2.1C, D). The shape, size, and relative position of bones, scutes, and scute sulci are important in identification.

Skull

Differences in dimensions of skulls, relative shape and size of individual skull bones, size and position of processes on bones, fossa (pl. fossae; depressions in bones) for connective tissue attachments, foramen (pl. foramina; holes) for passages of blood vessels and other tissues, construction of the inner ear, conditions of the palate, and development of masticatory surfaces have been used to separate North American turtle genera and major divisions within tortoises (see Bramble 1971, Auffenberg 1976, Gaffney 1979). Finding fossil turtle skulls, however, are rare events. This rarity may be related to the removal of turtle heads as one of the first actions in food-scavenging practices of vultures and other scavengers (D. Jackson, personal communication). This behavior is likely the basis for the lost-skull mystery that has plagued paleontologists studying fossil turtles for decades.

Other Bones

Skeletal bones, particularly the cervical and caudal vertebrae, pectoral girdles, humerus, and manus, are cooperating factors in the placements of fossil tortoises in a hierarchical classification. Most of these bones in tortoises are illustrated and reviewed in Auffenberg (1976); structural features of the cervical vertebrae, in Williams (1950b); and the carpel fusion and manus arrangement, Auffenberg (1966a, 1976), Bramble (1971, 1982), and Crumly (1994). Bramble's carpel depiction appears the most accurate of the three.

INTERPRETING THE FOSSIL RECORD

Paleontologists use different time scales to communicate events in Earth's history. The framing of these scales often uses the first appearance and/or final disappearance of fossil species. Index fossils could be plankton, mollusks, mammals, or other taxa that have well-constrained geologic histories. Some scales also incorporate major geologic or climatic events, such as glacial advances or sea level changes. The most common scale divides Earth history into eras, periods, and epochs (e.g., Cenozoic Era, Tertiary Period, and Pliocene Epoch). This global time-telling device, developed originally for fossil mollusk beds in Europe, is still popular, taught widely in classrooms, and used in scientific journals, books, magazines, and museum exhibit interpretations. A second device, the North American Land Mammal Age (abbreviated here as NALMA), is based on fossil mammal taxa and often appears in journal articles that may or may not include mammal subjects. Other fossil mammal–based measures are employed elsewhere in the world. The various scales are constantly being readjusted to accommodate new information; examples of more recent adjustments for the Cenozoic include Hulbert (2001) and Woodburne (2004); for the Plio-Pleistocene, Gibbard et al. (2010). The global and NALMA systems, along with real time in millions of years ago (Ma), are used in this paper to count time (fig. 2.2).

Tortoise systematics had remained relatively stable since Ernest E. Williams established the modern conceptual framework for all tortoises (Williams 1950a, 1952; Loveridge and Williams 1957), although the taxonomic arrangement within the gopher tortoise concept and the origins of this group has continued to generate controversy. I recommend that readers consult the dendrograms by Bramble (1971, 1983) and phylogenetic consensus trees by Crumly (1994), McCord (1997) (incorporating a stratocladistical component) (all three summarized in McCord 2002 and Reynoso and Montellano-Ballesteros 2004) (fig. 2.3) to view the range of available gopher tortoise phylogenies. Those taxa in the annotated list (below) marked with an asterisk are considered valid species of gopher tortoises in this treatment.

Hutchison (1996) proposed a different hypothesis for gopher tortoise systematics than those previously available. He established the subgenus *Oligopherus* for the fossils belonging to the *Gopherus laticuneus* group. He listed 15 character states that demonstrated the uniqueness of *laticuneus*. Extending this concept, Bramble and Hutchison (this volume) recommended the use of three gopher tortoise genera: *Oligopherus*, *Gopherus*, and *Xerobates*. I have used this approach here since this alignment allows for a measure of parity between various gopher tortoise units. This arrangement has not been specifically tested by a phylogenetic analysis.

Molecular data suggested that the split between *Xerobates* and *Gopherus* occurred more than 17 Ma (Lamb and Lydeard 1994); the split between the Sonoran and Mojave populations of desert tortoises was estimated at 5 Ma (Lamb and McLuckie 2002). The problem is that the fossil record for *Xerobates*, as currently portrayed, does not reflect its projected divergence times; the oldest known fossil *Xerobates* is estimated at about 500,000 years old (McCord 2002). If the Barstovian-aged *Testudo mohavetus* Merriam should represent *Xerobates* as suggested by DM Bramble (personal communication), then this fossil would help close the age gap for divergence.

PHYLOGENY

Family *Testudinidae* Batsch 1788, Subfamily *Xerobatinae* Gaffney and Meylan 1988

GENUS *OLIGOPHERUS* HUTCHISON 1996

Content: *laticuneus* and related fossil populations.

Field Characters: Late Eocene–Early Oligocene. Low-domed shell, shell bones moderately thin, epiplastron flattened with a very shallow excavation, gular shelf absent, gular elongate with toothed margins, entoplastral sculpture absent, cervical scute wide as long often wedge-shaped, vertebral strut to nuchal plate absent, xiphiplastral apices flange-like with toothed margins.

GENUS *GOPHERUS* RAFINESQUE 1832

Content: *atascosae, brevisternus, canyonensis, donlaloi, edae, flavomarginatus, hollandi, huecoensis, pansus, pertenuis, polyphemus, praecedens, vagus.*

Field Characters: Late Oligocene–Present. Low-domed shell, shell bones moderately to very thin, epiplastron strongly excavated, gular shelf prominent, gular rounded without prominent notching, entoplastral sculpture present, cervical scute wider than long usually rectangular, vertebral strut to nuchal plate present, xiphiplastral apices rounded or tapered but not toothed. Entoplastral sculptures and nuchal scars need to be substantiated for many species in the *Gopherus* group. The examination of these features requires exposed interior shell surfaces within the body cavity.

GENUS *XEROBATES* AGASSIZ 1857

Content: *agassizii, auffenbergi, berlandieri, lepidocephalus, mohavetus, morafkai.*

Field Characters: Middle Pleistocene (or Late Pliocene?) to Present. Higher-domed shell (particularly *G. berlandieri*), shell bones very thin, epiplastron strongly excavated, gular shelf pronounced, gular elongate with a prominent midline notch, entoplastral sculpture present but faint, cervical scute wider than long usually rectangular, vertebral strut to nuchal plate absent, xiphiplastral apices short and not toothed.

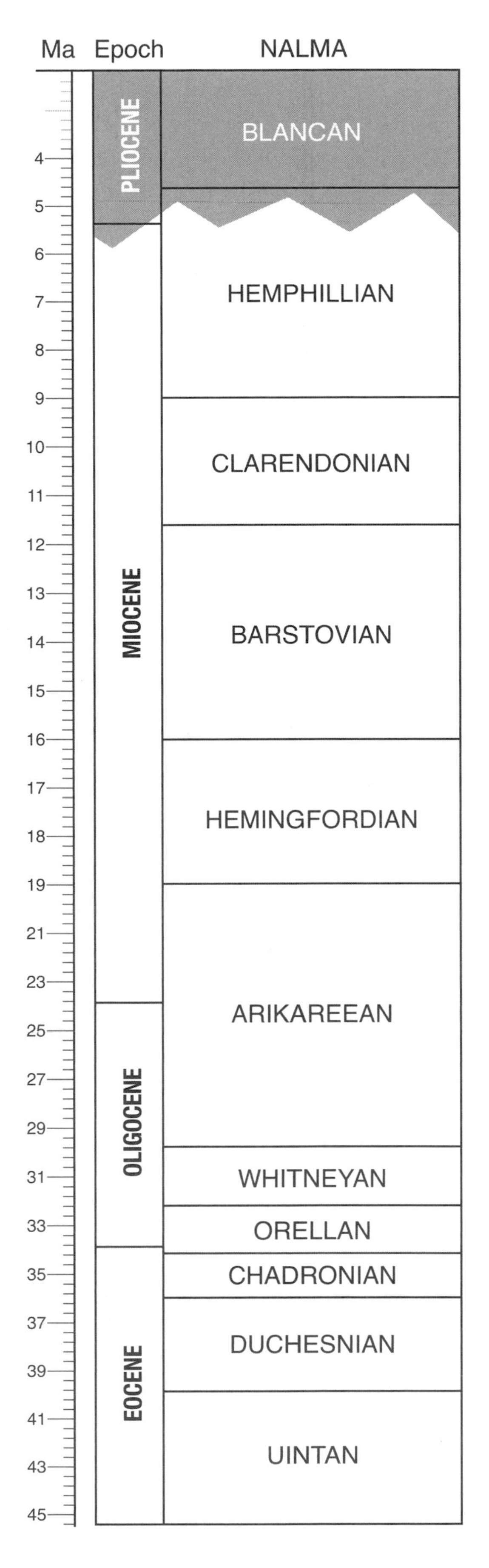

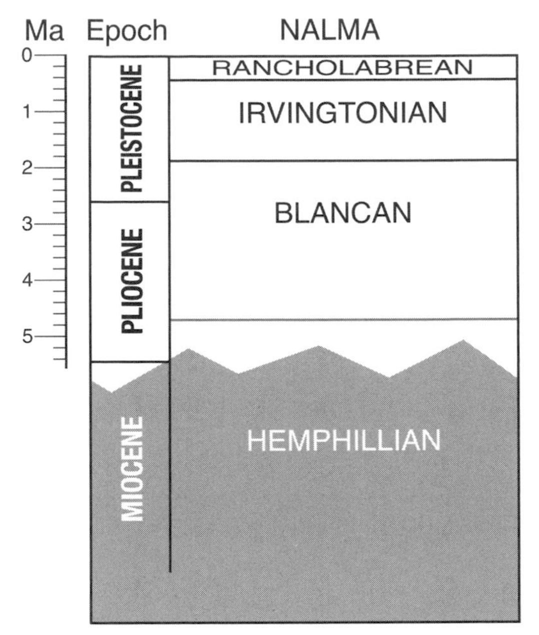

Fig. 2.2. Time scales. Traditional global scale (Epochs) and North American Land Mammal Ages (NALMA) presented in the context of real time in Ma (millions of years ago). Modified from Hulbert (2001), Woodburne (2004), and Gibbard et al. (2010).

UNASSIGNED OR QUESTIONABLE TAXA

Content: *brattstromi, copei, dehiscus, emiliae, hexagonata, laticaudata, milleri, neglectus, pargensis, undabunus.*

These taxa have been attributed to the genus *Gopherus* by Auffenberg (1974) and others, but their generic or specific status remains in question. This grouping includes species that either lack the necessary morphological information to assign them to one of the three known gopher genera or are candidates for placement in nongopher tortoise genera (e.g., *Stylemys* or *Hesperotestudo*). Rather than make an arbitrary decision of generic epithet, I have used the original description names for these fossils in this section.

ANNOTATED LISTING OF FOSSILS

A list of fossils attributable to *Gopherus* is provided below. Its composition shows a procession of taxonomic changes, as interpretations of gopher tortoise relationships evolved over the last 60 years.

Cataloguing North American fossil tortoises began with Hay (1908), who summarized the then-known menagerie of fossil species in his treatise *Fossil Turtles of North America*. He was followed by Williams (1950a, 1952) and Loveridge and Williams (1957) with their conceptual framework for tortoise systematics that is still in use today.

In 1974, Walter Auffenberg published his comprehensive catalogue of the fossil tortoises of the world, and it has become the starting point for all later studies. So, I have used it here as the entrance point for this review of fossil gopher

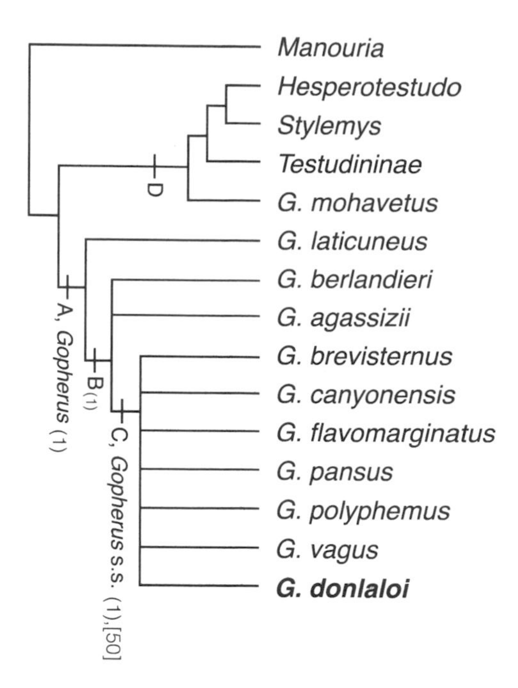

Fig. 2.3. Phylogenetic relationships of *Gopherus* (Reynoso and Montellano-Ballesteros 2004), showing *Manouria*, position of *G. mohavetus* in relation to *Stylemys, Hesperotestudo*, and Testudininae, and the basal position of *laticuneus* in regard to the *Xerobates* group and *Gopherus*. Bolded *G. donlaloi* = most recently described fossil species.

tortoises. Auffenberg listed 24 fossil tortoises in the genus *Gopherus*. An additional nine taxa supplement Auffenberg's original list, which brings the combined list attributable to *Gopherus* to 33 taxa. Some of the fossil names in the following compilation are no longer thought valid and have been placed in synonymy with other species or other genera, or are impossible to diagnose based on available materials. Taxa marked with an asterisk in the following list are considered valid members of gopher tortoise genera.

Fossil accounts are annotated briefly. Acronym and complete institutional names associated with type specimens are listed with first use in the following list; only the institutional acronym is employed afterward. The fossil ages and geologic units in this annotated list are mostly drawn from Auffenberg (1974), although in a few instances ages and geological names have changed in the interim. These newer interpretations are presented in Hunt (2002), Tedford et al. (2004), and Bell et al. (2004).

Genus *Oligopherus*

LATE EOCENE / EARLY OLIGOCENE; CHADRONIAN/ORELLAN/WHITNEYAN

**Oligopherus laticuneus* (Cope). *Testudo laticunea* Cope 1873. AMNH (American Museum of Natural History) 1160, Carapace, pelvis, and limb elements. Colorado: Head of Horsetail Creek, Weld County. Known from White River badlands of Colorado, Nebraska, South Dakota, and Wyoming.

Oligopherus praeextans (Lambe). *Testudo praeextans* Lambe 1913. NMC (National Museum of Canada) 8401. Shell. Wyoming: Sage Creek, Niobrara County. Listed as a synonym of *Gopherus laticuneus* by Hutchison (1996) and others, but not by Auffenberg (1974). See Gilmore (1946) for the best description of this tortoise.

Oligopherus quadratus (Cope). *Testudo quadrata* Cope 1884. AMNH 1149, Colorado: head of Horse Tail Creek, Weld County. Listed as *Geochelone quadrata* by Auffenberg (1974); as a synonym of *Gopherus laticuneus* by Hutchison (1996) and others.

Oligopherus thomsoni (Hay). *Testudo thomsoni* Hay 1908. AMNH 394, South Dakota: Corral Draw, Ziebach County, "lower Oreodon beds" (Orellan). Listed as *Geochelone thompsoni* (spelling error) by Auffenberg (1974); as a synonym of *Gopherus laticuneus* by Hutchison (1996) and others.

Genus *Gopherus*

MIDDLE–LATE PLEISTOCENE: IRVINGTONIAN/ RANCHOLABREAN/HOLOCENE

**Gopherus donlaloi* Reynoso and Montellano-Ballesteros 2004. Republic of Mexico: IGM (Institute de Geologia, Universidad Nacional Autonoma de Mexico) 6076. Shell, skull, and other bones. Municipio de Villagran, State of Tamaulipas. Close to *Gopherus polyphemus*.

**Gopherus flavomarginatus* Legler 1959. Living species. USNM (United States National Museum) 61253, mounted specimen. Republic of Mexico: "30–40 miles north from Lerdo," State of Durango. Fossils from the Cedazo local fauna, near Aguascaliente, State of Aguascaliente, Republic of Mexico (Mooser 1980). Other related fossils are listed from Rancholabrean sites in Arizona and Texas (McCord 1994, 2002) and New Mexico (gravel pits near Roswell) (RF obs); not *agassizii* as listed by Lucas and Morgan (1996).

**Gopherus polyphemus* (Daudin). *Testudo polyphemus* Daudin 1802. Living species. No designated type. Original location states "bords de la riviere Savanna et pres de l'Alatamah." Fossil *polyphemus* are reported from 52 Rancholabrean and Irvingtonian sites in Florida; six Rancholabrean sites in Georgia, Mississippi, and South Carolina; and 67 archaeological sites (including one Late Pleistocene site, 11,500 yrs BP) in Florida and Georgia (Franz and Quitmyer 2005).

Gopherus praecedens Hay 1916. FGS (Florida Geological Survey) 5463, left xiphiplastron. Florida: Vero Beach, St. Lucie County. Listed as a synonym of *G. polyphemus* by Auffenberg (1974).

Gopherus atascosae (Hay). *Testudo atascosae* Hay 1902. ANSP (Academy of Natural Sciences of Philadelphia), partial plastron. Texas: Atascosa County. Considered a synonym of *G. polyphemus* by Bramble (1971) and Reynoso and Montellano-Ballesteros (2004).

LATE PLIOCENE / EARLY PLEISTOCENE: LATE BLANCAN

**Gopherus canyonensis* (Johnston). *Bysmachelys canyonensis* Johnston 1937. PPHM (Panhandle Plains Historic Museum)

1534, skull, plastron, and partial appendular skeleton. Texas: North Cita (also spelled Ceta) Canyon, Randall County. Cita Canyon beds.

Gopherus huecoensis Strain 1966. TMM 40240-27, plastron and appendular skeleton. Texas: Madden Arroyo, Hedspeth County. Fort Hancock Formation. Considered a synonym of *Gopherus flavomarginatus* by Auffenberg (1974), McCord (2002), and others.

**Gopherus pertenuis* Cope 1892. TMM 40287-32, partial shell (now shattered). Texas: near Mount Blanco, Crosby County. Considered close to *G. canyonensis* and *G. laticaudatus* by Auffenberg (1974). The type specimen in the TMM collection has been shattered into small fragments, which no longer provides any useful information.

MIDDLE MIOCENE: BARSTOVIAN

**Gopherus pansus* (Hay). *Testudo pansa* Hay 1908. AMNH 5869, shell and pelvis. Colorado: north of Sterling, Weld County. Pawnee Creek Formation. Considered close to *brattstromi* (as *depressus*) and *mohavetus* by Brattstrom (1961). Listed in *Gopherus,* but with uncertain relationships by Reynoso and Montellano-Ballesteros (2004). Placed in the new definition of *Gopherus* (strong circumstantial evidence for the presence of the vertebral strut, DM Bramble personal communication).

**Gopherus vagus* (Hay). *Testudo vaga* Hay 1908. AMNH 1327, parts of three individuals. Wyoming: near Laramie Peak, Albany County. "Deep River" beds. Placed in *Geochelone (Hesperotestudo)* by Auffenberg (1974). Listed as *Gopherus* by Bramble (1971), Crumly (1994), McCord (2002); kept in *Gopherus,* but with uncertain relationships by Reynoso and Montellano-Ballesteros (2004). Placed in the new definition of *Gopherus* (strong circumstantial evidence for the presence of the vertebral strut, DM Bramble personal communication).

EARLY MIOCENE: LATE ARIKAREEAN

**Gopherus brevisternus* (Loomis). *Testudo brevisterna* Loomis 1909. AC (Amherst College) 2006, nearly complete skeleton. Wyoming: Muddy Creek, Laramie County. Upper Harrison beds. Hunt (2002) proposed a replacement name, Anderson Ranch Formation, for the Upper Harrison beds. This unit lies above the Harrison Formation and below the Hemingfordian-aged Runningwater Formation (Tedford et al. 2004).

**Gopherus edae* (Hay). *Testudo edae* Hay 1907. CM (Carnegie Museum) 1535, shell. Nebraska: near Running Water Creek, Sioux County. Harrison Formation. Best description in Hay (1908).

Gopherus hollandi (Hay). *Testudo hollandi* Hay 1908. CM 1561, shell. Nebraska: near Running Water Creek, Sioux County. Harrison Formation. Listed as a synonym of *Gopherus edae* by Bramble (1971).

Genus *Xerobates*

LATE PLEISTOCENE: RANCHOLABREAN/HOLOCENE

**Xerobates agassizii* Cooper 1861. Living species. (Mojave populations of desert tortoises). CSGS (California State Geological Survey). Three young specimens. "Mountains of California, near Fort Mojave." USNM 7888 (listed as cotype). Two syntypes deposited in CSGS (California State Geological Survey) lost. Type Locality restricted to "California, San Bernardino County; Mountains of California, near Fort Mojave; Soda Valley (very approximately 35° 6′ N, 116° 6′ W), by Murphy et al. 2011. More than 20 fossil sites in Arizona, California, Nevada, Texas, with oldest record 38,000 years BP (summarized by McCord 1994, 2002). Fossils from gravel pits near Roswell, New Mexico, probably late Pleistocene in age, listed as *agassizii,* by Lucas and Morgan (1996); now considered close to *flavomarginatus* (RF obs). Wisconsin and Holocene cave records from New Mexico summarized in Harris (1993). The fossil records from Arizona, New Mexico, and Texas presumably represent the Sonoran species.

Xerobates lepidocephalus (Ottley and Velazquez Solis 1989). Living tortoise. BYU (Brigham Young University) 39706. Sierra San Vincente region, approx. 1.5 km north of the Buena Mujer Dam, 20 km (by air) south of La Paz, Baja California Sur, Mexico (Ottley and Velazquez Solis 1989). Listed as synonym of *X. agassizii* by Crumly and Grismer (1994). No fossil record.

**Xerobates morafkai* (Murphy, Berry, Edwards, Leviton, Lathrop, Riedle 2011). *Gopherus morafkai* Murphy, Berry, Edwards, Leviton, Lathrop, Riedle 2011. Living species. (Sonoran populations of desert tortoises). CAS (California Academy of Science) 33867. Juvenile, Tucson, Pima County, Arizona. Fossil record of *X. morafkai* has not been separated from that of *X. agassizii.*

**Xerobates berlandieri* Agassiz 1857. Living species. Two syntypes, USNM 60 (2), collected by JL Berlandier. Type locality not specified. Not listed as a fossil by Auffenberg (1974). *X. auffenbergi* from the Irvingtonian of Aguascaliente (Mooser 1972) and a robust *berlandieri* from the late Pleistocene at Desemboque, Sonora, Mexico, are the only purported fossil occurrences of *berlandieri* (Lamb et al. 1989, McCord 2002).

EARLY/MIDDLE PLEISTOCENE: LATE BLANCAN OR IRVINGTONIAN

**Xerobates auffenbergi* (Mooser 1972). *Gopherus auffenbergi* Mooser 1972. New combination. Republic of Mexico: FC 500 (paratypes FC 501–504), Cedazo local fauna, Arroyo Cedazo and Arroyo San Francisco, few km southeast of the City of Aguascaliente, State of Aguascaliente. Cedazo local fauna is exposed in the Tacubaya Formation (Mooser and Dalquest 1975). Considered a synonym of *Xerobates berlandieri* (Crumly 1994, McCord 2002). Listed by McCord (2002) as the oldest known occurrence for the *Xerobates* lineage, considered to be 500,000 years old, but could be older, possibly as much as two million years old (Late Blancan) (Bell et al. 2004). The specific age of this fossil species and the accompanying fauna needs to be verified.

MIDDLE MIOCENE: BARSTOVIAN

**Xerobates mohavetus* (Merriam). *Testudo mohavense* Merriam 1919. UC (University of California) 21575, shell. California: Barstow syncline, Mojave Desert, San Bernardino County. Barstow beds. Spelling changed to *mohavetus* by Des Lauriers

(1965). Removed from the genus *Gopherus,* placed near *Stylemys, Hesperotestudo,* and Testudines, by Reynoso and Montellano-Ballesteros (2004), but its generic affinities were not clarified. I follow DM Bramble (personal communication) and include *mohavetus* in *Xerobates.*

Unresolved Taxa

MIDDLE PLEISTOCENE: IRVINGTONIAN

Testudo hexagonata Cope 1893. TMM 41412-3, partial shell (now in pieces). Texas: Equus beds on Rock Creek, Tule Canyon, Briscoe County. Tule Formation, Rock Creek local fauna. See Auffenberg (1962) for the best description of this fossil species. Westgate (1989) reported a mass occurrence of specimens in the late Pleistocene Beaumont Formation, Willacy County, Texas. *G. hexagonatus* is considered *nomum vanum* (empty name) by Reynoso and Montellano-Ballesteros (2004) (see Mones 1989 for discussion of term). The status of this gopher tortoise-like species needs to be verified.

Testudo laticaudata Cope 1893. TMM (Texas Memorial Museum) 41412-2, epiplastron and xiphiplastron. Texas: Equus beds on Rock Creek, Tule Canyon, Briscoe County. Tule Formation, Rock Creek local fauna. Considered a synonym of Gopherus *hexagonatus* by Auffenberg (1974). Listed as *nomem vanum* by Reynoso and Montellano-Ballesteros (2004).

LATE PLIOCENE / EARLY PLEISTOCENE: LATE BLANCAN OR IRVINGTONIAN

Gopherus pargensis Mooser 1980. TMM 41536-29, left side of posterior carapace. Republic of Mexico: Cedazo local fauna, near Aguascaliente, State of Aguascaliente (Mooser 1980). The Cedazo lf, listed as Irvingtonian by Mooser (1972). Considered mixed faunas, including Rancholabrean, Irvingtonian, and Blancan fossil components (Bell et al. 2004). The specific age of this fossil tortoise needs to be established.

MIDDLE MIOCENE: BARSTOVIAN

Gopherus brattstromi Auffenberg 1974. Substitute name for *G. depressus* Brattstrom 1961. CIT (California Institute of Technology) 498/5133, shell. California: Tchachapi Mountains, Kern County. Bopesta Formation, Cache Peak fauna. Closely related to *G. mohavetus* and *G. pansa,* possibly derived from a primitive form, such as *G. vagus* (Brattstrom 1961). Placed in *G. mohavetus* by Reynoso and Montellano-Ballesteros (2004).

Testudo copei Koerner 1940. YPM (Yale-Peabody Museum), shell. Montana: Meagher County. Deep River Formation, listed as early Barstovian by Tedford (2004). Close to *G. emiliae.* Placed in *Stylemys* by Bramble (1971).

Gopherus dehiscus Des Lauriers 1965. LACM (Los Angeles County Museum) 400(26)/5178, internal cast of shell. California: Cajon Pass, west end of Cajon Valley, San Bernadino County. Placed in *Hesperotestudo* by Bramble (1971), but retained in *Gopherus* by Auffenberg (1974) without comment.

Testudo milleri Brattstrom 1961. UCMP (University of California Museum Paleontology) 21574. California: Barstow syncline, Mojave Desert, San Bernardino County. Transferred to *Geochelone,* subgenus *Caudochelys,* by Auffenberg (1974). Listed as a synonym of *Gopherus mohavetus* by Reynoso and Montellano-Ballesteros (2004).

EARLY MIOCENE: LATE ARIKAREEAN

Testudo undabuna Loomis 1909. AC, shell. Wyoming: Muddy Creek, Laramie County. Upper Harrison Beds (= Anderson Farm Formation of Hunt 2002). Listed as *Stylemys* by Bramble (1971), not mentioned by later researchers.

LATE OLIGOCENE: EARLY ARIKAREEAN

Testudo emiliae Hay 1908. AMNH 6135, shell. South Dakota: Porcupine Creek. Lower Rosebud Formation. Placed in *Stylemys* by Bramble (1971).

Gopherus neglectus Brattstrom 1961. CIT 126/1672, shell. California: Key Quarry, Ventura County. Upper Sespe Formation. Listed as Whitneyan by Auffenberg (1974), but now considered Arikareean. Probably close to *O. praeextans* (Brattstrom 1961). Placed in *Stylemys* by Bramble (1971).

NEW FOSSIL DISCOVERIES

Recent inventories of North American museum collections revealed a wealth of unstudied gopher tortoise fossils from Arizona, Florida, Nebraska, New Mexico, Oklahoma, South Carolina, and Texas. At least some of them represent new taxa, and some will help fill in unexplained time gaps in the fossil record. The following list offers a few examples of the unstudied diversity from Florida and South Carolina. Many of these new specimens were not available to Walter Auffenberg in the 1970s. This unexplored diversity also shows the need for renewed participation in fieldwork throughout the southern United States, and particularly Mexico, in the quest for unique tortoise taxa. New discoveries in collections and in the field will challenge many commonly held paradigms concerning informative character states, phylogenies, and ideas of geographic origins.

LATE OLIGOCENE: EARLY ARIKAREEAN

Gopherus (South Carolina) species A. ChM (Charleston Museum) VP-7180, Charleston County, Dorchester Creek. Chandler Bridge Formation. This fossil is represented by a complete anterior carapace (fig. 2.5). The two most remarkable aspects presented by this specimen are its prominent vertebral strut scar on the back of the nuchal plate (only in *Gopherus*) and its early age and provenance. Its early Arikareean age appears not to be in question (A. Sanders personal observation). This fossil may challenge the geographic origins and timing of the *Gopherus* group.

EARLY MIOCENE: EARLY HEMINGFORDIAN

Gopherus (Florida) species B. UF (Florida Museum of Natural History), MCZ (Harvard Museum of Comparative Zoology), Thomas Farm, near Bell, Gilchrist County. A second tortoise species was recently discovered among samples of the tortoise *Hesperotestudo tedwhitei* (Williams) from Thomas Farm.

These fossils are the earliest known occurrences of *Gopherus* in Florida. They represent an undescribed gopher tortoise that lacks the vertebral strut scar, which clearly distinguishes the Thomas Farm species from the earlier strut-bearing *Gopherus* species A from South Carolina.

EARLY PLIOCENE: LATE HEMPHILLIAN

Gopherus (Florida) species C. UF, Polk County, Bone Valley, Whitten Creek local fauna. Very large species, known from a single anterior lobe of the plastron. This is the largest known gopher tortoise from the Southeast.

LATE PLIOCENE: LATE BLANCAN

Gopherus (Florida) species D. UF, Citrus County, Inglis IA and IC. A miniature species related to *G. polyphemus* (Franz and Quitmyer 2005).

THE MORPHOLOGICAL EVOLUTION OF BURROWING

Extant gopher tortoises use soil manipulation to create protective spaces to escape harsh surface conditions and reduce predation. *Xerobates berlandieri* and *X. morafkai* construct pallets, often associated with cacti or other armored plants and retreats in rocky outcrops or erosional cavities in gullies. *Xerobates agassizii, Gopherus flavomarginatus,* and *G. polyphemus* excavate tunnel-like structures that they use as homes on a regular basis. Their front limbs are the primary apparatus for digging these retreats.

Tunnels are usually single tubes, wide enough for tortoises to turn around anywhere along their length. Soils dug from these burrows are pushed to the surface by the bodies of the tortoises, as they move toward the surface. This soil accumulates around the outside of the opening to the burrow in a wedge-shaped mound, referred to as the apron.

The morphological evolution of gopher tortoises is linked to its burrowing habitus. Structural modifications of the skeleton and associated muscles provide tortoises with greater efficiency for digging. The shells of burrowing gopher tortoises are more flattened (less domed); anterior shell opening larger to accommodate the movement of the humerus when digging; shell bones, thinner; heads, more robust; eyes, more lateral and flush with the skin; the external nares, smaller; and tympanic membranes, covered with small scales (Bramble 1971 personal observation). The head and anterior lobe of the plastron (including the gular projection) and the extension of the elephantine hind limbs serve to hold the tortoise in position during excavation. The front limbs, one at a time, dig and move the soil. A bony strut formed by an extension of the neural spine of the first shell vertebra attaches to the under-surface of the nuchal plate (seen as a substantial bone scar on the nuchal in fossils) (fig. 2.4). The strut helps stabilize the position of the head and neck during excavation. The strut is unique to advanced burrowers in *polyphemus, flavomarginatus,* and closely related fossil species. The front limbs include subtle modifications of the humerus, fore bones, and carpal bones (illustrations in Bramble 1982). The manus is constructed to form a very broad short hand with large flat nails that becomes a highly efficient paddle-like "earth mover" (Bramble 1971). The interior surface of the entoplastron in both *Gopherus* and *Xerobates* has a distinctive interclavicular sculpture that consists of a pair of divergent brows, a strong midline keel that extends as a blade onto the hyoplastron, and a pair of shallow depressions on either side of the keel (fig. 2.4). Bramble (1971) postulated that this sculpture facilitated buttressing of the scapula within the scapula-interclavicular joint to compensate for the resistance of the soil that the front limb encountered during the power stroke. This sculpture, reminiscent of a bird face (see fig. 2.4), is found also in several nongopher tortoise genera, but the pattern of these sculptures is different for each group (see Franz and Franz 2009). The strut and entoplastral sculpture are absent in *Oligopherus* and unreported for other early gopher tortoises.

EXTINCTIONS AND THE LOSS OF TORTOISE DIVERSITY

Following their arrival, invading *Hadrianus*-like tortoises spread across the North American continent. Their fossils are reported from Alabama, New Mexico, Utah, and Wyoming (Bramble 1971, Auffenberg 1974). Modern genera—*Stylemys, Gopherus* (and relatives), and *Hesperotestudo*—evolved from them. None of these tortoises penetrated into South America. *Stylemys* became extinct in the middle Miocene. *Hesperotestudo* disappeared at the end of the Pleistocene. Gopher tortoises managed to survive into the modern age, but their ranges became contracted and populations allopatric. What happened? Changing climates? Changing sea levels? Disease? Other more subtle, undetected environmental parameters (in the case of Hesperotestudo)? Invading peoples? Or some combinations of these factors?

CONCLUSIONS

When conditions began drying out, following shifts in worldwide climates during the middle to late Eocene, lush forests gave way to drier woodlands, savannas, prairies, and possibly deserts. Tortoises found these emerging conditions favorable to their lifestyle, and they flourished and radiated. The genera *Stylemys, Oligopherus,* and *Hesterotestudo* all appeared about this time. This radiation of genera spawned multispecies groupings of tortoises. Initial groups consisted of species pairs (e.g., *Stylemys nebrascensis* and *Oligopherus laticuneus* in the late Eocene White River deposits in the West), and later groupings expanded into trios, which usually incorporated *Gopherus* and/or *Xerobates,* a small *Hesperotestudo* (turgida or incisa groups in the subgenus *Hesperotestudo*), and a large *Hesperotestudo* (in the subgenus *Caudochelys*) (Franz and Quitmyer 2005). For unknown reasons, large *Hesperotestudo* disappeared from tortoise groupings in the Southwest during the Irvingtonian; however, gophers and the small *Hesperotestudo* persisted there until the end of the Pleistocene. The trios existed

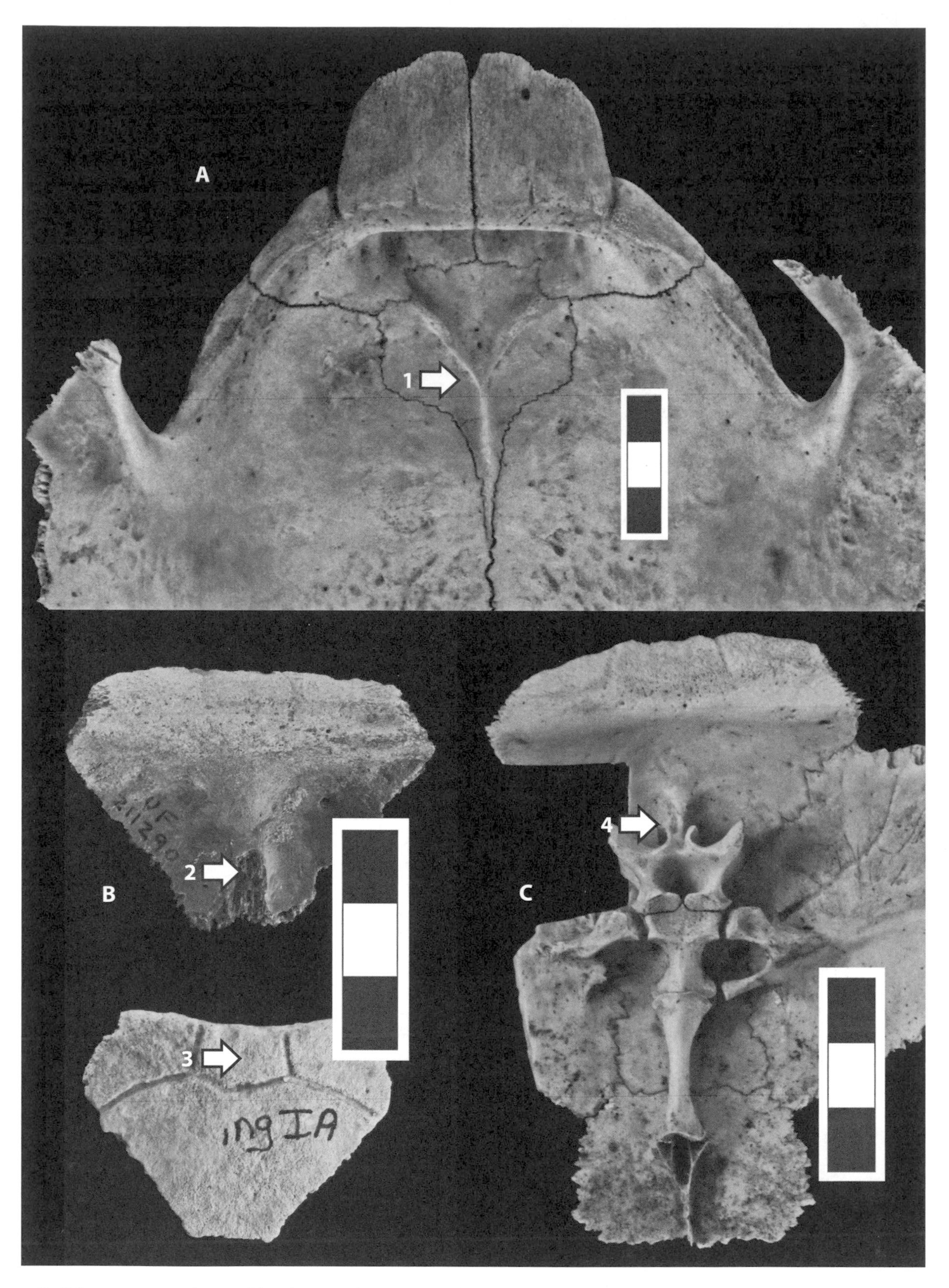

Fig. 2.4. Shell features. Internal view of the plastron of a recent *Gopherus polyphemus* showing: *A*, interclavicular sculpture (bird face) on the entoplastron *(1)*. *B*, vertebral strut scar *(2)* on the nuchal plate of a miniature *G. polyphemus* fossil from Inglis, Florida, and the wide cervical scute *(3)* typical in *Gopherus*. *C*, underside showing the bony vertebral strut from the first dorsal vertebra in a recent specimen of *G. polyphemus (4)*. Scale = 3 cm.

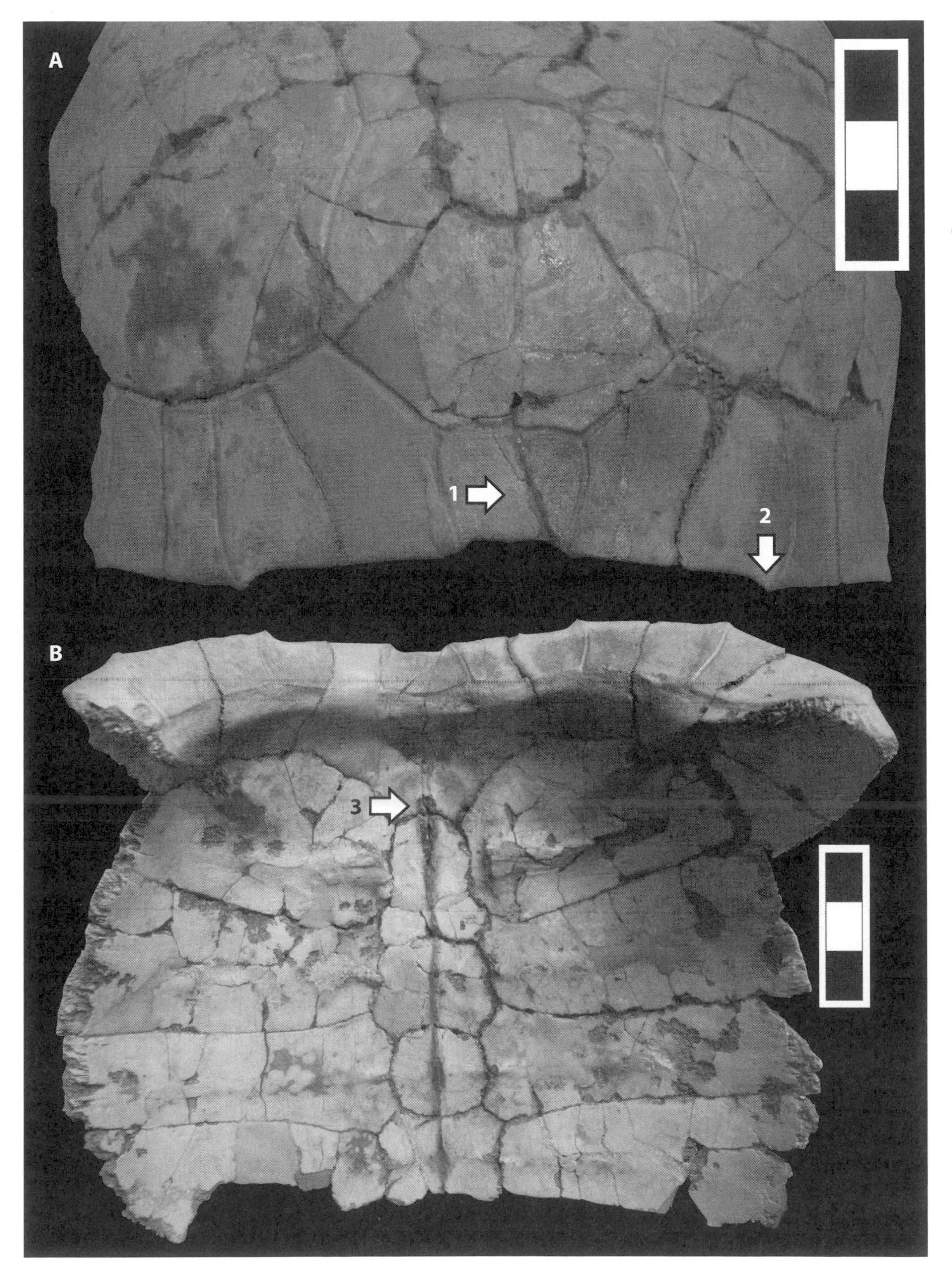

Fig. 2.5. New fossil *Gopherus* (A) from the Chandler Bridge Formation, Charleston, South Carolina. *A,* dorsal view of carapace of this 27-million-year-old tortoise, showing the wide cervical scute *(1)* and marginal sulcal spurs *(2)*, both definitive features of *Gopherus. B,* internal view of carapace, showing the vertebral strut scar on the distal edge of the nuchal plate, a definitive feature of advanced gopher tortoises in the *polyphemus* group *(3)*. Scale = 3 cm.

continually in the southeastern United States, as far west as eastern Texas, until the end of the Pleistocene, when all *Hesperotestudo* went extinct. The gopher tortoises *Gopherus* and *Xerobates* managed to survive into the modern age, but their ranges became contracted and populations allopatric.

The presence of multispecies groupings of tortoises in North America begs certain ecological questions. How were pairs or trios of tortoises able to coexist in the same geographic space and at the same time? Were the tortoise species occupying closely associated, but different, habitats, within a small area? Did they compete for food, protective night shelters, shaded daytime retreats, and nesting areas? How were their populations structured? Were body size and burrowing behaviors contributing factors to their successful co-occupation? Did the multispecies groupings partition the available forage? Did they compete with associated mammalian herds for graze? What effects did they have on local soils? Some of these questions can be resolved from the fossils themselves through examining death assemblages, associated fossil plant and animal taxa, evidence of predation (tooth marks on bone), and stable isotope analysis of the fossils to determine food types. Creative paleontologists continue to find new ways to extract information from these fossil accumulations.

For fossilization to occur, carcasses need to find their way into an environment suitable to allow preservation. These are often wind-blown accumulations, water bodies, or natural traps (e.g., sinkholes or caves) where carcasses can collect over time. For long-term preservation, carcasses need to be protected from weather and scavengers and be accessible for quick burial by appropriate geological processes. Locations of vertebrate fossil sites are unpredictable and are not common features across landscapes. Sites, even those in close proximity to one another, often sample different segments of time and have very different fossil faunas. Through them, we learn about prehistoric environments and the organisms that inhabited them. As a result, each time fossil sites are discovered they deserve the utmost care in excavations, with the fossils accessible for study and protected in perpetuity.

Acknowledgments

I thank Dennis Bramble, C. Kenneth Dodd, Gary S. Morgan, and David Steadman for their critical reviews of this manuscript; Jason Bourque for preparation of the tortoise shell diagrams; Nancy Albury for specimen photography; and Melanie Wegner for photo-conversions, figure layouts, and chart production. Permission was given by VH Reynoso to use figure 8 from his 2004 paper in the *Journal of Vertebrate Paleontology.*

3 Systematics of Extant North American Tortoises

ROBERT W. MURPHY

Systematics concerns the phylogeny and taxonomy of species, and usually the drivers of their evolution. The field spans from microevolutionary relationships, typically maternal genealogies, to macroevolutionary, interspecific associations, and biogeography. In terms of microevolutionary relationships, the five species of North American tortoises of the genus *Gopherus* receive unequal coverage. Whereas much data exist for *G. polyphemus* and *G. agassizii,* precious little exists for *G. berlandieri* and *G. flavomarginatus.* Moderate amounts of data are available for the most widespread species, *G. morafkai,* but only within the United States.

Two seemingly opposing camps pursue systematics: morphology and molecules (Hillis 1987). Nowadays, arguments over the superiority of one type of data over the other seem to involve problems of data-coding and limitations of alternative states (e.g., Murphy and Doyle 1998). If DNA underlies the morphology, as it should in systematics, then the dichotomy between morphology and molecules is an artificial construct based on the types of data systematists prefer to gather and analyze. This review primarily considers the contributions of molecular data to revealing the macro- and microevolutionary associations of the five to six species. Bramble and Hutchison (chapter 1 of the present volume) and Franz (chapter 2) present morphological and paleontological data, respectively.

Herein, I apply the evolutionary species concept of Simpson (1961; Wiley 1978). Monophyly is not required because introgressive hybridization results in species being "non-monophyletic." At a microevolutionary level, all sexually reproducing individuals are polyphyletic in having different paternal and maternal histories (Murphy and Méndez de la Cruz 2010). The microevolutionary null hypothesis is panmixia and genetic recombination and gene sorting quickly results in an intractable network of gene flow. In tortoises, we can only track the maternal history using mtDNA and I refer to these as genealogies. I use the term "phylogeny" exclusively for macroevolutionary relationships, where the null hypothesis is the absence of gene flow.

Nomenclature

Zoological nomenclature starts with the 10th edition of *Systema Naturae* (Linnaeus 1758), yet for *Gopherus polyphemus,* a review of older literature is required. Errors constitute a Hercules' knot. Herein, I base taxonomic determinations on the International Commission of Zoological Nomenclature (ICZN 1999). Fritz and Havaš (2007) provide an accurate synonomy, except for date of publication for *Xerobates agassizii* by Cooper, which is 1861 (Murphy et al. 2011). My review avoids abstracts published online.

EVOLUTIONARY RELATIONSHIPS: MICROEVOLUTION

It has taken almost 200 years to disentangle the history of the species (fig. 3.1). Errors arose from the start, and some are quite recent. Below, I review the species in their order of description.

Gopherus polyphemus (Daudin 1803) (gopher, gopher tortoise)

The taxonomy of *G. polyphemus* was once trapped "in a whirlpool of conflicting opinion and uncertainty" (True 1881: 434). The species first appeared as an inaccurate figure in Seba (1734–1765) and named *"Testudo tessellata minor carolininia."* The image, discovered by Linnaeus and reported in his 12th ed. of *Systema Naturae* (1766), was credited to *Testudo carolina.* Creating an error, Gmelin (1788: 1041) expanded on Linnaeus' description of *T. carolina* and in doing so diagnosed *Gopherus polyphemus* and not *Terrapene carolina.* This error was perpetuated by Le Conte (1829) and others. Schweigger (1812) placed the tortoise in the genus *Emys,* yet many of the early researchers assigned it to the genus *Testudo* (e.g., Gray 1831, 1855;

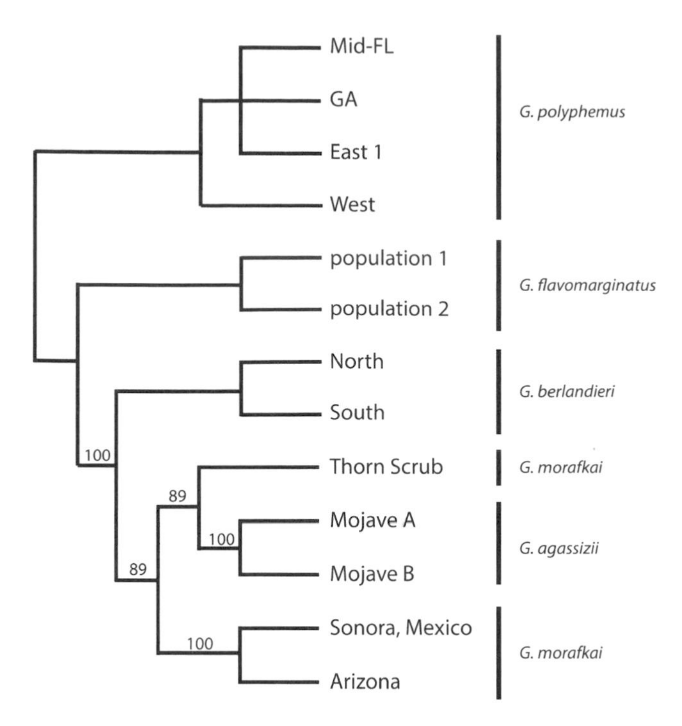

Fig. 3.1. Composite matrilineal genealogy of the genus *Gopherus*. Relationships for the *G. agassizii* species complex are from Edwards et al. (in press), *G. polyphemus* from Osentoski and Lamb (1995) and Clostio et al. (2012), *G. flavomarginatus* from Ureña-Aranda and de los Monteros (2012); and *G. berlandieri* from Fujii and Forstner (2010), assuming the nuclear DNA structure corresponds to matrilineal associations.

Holbrook 1836, 1842; Cope 1875). Apparently unaware of the true identity of *G. polyphemus* (Daudin 1803), Gray (1844) described *Testudo gopher* for the same species and, thus, created a junior synonym.

Agassiz (1857) erected a new genus for the tortoise, *Xerobates,* and Cope (1875), Gray (1870), Cooper (1861), and others followed this determination. Subsequently, Stejneger (1893) pointed out the existence of an older generic name, *Gopherus* Rafinesque 1832.

Today, *G. polyphemus* is a keystone species for many southern regions of the United States east of the Mississippi River (Auffenberg and Franz 1982; Kushlan and Mazzotti 1984; Diemer 1986, 1992a; Breininger et al. 1994). Historical female dispersion—the matrilineal genealogy (phylogeography)—and genetic structuring are reasonably well documented. Employing restriction length fragment polymorphisms (RFLPs), Osentoski and Lamb (1995) detected three major matrilines termed the Western, Eastern, and Mid-Florida assemblages (fig. 3.1). The former is the sister-group of the latter two. Their difference in terms of RFLPs hovers around 1%. Nucleotide sequence difference based on the mtDNA Control Region between the Eastern and Western lineages averages about 2.1% to 2.3% (Osentoski and Lamb 1995, Clostio et al. 2012), which is typical of intraspecific divergence. Osentoski and Lamb (1995) date this divergence to the Pleistocene, about 1.3 Ma yet Clostio et al. (2012) date it to 3–5 Ma. The latter estimate seems inaccurate because it does not correspond to associated Plio-Pleistocene events (Auffenberg and Milstead 1965, Osentoski and Lamb 1995); the clock-calibration needs to be evaluated. Regardless, the Apalachicola River seems to be an important determinant of matrilineal dispersal. The origin of the Mid-Florida assemblage may be a consequence of a historic ridge-island effect owing to the Brooksville Ridge. A novel matriline also occurs in western Georgia (Clostio et al. 2012; fig. 3.1). Much localized divergence occurs within the eastern lineage. Individuals from divergent lineages occur sympatrically, and this may be due to either naturally occurring parapatry or human translocations of tortoises.

Regarding nuclear gene assessments, Schwartz and Karl (2005) use nine polymorphic microsatellite loci to assess gene flow within southern Georgia and peninsular Florida. Their analyses discern eight breeding units, but they did not sample the Florida panhandle, Louisiana, Alabama, Mississippi, or South Carolina, i.e., the Western assemblage. Whereas Osentoski and Lamb (1995) define three assemblages using RFLP data within Florida, Schwartz and Karl (2005) identify five major units using microsatellites; different sampling strategies preclude a geographic comparison. Sinclair et al. (2010) and Richter et al. (2011) report the absence of genetic substructuring with a 45 km distance within central Florida and Mississippi, respectively. Genetic divergence involves greater distances. The only range-wide comparison (Clostio et al. 2012) obtained five genetic units with limited gene flow between them, especially across the Apalachicola River. Most genetic assessments (Osentoski and Lamb 1995; Schwartz and Karl 2005, 2008; Sinclair et al. 2010; Ennen et al. 2010; Tuberville et al. 2008; Richter et al. 2011; Ennen et al. 2011; Clostio et al. 2012) suggest gene flow among the populations but highly restricted levels of admixture (Clostio et al. 2012), which indicates that *G. polyphemus* may be speciating. This possibility requires further work. Translocations and relocations may be obscuring the natural historical patterns (Schwartz and Karl 2005, Clostio et al. 2012).

Gopherus berlandieri (Agassiz 1857) (Berlandier's tortoise, Texas tortoise)

The species has a larger distribution in Mexico and, thus, the common name "Texas tortoise" does not seem appropriate. Agassiz (1857) did not suggest a common name. This tortoise occurs in a suite of low-density, disjunct populations. In Texas, the species is recorded from Val Verde County eastward to Matagorda County and southward into Tamaulipas, northern and eastern Nuevo Leon, northeastern San Luis Potosi and northern Coahuila, Mexico (Rose and Judd 1989).

Fujii and Forstner (2010) provide the only genetic assessment of population structure using 138 individuals using eight polymorphic microsatellite loci from Texas; they do not assess Mexican tortoises. Assignment tests resolve weak northern and southern breeding units whose distributions roughly correspond to southern Duval County. Most of the variation, 85.6%, occurs within groups; only 9.1% occurs between groups. Significant isolation-by-distance occurs. Recent migration in a few individuals documents ongoing gene flow between the two units. The absence of Mexican tortoises and mtDNA analyses precludes an evaluation of female dispersal and biogeography.

Gopherus agassizii (Cooper 1861) (gopher, Agassiz's desert tortoise, Agassiz land-tortoise, desert tortoise; Mojave [Mohave] desert tortoise)

The taxonomic history of *G. agassizii* is steeped in errors and controversy far more so than that of all other "species." Errors involve fundamental data points, such as date of publication, type-locality, fates of the cotypes, common name, and the validity of *G. lepidocephalus* Ottley et Velázques Solis 1989. Murphy et al. (2011) resolve these conundrums. The publication date is 1861, not 1863 and the type-locality is Soda Playa in California, not Arizona. Finally, *G. lepidocephalus* is neither native to Baja California nor has an origin from Tiburon Island, Sonora, Mexico. Its maternal ancestor is from the Mojave Desert but the population in Baja California Sur might consist of hybrids between *G. agassizii* and *G. morafkai*, because both matrilines occur in the region (Murphy et al. 2011). Regardless, this invalid taxon is a junior synonym of *G. agassizii* and not *G. morafkai*.

Gopherus agassizii has only 30% of its former range because of the description of *G. morafkai*. The historical associations of *G. agassizii* are researched more so than for any other species. This tortoise has major matrilines A and B (Murphy et al. 2007). Lineage A occurs throughout the range of the species except in the Northeastern Mojave Recovery Unit. In contrast, Lineage B occurs primarily in the Northeastern Mojave Recovery Unit but has occasional occurrences elsewhere. Lineage B has two sublineages (Lamb et al. 1989), one of which occurs only in extreme northeastern parts of the Mojave Desert. The low levels of sequence divergence and widespread co-occurrences of haplogroups preclude the occurrence of multiple species. The broad distribution of some haplotypes may reflect either a natural occurrence or the documented historical mass relocations of thousands of tortoises (Murphy et al. 2007).

Rainboth et al. (1989) report significant nuclear DNA (allozyme) differentiation between two localities, Kramer Hills (western Mojave Desert) and Chemehuevi Valley (eastern Mojave Desert) and reject the null hypothesis of a single randomly breeding unit. They conclude that because of the high level of genetic similarity translocations between these populations would have a minimal effect on genetic structuring. A polymorphic albumin-like general protein is polymorphic in most of the Mojave Desert, yet monomorphic for one allele in extreme northwestern Arizona and adjacent Utah for the alternative allele in *G. morafkai* (Glen et al. 1990). The absence of an outgroup precludes an evolutionary interpretation to this pattern.

Britton et al. (1997) evaluate nuclear and mitochondrial genes and morphology to assess their correspondence to proposed desert wildlife management areas for tortoises in Nevada and Utah only. The nuclear gene assessment is limited to a single allozymic locus and the small geographic sampling limits conclusions about species diversity.

Berry et al. (2002a) emphasize the necessity of range-wide assessments of *G. agassizii* using nuclear gene markers to determine the number of species that might be involved in the complex (Lamb et al. 1989, Lamb and Lydeard 1994) and to provide genetic assessments for conservation. In developing a suite of microsatellites, Edwards et al. (2003) facilitate such assessments.

Employing microsatellites, Murphy et al. (2007) report substantial genetic substructuring that corresponds to montane barriers to dispersal. Hagerty and Tracy (2010) also report genetic structuring, but propose different breeding units in the western Mojave Desert, a difference they attribute to sampling strategies. Recently, Hagerty et al. (2011) synthesize the microsatellites with ecological and geological data and reach an assessment that is virtually identical to that of Murphy et al. (2007). These assessments neither reveal insights into dispersal history nor suggest the existence of multiple species within *G. agassizii*. Finally, Latch et al. (2011) discover weak cryptic genetic patterning on a local, fine scale driven by both slope and anthropogenic roads.

Gopherus flavomarginatus Legler 1959 (Bolsón tortoise, Mexican giant tortoise, yellow-bordered tortoise)

Occurring in the Chihuahuan Desert of north-central Mexico, *G. flavomarginatus* has the smallest distribution of all *Gopherus*, the least amount of genetic investigation, and little genetic variability. Morafka et al. (1994) analyze allozymes within and between two populations, including one captive population from a different locality. The two populations are genetically indistinguishable. Ureña-Aranda and Espinosa (2012) report two matrilines among 76 individuals, and 95% of the variation occurs among populations, rather than within. A fine-scale analysis of genetic substructuring using microsatellites is desirable.

Gopherus morafkai Murphy, Berry, Edwards, Leviton, Lathrop et Riedle 2011 (Morafka's desert tortoise, Sonoran desert tortoise)

This species is the object of much attention (Van Devender 2002a). Lamb et al. (1989) and Lamb and Lydeard (1994) resolved two major matrilines whose patterns tend to have a north to south distribution (Lamb and McLuckie 2002). Sympatric occurrences of the divergent lineages exist from northern Sinaloa to near Hermosillo at least in central Sonora, Mexico (Edwards et al., unpublished data). Within the northern lineage, no geographic pattern appears to exist. Edwards et al. (2004a: 496) note "a low degree of nucleotide diversity and polymorphism among mitochondrial DNA sequences." Further, Edwards et al. (2010) report the existence of two additional lineages whose area of origin remains unknown.

A nuclear genes assessment exists for Arizona only (Edwards et al. 2004a), which is a small portion of the species' range. The "isolated" montane populations appear to have exchanged genes recently. Genetic structuring is limited to isolation-by-distance and this may be due to both long-distance dispersion of tortoises between mountains, sperm-storage for extremely long periods of time (Palmer et al. 1998, Murphy et al. 2007), and/or high levels of polyandry, assuming that *G. morafkai* and *G. agassizii* (Davy et al. 2011) have identical reproductive strategies. Further work throughout the distribution of the species is underway by Edwards and colleagues. *Gopherus morafkai* may be two species (Lamb et al. 1989; Berry et al. 2002; Murphy et al. 2007, 2011; Edwards et al., 2013).

Fujii and Forstner (2010) compare their microsatellite analyses for *G. berlandieri* to that for *G. morafkai* (Edwards et al. 2004a), albeit as *G. agassizii*. *Gopherus morafkai* appears to have greater genetic diversity yet less population structure than *G. berlandieri*. Thus, *G. morafkai* may have a greater extent of gene flow among localities than *G. berlandieri* (Fujii and Forstner 2010). Range-wide sampling of both species may confirm this observation.

EVOLUTIONARY RELATIONSHIPS: MACROEVOLUTION

Spinks et al. (2004) report that *Manouria* is the sister group of *Gopherus* plus *Indotestudo*, but this set of relationships is unsupported by bootstrapping and decay indices. Le et al. (2006), in a massive study of 65% of species of tortoises, resolve *Manouria* and *Gopherus* as sister genera with moderate support; *Indotestudo* is strongly excluded as the sister group of *Gopherus*. Thomson and Shaffer (2010) also obtained the latter result, albeit without reporting support values. Regardless, it seems certain that *Manouria* and *Gopherus* are sister-genera.

Gopherus contains five extant species, although a sixth species may exist. Lamb et al. (1989) provide the first molecular assessment of interrelationships of all five species and report two major lineages, one comprised of *G. polyphemus* and *G. flavomarginatus* and the other consisting of *G. agassizii* (including *G. morafkai*) and *G. berlandieri*. Unfortunately, the study suffers from four problems: divergent RFLP fragments (Lamb and Lydeard 1994); unmapped RFLPs that might not be homologous fragments (Murphy et al. 2007); presence/absence coding of data that leads to the clustering of parallel losses (Murphy 1993; Murphy and Lovejoy 1998); and the unrooted network. The data for *G. agassizii* are further evaluated by Nichols (1989). Notwithstanding, one result is surprising: *G. morafkai* clusters with *G. berlandieri*, rather than with *G. agassizii*, including the Sinaloan lineage.

Lamb and Lydeard (1994) reassess relationships using mtDNA cytochrome b sequences; they rooted the tree using Manouria. Again, they resolve two major lineages. All the samples of *G. agassizii* cluster together and *G. berlandieri* forms their sister-group. Within *G. agassizii*, the Sinaloan lineage clusters with tortoises from the Mojave Desert, and their sister group is *G. morafkai*. Neighboring geographic lineages are not sister-groups. This curious arrangement receives a high bootstrap value (BS = 89) in an expanded analysis (Edwards et al., 2013; fig. 3.1).

Morafka et al. (1994) provide an allozyme study of *Gopherus* and report similar relationships to those of Lamb and Lydeard (1994), although samples of the *G. agassizii* group are from Arizona (presumably *G. morafkai*). Their report of a lack of genetic differentiation between *G. agassizii* and *G. berlandieri* is compromised by employing blood samples for *G. agassizii* from captive tortoises, which consist of hybrids (Edwards et al. 2010). Murphy et al. (2007) evaluate matrilineal relationships within the *G. agassizii* group using *G. flavomarginatus* as the outgroup. They resolve *G. berlandieri* as the sister group to *G. agassizii* plus *G. morafkai* (as *G. agassizii*). Unfortunately, they do not sample the Sinaloan lineage. Regardless, *G. agassizii* and *G. berlandieri* appear to be sister taxa. Thomson and Shaffer (2010) provide a mega-analysis of turtle relationships and resolve two primary lineages within *Gopherus* with high bootstrap support (>95%): one clade contains *G. polyphemus* and *G. flavomarginatus* and the other *G. agassizii* and *G. berlandieri*. They do not provide a resolution for the three matrilines of the *G. agassizii* group, one being *G. morafkai*, and their contribution does not note which lineage they analyze from the *G. agassizii* group.

Le et al. (2006) analyze only *G. polyphemus* and *G. agassizii*. By default, they are resolved as sister taxa. Reynoso and Montellano-Ballesteros (2004) hypothesize phylogenetic relationships by combining Crumly's (1993) morphological data with their own. This approach treats all lineages of the *G. agassizii* group (including *G. morafkai*) as a single taxon. Whereas they resolve *G. flavomarginatus* and *G. polyphemus* as extant sister lineages, *G. agassizii* and *G. berlandieri* have unresolved relationships near the base of the tree, which is compatible with their being sister lineages. Their assumption of *G. agassizii* being a single species may have compromised their results with respect to the relationships of *G. berlandieri*, *G. agassizii*, and *G. morafkai*.

Reynoso and Montellano-Ballesteros (2004) provided a fossil-based synthesis of the biogeography of the species group. *Gopherus* lived in the central U.S. during the Oligocene. The two major lineages formed in the Miocene. In Plio-Pleistocene times, the ancestor of *G. agassizii–G. berlandieri* occurred from California eastward to New Mexico. Subsequently, the lineage becoming *G. berlandieri* dispersed into Texas and adjacent Mexico, where it displaced congeners that belonged to the *G. polyphemus–G. flavomarginatus* group. Although the explanation corresponds to the fossil record, some aspects strongly conflict with molecular dating. The split between *G. agassizii* and *G. morafkai* is dated at 6 Ma (Lamb and Lydeard 1994), and this occurred after these two species separated from *G. berlandieri*. Whereas Reynoso and Montellano-Ballesteros (2004) suggest a Pleistocene to Recent separation of the *G. agassizii* group from *G. berlandieri*, molecular data date this event before 6 Ma, the Late Miocene. The different scenarios rest upon the assumptions that the mitochondrial DNA of tortoises evolves at a very slow rate and the fossil record is sufficient to date historical events; both assumptions cannot be true. Either the mitochondrial gene used to date the tree (cytochrome b) is not evolving as slowly as thought (the calibration rate for the gene is too low), the fossils tortoises are dead-end lineages not in line to extant species, or the dating of the fossils is inaccurate. One fossil-based scenario may be in error. The split between *G. agassizii* and *G. morafkai* dates to the formation of the Colorado River between Arizona and Colorado, which itself occurred 5.3 Ma (Dorsey 2012), and this divergence occurred after the split of these species from *G. berlandieri*.

HYBRIDS

Auffenberg (1976) cites the possibility of hybrids between *Gopherus agassizii* and *G. berlandieri* (Woodbury 1952, Mertens 1964) but notes the lack of confirmation of parentage in one case and unknown viability of the offspring. Parentage of hybrids is now documented for many individuals, both in nature (McLuckie et al. 1999) and ex situ (Edwards et al. 2010). Viable offspring seem certain between combinations of *G. agassizii* with both *G. morafkai* and *G. berlandieri*, as well as between *G. berlandieri* and *G. polyphemus* (Edwards personal communication). No data exist on whether or not all possible combinations of *Gopherus* can produce viable hybrids. Further, no analyses document the extent of backcrossing for hybrids, if any, both in situ and ex situ.

CONCLUSIONS

The extent of investigation varies substantially among the species. Whereas some species enjoy rather extensive attention—e.g., *Gopherus agassizii*—other widespread species have hardly been investigated, including *G. berlandieri* and *G. morafkai*. With the exception of *G. flavomarginatus*, little genetic work exists on Mexican tortoises and to the extent that *G. morafkai* may be a composite of two species. Patterns of matrilineal dispersion are largely undocumented for both *G. berlandieri* and *G. morafkai* and much work remains for *G. polyphemus*.

Available macroevolutionary analyses converge on defining only three of the five or six potential nodes of the tree. First, *Gopherus* appears to be a monophyletic genus, and its sister group is the Asian genus *Manouria*. Second, *G. polyphemus* and *G. flavomarginatus* are sister taxa. And third, *G. berlandieri* is the sister taxon of the *G. agassizii* group. The historical relationships within the *G. agassizii* group remain to be resolved confidently. Further, the controversial biogeographic history of *Gopherus* requires further investigation.

The review serves to identify gaps in our knowledge, of which there are many. These involve both micro- and macroevolutionary levels.

Microevolutionary Questions

Gopherus polyphemus would benefit from additional mtDNA sequencing, including DNA barcoding (Hebert et al. 2003, 2009) and broader geographic sampling. Microsatellite DNA studies should use a greater number of loci, and loci developed for *G. agassizii* (Hagerty et al. 2011; Edwards et al. 2003, 2011) might work on *G. polyphemus*. Such assessments would facilitate conservation, identify translocated and relocated individuals, determine the source populations of illegally collected individuals, and resolve taxonomic uncertainties.

Gopherus berlandieri needs much further investigation, especially comparisons involving Texas and Mexico. Such work should involve mtDNA sequences as well as the incorporation of additional microsatellites.

Gopherus flavomarginatus has the smallest range and the least amount of genetic variation of all the species. Genotyping of individual tortoises might serve to maintain the existing variation.

Gopherus agassizii may be the only population with adequate genetic analyses. Little can be said about matrilineal dispersion because the species only has two or three major lineages. A synthesis of nuclear gene assessments using samples from Murphy et al. (2007) and Hagerty and Tracy (2010) is highly desirable.

Gopherus morafkai may be Pandora's box with respect to *Gopherus*. Essential genetic assessments are currently underway. Fundamental questions range from deciphering the matrilineal history and resolving the possibility of two species, to ecological genetics and the role hybridization might play in evolution and adaptation.

Hybrids in northern Arizona can provide data critical to understanding how hybridization might shape speciation within the *G. agassizii* group. Investigations should synthesize mtDNA sequence data with nuclear gene assessments.

Macroevolutionary Questions

The matrilineal relationships within the *G. agassizii* group are inadequately resolved. Genetic assessments should use a broader array of mitochondrial genes, and involve wide geographic sampling and DNA barcoding. The molecular clock estimates need to be validated using more sequence data and statistical approaches, as available in BEAST (Drummond and Rambaut 2007). Nuclear gene assessments are required to corroborate the mtDNA tree.

Acknowledgments

This research was supported by the Natural Sciences and Engineering Research Council Discovery Grant A3148. Jack Sites Jr., Trip Lamb, Mike Forstner, and Taylor Edwards provided literature and answers to my inquiries while I was working in China. Amy Lathrop prepared figure 3.1.

4 Thermoregulation and Energetics of North American Tortoises

JAMES R. SPOTILA
THOMAS A. RADZIO
MICHAEL P. O'CONNOR

The basic thermal biology of tortoises has been known for more than 120 years. Hubbard (1893) described the summer foraging of *Gopherus polyphemus* in Florida and noted that they emerged in midday between 1100 and 1400 at temperatures around 33°C, and foraged for an hour before retreating to their burrows "in the moist, cool sand." In winter, "the gopher very rarely quits its burrow, and comes forth to feed only on the very hottest days at noon." Hubbard also described the burrow and reported that temperatures were stable, generally remaining above 23°C in winter and below 26°C in summer. Cowles and Bogert (1944) quantified thermal requirements of desert reptiles; and in the 1960s and 1970s, a new wave of studies in physiological ecology examined thermoregulation in tortoises (e.g., McGinnis and Voigt 1971, Voigt 1975, Voigt and Johnson 1977, Douglass and Layne 1978) and other reptiles. One hundred years after Hubbard, Zimmerman et al. (1994) quantified thermoregulation of *G. agassizii* and described the relationship between operative temperature (T_e), body temperature (T_b), and microhabitat utilization.

BIOPHYSICAL ECOLOGY

It is the interaction of the physical aspects of the environment and the size and physical characteristics of the tortoise that determine its thermal ecology and thermoregulation. All tortoises have the same basic shape, shell type, and thermal characteristics. Desert tortoises (the literature has heretofore not recognized the two proposed species as different, and the life-history information presented here largely amalgamates data and analyses from desert tortoises in all southwestern deserts of North America) live in hot deserts and have a rounder and more dome-shaped shell. *Gopherus polyphemus* lives in eastern pine-oak forests, beach scrub, oak hammocks, or pine forests, and has a flatter shell. Desert tortoises also live in Sinaloan thornshrub and deciduous forest in Mexico, where they encounter lower temperatures in summer and higher temperatures in winter than relatives in the Mojave and Sonoran Deserts (Germano et al. 1994). In Sinaloan ecosystems, *G. morafkai* may experience similar thermoregulatory environments to *G. polyphemus* in Florida. *Gopherus berlandieri* is intermediate in size and lives in habitats ranging from near-desert to brush grasslands. *Gopherus flavomarginatus* is the largest North American tortoise (as large as 400 mm carapace length), but has the most restricted range, limited to a small area in north-central Mexico. It lives in semidesert or dry savanna up to 1400 m. All tortoises face similar physical constraints, avoiding heat during the middle of the day in the hottest portion of the year and retreating to burrows or surface refuges to survive both heat and cold.

Tortoises control T_b by balancing heat absorbed from the environment in the form of solar and thermal radiation, with heat loss and gain from conduction with the substrate and convection (wind) from the atmosphere, and heat loss from evaporation. Porter and Gates (1969) and Spotila et al. (1972) discuss these processes. Different-sized tortoises are more or less affected by these avenues of heat exchange, because animal mass buffers transients in heat flux. For example, hatchling tortoises are active above ground on warm sunny days in winter because they heat up quickly. They are constrained from operating on the surface during midday on sunny summer days, however, because they will rapidly overheat. At the opposite extreme, large adult *Gopherus agassizzi* extend surface activity time by anticipating retreat to cooler microclimates in the morning and by exploiting their thermal inertia to dampen heating rates in the afternoon (Zimmerman et al. 1994). Little is known about thermoregulation by *G. flavomarginatus* in nature, but we expect that large adults exploit body size to buffer heat gain and loss (Bonin et al. 2006). At higher elevations, individuals can avoid the heat stress typical of hot, dry desert areas at the foot of Cerro San Ignacio in Durango.

By measuring microclimatic variables and using models of tortoises painted to have the same solar absorptivity (68.0–73.2%) as living tortoises, Zimmerman et al. (1994) determined the role of heat energy exchange in the thermal ecology of tortoises in the Mojave Desert near Las Vegas, Nevada. They constructed aluminum models of tortoises and used them to measure T_e of the environment. Bakken and Gates (1975) defined T_e as the temperature of a mass-less animal with external heat exchange characteristics of the animal. It is equivalent to the temperature that an animal would attain if left indefinitely in a static thermal environment similar to the one it currently occupies (Bakken 1981b, Bakken et al. 1985). Operative temperature is also used to characterize an animal's thermal environment, and is estimated using computational heat transfer models and microclimate data or, more commonly, a physical model of an animal that replicates external heat exchanges of the animal in the environment of interest (Grant and Dunham 1988; Huey et al. 1989; Bakken 1989, 1992). Although they are widely used for studies of small reptiles such as lizards, T_e models are not straightforward to use, especially for larger animals. Use of such models is discussed below, after discussion of tortoise heat exchange.

Zimmerman et al. (1994) took into account both theoretical and practical considerations in their use of T_e models for studies of large reptiles (O'Connor 2000, O'Connor et al. 2000, Dzialowski and O'Connor 2001). They constructed thick-walled models (see below) and placed them in typical tortoise microhabitats, including open areas exposed to full sun, partially protected areas—such as under a bush or in a pallet (shallow depression in ground made by a tortoise, typically at the base of a shrub)—on various slopes with different exposures to sky, in burrows, and in a caliche den. They measured T_b of tortoises with radio transmitters.

Air, ground, and operative temperatures coincided with daily and monthly patterns of incident solar radiation. Variation in T_b primarily was related to microhabitat selection, especially use of burrows. During July–October, tortoises in burrows had lower T_b than tortoises on the surface, in the morning. Tortoises on the surface selected microhabitats to avoid extreme temperatures. They used shaded locations until conditions became too hot and then retreated to burrows. At midday, tortoises remained in burrows, and maintained cooler body temperatures than operative temperatures on the surface, which exceeded 40°C from late April to October. Lethal mean surface T_es (above 50°C) were common (fig. 4.1). Tortoises could be active from 0800 to 1700 in April and 0800 to 1700 in October, but were constrained to a bimodal activity pattern in the intervening months.

In the morning during the summer, free-ranging desert tortoises were active on the surface, and T_b tracked T_e measured with tortoise models. As T_e approached 40°C and T_b reached about 35°C, tortoises retreated to burrows or caliche dens, and T_b decreased and stabilized. After equilibrating to burrow temperature, T_b remained stable until tortoises emerged during late afternoon or evening for another round of activity. After dark, body temperatures either tracked T_e or remained above T_e, depending upon whether they remained on the surface or retreated to burrows (fig. 4.2). Hillard (1996) found that T_e was a useful predictor of daily and seasonal

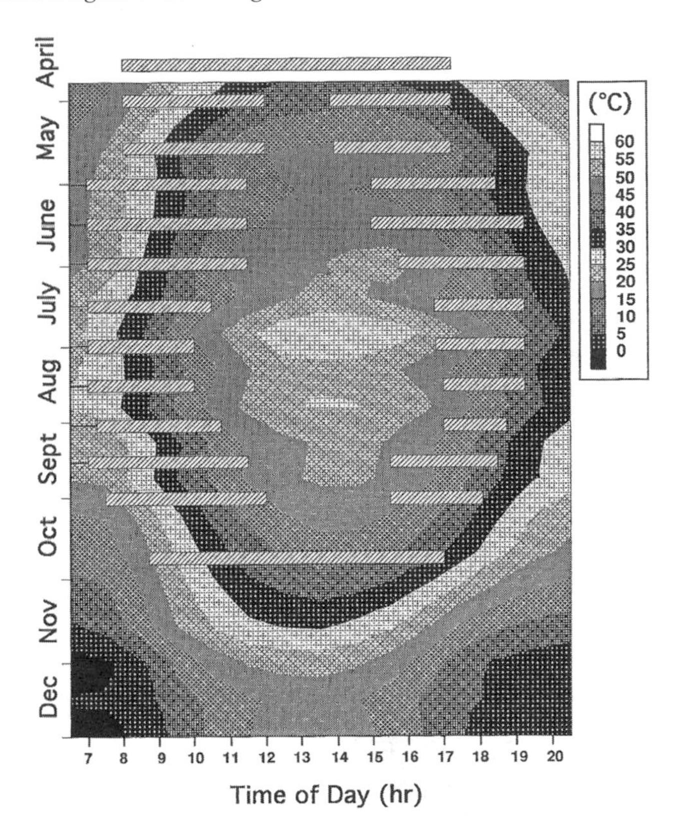

Fig. 4.1. Daily and monthly potential activity times (horizontal bars) of *Gopherus agassizii* superimposed on mean surface operative temperatures at the Desert Tortoise Conservation center near Las Vegas, Nevada. Reproduced from Zimmerman et al. (1994) with permission.

population activity of juvenile desert tortoises and was able to predict potential time that juveniles could spend above ground to forage.

The use of T_e models is simple for small ectotherms like lizards because the physical model is a thin copper shell painted to have the same solar absorptivity as the real lizard (Clusella Trullas et al. 2009). The model has the appropriate characteristics of minimal mass, heat capacity, and thermal inertia that are needed to integrate heat flows rapidly and provide nearly instantaneous measures of T_e and a single temperature of physiological importance. Unfortunately, many investigators have been less than rigorous in their use of T_e models. They have used them to substitute improperly for T_b, not considered mass, filled them with fluids and even ground meat to mimic body mass, used paints of incorrect absorptivity, and made other errors, such that T_es were inaccurate. Investigators should review the theoretical literature on T_e before using T_e models in their studies. It is even more difficult to use T_e models for large animals such as tortoises. In large T_e models, the longer distances and heterogeneous heat transfer rates between different parts of the body can lead to different estimates of T_e, depending upon where temperature is measured in the model (Bakken 1992). When anatomical variations in heat transfer are important, temperature variation

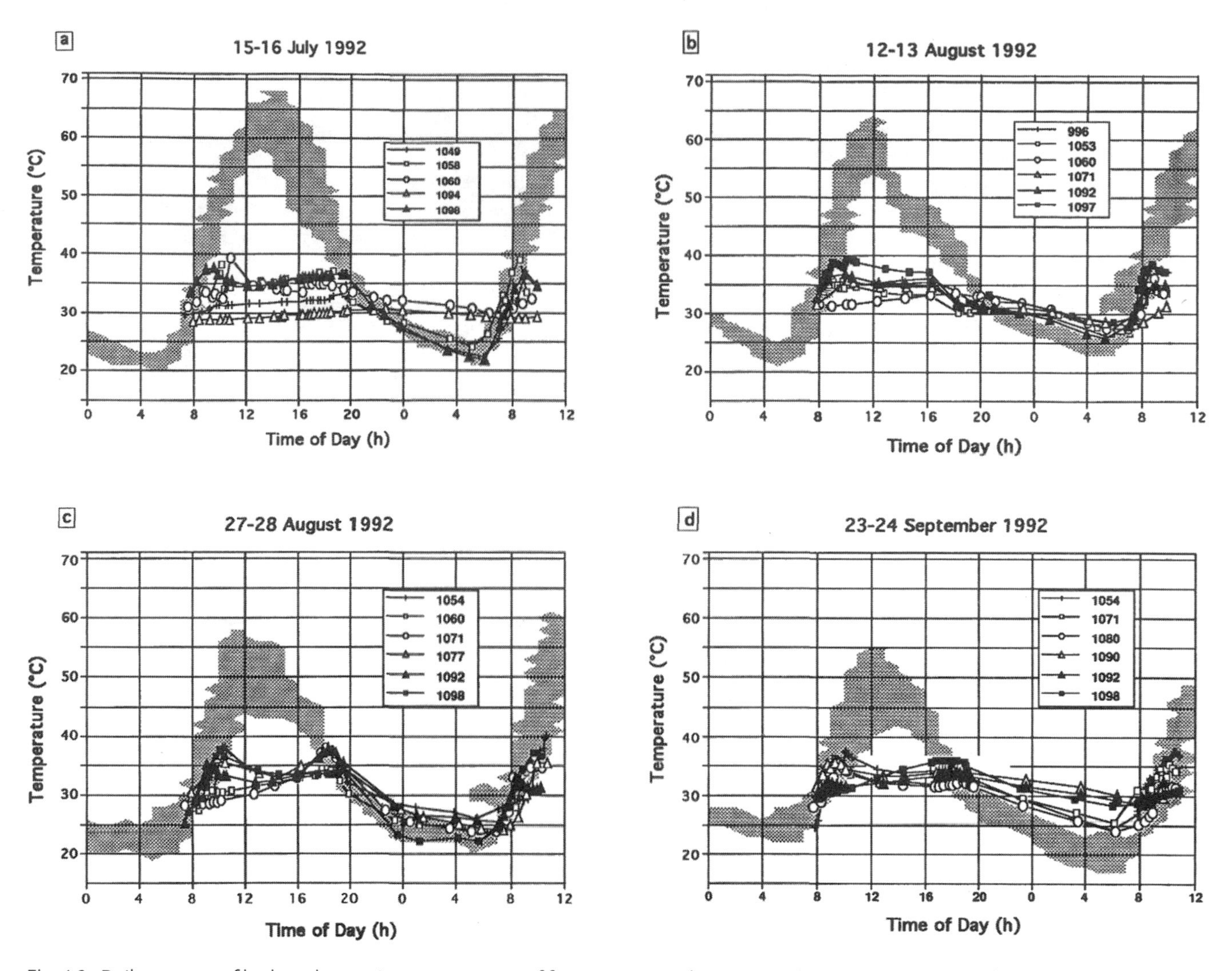

Fig. 4.2. Daily patterns of body and operative temperatures of free-ranging *Gopherus agassizii* near Las Vegas, Nevada. Each symbol represents the T_b, measured with a radio transmitter, of a different tortoise. Ranges of T_e, indicated by shading, were measured with T_e models of tortoises. Reproduced from Zimmerman et al. (1994) with permission.

among parts of the model is informative (Bakken 1981a), but when T_e is used to study an animal's ecology, such variation can obscure patterns of T_e. O'Connor et al. (2000) analyzed regional variation in temperatures possible in T_e models of tortoises and found that thick-walled models kept regional temperature differences smaller than thin-walled models and increased thermal inertia of models. Time constants of models always were faster than those of real animals and allowed models to sample the range of T_es available to the animal. O'Connor (2000) provided algorithms to extract estimates of instantaneous T_e from model temperatures and a deconvolution method to use when wind speeds were constant. An iterative method is needed when wind speed varies. Thus, thick walls can improve T_e estimates for tortoises and other moderate- to large-sized reptiles if appropriate theoretical and practical concerns are considered.

Individuals of *G. flavomarginatus* emerge from their burrows at night in response to rising deep soil temperature (>34°C between 2300 and 0100), caused by thermal lag in the soil. By 0700, soil temperature at a depth of 15 cm is still above 31°C, while temperature of an adult tortoise (7.3 kg) is below 30°C when it emerges to forage (Adest et al. 1988). The large thermal mass of *G. flavomarginatus* allows them to dump heat at night like a camel (Schmidt-Nielsen 1964), such that their body temperature is lower in the morning than if they had stayed in their burrows all night. By starting at reduced body temperatures, *G. flavomarginatus* can forage for a longer period of time before retreating to burrows.

THERMOREGULATION

As might be expected, *Gopherus* spp. have the highest temperature tolerances of North American turtles. *Gopherus polyphemus* has a CTM (Critical Thermal Maximum) of 43.9°C, *G. berlandieri* a CTM of 42.8°C, and *G. agassizii* a CTM of 43.1°C (Hutchison et al. 1966). Judd and Rose (1977) reported a CTM of 43.7°C for one *G. berlandieri*. Body temperatures of active individuals are similar among species. *Gopherus polyphemus* had a mean T_b of 34.7°C (31.7–38.3°C) in summer in southern Florida (Douglass and Layne 1978) and a range of T_b of 15° to 35° C from March to November in Mississippi (Anderson 2001). *Gopherus agassizii* had a mean T_b of 32.3°C (20.8°–38.0°C) and 34.7° C (27.5°–38.3°C) in May and July, respectively, in the Mojave Desert of California (McGinnis and Voigt 1971). Active *G. agassizii* had a range of mean T_b of 34.0 to 35.8°C in summer at Las Vegas, Nevada (Zimmerman et al. 1994). Brattstrom (1965) reported that T_b ranged from 19.0 to 37.8°C for active desert tortoises (place and season not reported). One study of *G. berlandieri* showed two individuals to have T_bs of 32.2° and 33.1°C in June (Voigt and Johnson 1976), and a larger study showed mean T_b to be 31.1°C (24.1° to 35.6°C) in spring, 33.1°C (28.0° to 39.0°C) in summer, and 32.9°C (31.0° to 35.0°C) in fall (Judd and Rose 1977).

Gopherus spp. have distinct unimodal activity patterns in spring and early summer and shift to bimodal activity patterns in the heat of midsummer (Auffenberg and Weaver 1969, Rose and Judd 1975, Judd and Rose 1977, Bury and Smith 1986, Zimmerman et al. 1994). Tortoises hibernate in winter and survive freezing temperatures by supercooling to as low as –5.25°C (Lowe et al. 1971). *Gopherus polyphemus* in Florida are active in winter on warm days when air temperatures are above 21°C (Douglas and Layne 1978). An adult female basked outside her burrow in Florida in winter at 18°C to 23°C (Diemer 1992a).

All *Gopherus* spp. use burrows as retreats from extreme temperatures and could not survive extreme temperatures of summer and winter without them. Ruby et al. (1994) reported that *G. agassizii* in enclosures near Las Vegas used water from sprinkler systems to thermoregulate and remained active after sprinklers shut off, until their carapaces dried and they began to heat rapidly. They then "ran" to their burrows to avoid overheating. Use of burrows has been described in *G. polyphemus* by Douglass and Layne (1978), Diemer (1992a), and Anderson (2001); and in desert tortoises by McGinnis and Voigt (1971), Bulova (1994, 2002), Ruby et al. (1994), and Zimmerman et al. (1994). *Gopherus berlandieri* uses both burrows and surface pallets to avoid extreme temperatures (Rose and Judd 1975, Voigt and Johnson 1976, Judd and Rose 1977). *Gopherus flavomarginatus* also uses burrows as thermoregulatory refuges (Adest et al. 1988). Bulova (2002) described patterns of burrow use by desert tortoises and, by combining field behavior and microclimate measurements with biophysical models, assessed the value of burrows as thermal refuges. She found that burrows not only provided thermal refuges, but also were important in water conservation, because humidity was significantly higher and T_b and predicted evaporative water loss lower inside burrows than on surface (fig. 4.3). Greater burrow length and smaller entrances were correlated with higher burrow humidity.

Burrows are especially important for hatchling and juvenile tortoises for thermoregulation as well as avoiding predation. Hatchling *G. polyphemus* remain inside burrows of adults,

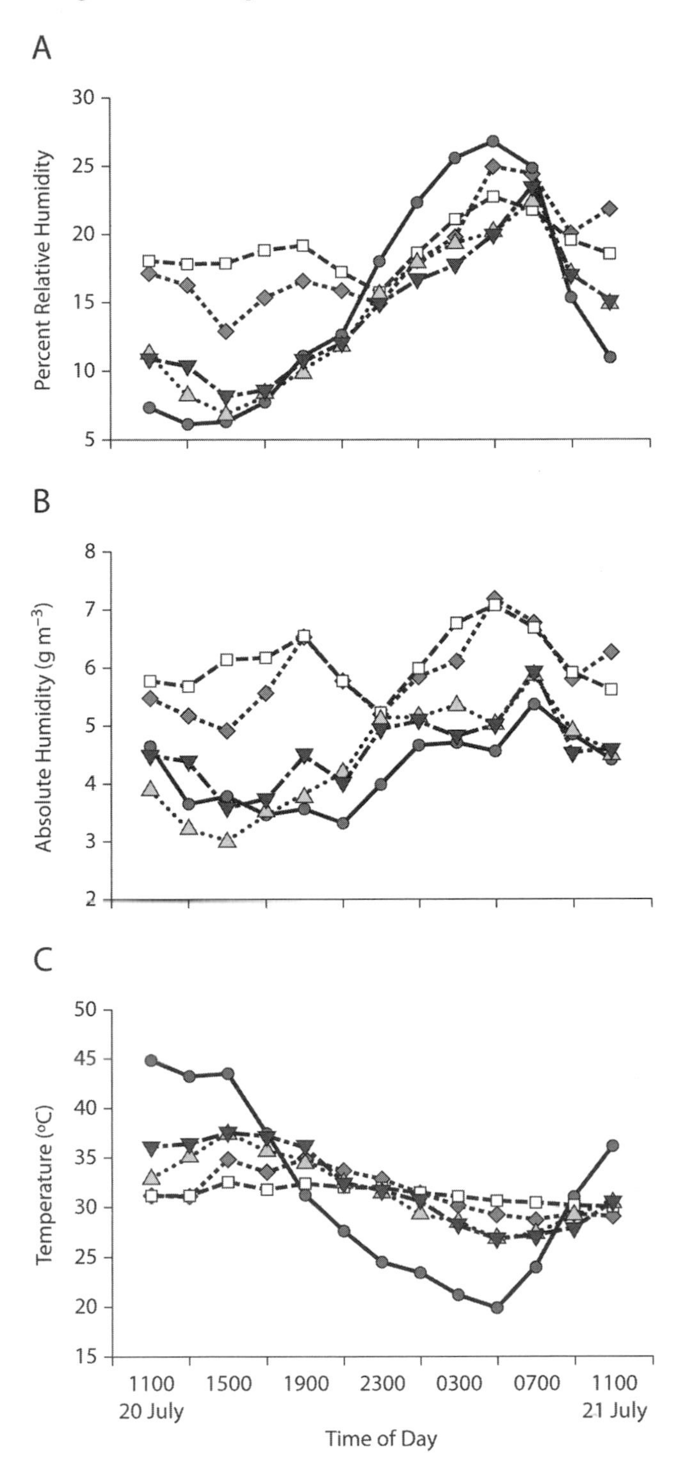

Fig. 4.3. Burrow and surface microclimate over 24 h for *Gopherus agassizii* near Las Vegas, Nevada. Measurements were made inside and on the surface near four burrows. Closed circles and solid lines represent the surface, and open squares, diamonds, triangles, and dashed and dotted lines represent four different burrows. *A*, percent relative humidity; *B*, absolute humidity (gm^{-3}); *C*, temperature (°C). Reproduced from Bulova (2002) with permission from Elsevier.

dig their own burrows, or remain buried under sand or litter during their first winter (Douglass 1978). During their first year, they spend large amounts of time in burrows and in pallets under vegetation (Butler et al. 1995, Pike 2006). Juvenile *G. agassizii* dig burrows under shrubs (80% of burrows) and primarily under large shrubs with more protection from predators and a more stable microclimate (Wilson et al. 1999a). *Gopherus flavomarginatus* hatchlings excavated multiple burrow and pallet sites during the rainy season and moved among them, but in the dry season, occupied only one burrow. Most burrows and pallets were in cactus and shrub microhabitats, rather than grass microhabitat (Tom 1994).

PHYSIOLOGICAL ADAPTATIONS

Gopherus spp. heat faster than they cool, as do most reptiles (Voigt 1975). There are large differences in T_b and head temperature of *G. berlandieri* during heating and cooling (Voigt and Johnson 1977). When extended, the head heats and cools faster than the body; but when retracted, it heats and cools slower than the body, because the body insulates the head. O'Connor (1999) developed a mathematical model of heat exchange in reptiles to investigate the role of blood flow in controlling rates of heating and cooling in complex thermal environments. Mass, blood flow, and shuttling behavior affect rates of heating and cooling. For animals experiencing changing T_e, deep T_bs depend upon body mass, range of T_e, and rapidity with which T_e varies. Potential effects of blood flow on T_bs depend on the same factors. A 20 g lizard (*Cnemidophorus* sp.) would need to move every 3 min to maintain a constant T_b range between 33° and 40°C, while a desert tortoise would only need to move every 90 min to maintain the same T_b range, and would have more latitude in behavior by adjusting that period via blood flow.

Although burrows provide stable temperatures and high humidity, they also create environments characterized by both hypoxia (low level of oxygen in the air) and hypercarbia (high level of carbon dioxide in the air). In burrows in sandy soils, there is no significant physiological challenge for *G. polyphemus* (O_2 = 20.14%, CO_2 = 0.78%) (Ultsch and Anderson 1986). In clayey soils, however, burrows exhibit moderate hypoxia (18.19%) and levels of CO_2 high enough (2.35% to 6.77%) to be of physiological importance. Burrows in clayey soils tend to be shorter (3.2 m) than those in sandy soils (4.1 m), in part because of this physiological limitation (Ultsch and Anderson 1986). *Gopherus polyphemus* is better able to tolerate hypoxia than the Eastern Box Turtle (*Terrapene carolina*) and its responses to hypercarbia are less marked (Ultsch and Anderson 1988). Although it does not live in burrows, *T. carolina* does hibernate in water, and thus may experience alveolar hypercapnia (high level of carbon dioxide in the blood). Standard metabolic rates of *G. polyphemus* and *T. carolina* are about as expected for turtles of their body sizes. Critical oxygen tension (P_c) of *G. polyphemus* (10–11 mm Hg) is lower than that of *T. carolina* (26 mm Hg). The P_c of both species is unaffected by hypercarbia (6% CO_2), but in moderate hypoxia (15% O_2), 45 min of hypercarbia causes an increase in oxygen consumption of 72% in *G. polyphemus* and 206% in *T. carolina*. After 12–18 h, oxygen consumption of *T. carolina* returns to normal, but remains 22% elevated in *G. polyphemus*. Both species hyperventilate in hypercarbia and hypoxia. Both species survived a 24 h exposure to 6.3% CO_2 with no apparent ill effects. Thus, both species are equally well adapted to hypercarbia (Ultsch and Anderson 1988).

Physiological indicators of stress include heat shock proteins and other biomarkers (Yu et al.1998). No studies of the physiological response of tortoises to heat stress have been undertaken, however. Tortoises exposed to drought may reduce time on the surface not only because of heat stress directly, but also because of osmotic stress. Exposure to drought causes an increase in osmolality of both bladder urine and blood in desert tortoises, because of increases in plasma sodium, chloride, and urea concentrations (Peterson 1996a). Relatively short-term drought has been associated with severe die-offs of desert tortoises (Longshore et al. 2003). O'Connor et al. (1994) examined potential hematological and biochemical indicators of stress in free-ranging and captive desert tortoises. There were significant increases in plasma electrolyte and urea nitrogen concentrations and white blood cell counts in water-stressed animals; however, no generalized stress response was detected. High sodium, chloride, bicarbonate, and urea nitrogen concentrations suggested water stress in captive animals that did not receive supplemental water, and rising electrolyte concentration suggested negative water balance and hydration stress late in the active season in both captive and free-ranging tortoises. Glucose levels in blood decreased through the course of the activity season, suggesting that energy reserves were decreasing. Blood panels are unlikely to identify stress in randomly sampled individual tortoises. A more useful approach would be to identify specific stressors thought to affect tortoises and to identify markers for those stressors.

ENERGETICS

The study of energetics has been enhanced by the use of doubly labeled water (water containing enriched amounts of 18O and 3H or 2H) to measure flux of water and metabolic rate of animals (Lifson and McClintock 1966). Differential loss of these isotopes by evaporation (H_2O) and metabolism (H_2O and CO_2) allow measurement of metabolism under actual field conditions. The use of nonradioactive 2H, coupled with explanations and detailed instructions for its use (Speakman 1997), have greatly facilitated use of this technique. Because desert tortoises experience more irregular rainfall and more limited resources in the Mojave Desert than in the Sonoran Desert of Arizona they have different growth strategies and longevity (Curtin et al., 2009; chapter 10). By using doubly labeled water, Nagy and colleagues, in several papers, elucidated the metabolic rates of desert tortoises in the field and the role of climate and food supply in their reproductive and annual energy budgets. By combining use of doubly labeled water, radio telemetry, and field observations, Nagy and Medica (1986) measured seasonal changes in water balance, energetics, food consumption, daily behavior, diet, osmoregulation, and body mass of *Gopherus agassizii* in Nevada. In spring, tortoises gained mass while eating succulent annual plants, but energy intake was less than required to meet energy expenses. As food plants dried in late spring, tortoises entered positive energy balance while eating grasses, but body masses declined as a result of negative water balance. During summer, tortoises emerged from aestivation in burrows to drink and feed after rainfall, and their energy and water balances fluctuated until they entered hibernation. Tortoises tolerated large imbalances in water, energy, and salt budgets during the active season, while exploiting resources only available periodically, resulting in an annual energy surplus.

Peterson (1996a), using doubly labeled water, found that field metabolic rates and feeding rates of *G. agassizii* were highly variable, with differences between seasons within populations, between years, and between populations. These differences were related to differences in rainfall between years and sites. Also using doubly labeled water to measure field metabolic rates, as well as obtaining data on lipid and nonlipid dry mass, Henen (1997) determined how female desert tortoises reproduce every year despite variability in winter rainfall and food availability. Females produce eggs in years with low levels of winter annual plants by "relaxing" their control of energy and water homeostasis, and tolerating large deficits and surpluses in body matter composition on seasonal, annual, and longer time scales. They increased body energy content before winter and used the reserve in the following spring to produce eggs. Nitrogen or crude protein appears to be the primary limiting resource for producing eggs. By reducing metabolic rates 90%, female desert tortoises conserved enough energy to produce eggs during extreme drought conditions (Henen 2002b). Long-term chronic impacts of climate upon the diet and nutrition of female desert tortoises can decrease fecundity and cause local extinctions (Henen et al. 1998, Henen 2002a). For example, local drought over a period of three years resulted in decline of a desert tortoise population in a small area, while a population at a nearby site that received more rain remained stable (Longshore et al. 2003). Thus, a desert tortoise population at a given site is dependent upon local precipitation and annual biomass production for its survival and reproduction and that cannot be assessed by average precipitation over a regional area. Relatively short-term drought (annual or longer) at a local site can cause severe reductions in tortoise survival at that site. Differences in precipitation at different sites in the Mojave Desert may be responsible for the phenomenon of localized die-offs observed there.

Young desert tortoises use energy and water at rates similar to those expected for desert reptiles of their size, and they can conserve water and energy during dry seasons (Nagy et al. 1997). Temperature, rainfall, and the presence of green, succulent plant food affect rates of physiological processes within and between seasons and between years. Metabolic rates peak in late spring, and rates of water intake are highest when green annual plants or rain is available. During hibernation, rates of metabolism and water loss are low, and young tortoises lose little body mass. Dry periods in summer are most stressful, as young tortoises lose water even though they retreat into their burrows. Dry years pose a special threat to the survival of neonate and one-year-old desert tortoises.

Daily energy expenditure measured with doubly labeled water is greater in *G. polyphemus* than in desert tortoises (Jodice et al. 2006) and box turtles (Penick et al. 2002). It would be very informative to extend use of doubly labeled water to investigate energetics of other species such as the *G. flavomarginatus* and *G. berlandieri*, as well as other populations of desert tortoises in habitats such as Sinaloa thornshrub and deciduous forests.

CLIMATE CHANGE

Climate change will be a dominant feature of the Earth's physical environment in the 21st century. Heating of the planet as a result of anthropogenic production of CO_2 and other greenhouse gases changes climates and causes shifts in biota (Kolbert 2006, Dillon et al. 2010). Extinction and changes in ranges are predicted for both amphibians and reptiles (Milanovich et al. 2010, Sinervo et al. 2010). Potential impacts on tortoises are unknown. It is clear from studies on the role of rainfall and temperature on energetics of *Gopherus agassizii* that changes in climate of the Mojave Desert have large effects on reproduction and survival of individual tortoises and extinction of local populations. Climate models indicate that extreme-temperature days will increase in southern California in the 21st century and that June–August temperatures will increase from 3° to 6°C in the Mojave Desert (Cayan et al. 2008). Increases of that magnitude will increase metabolic demands on tortoises and affect production of vegetation. While overall mean precipitation may remain similar to historic levels (1960–1990), extreme rainfall events and droughts will increase in frequency. Since local droughts have large negative effects on viability of local tortoise populations (Longshore et al. 2003) there may be major consequences for tortoise populations.

Given predictions for effects of climate change on lizards in Mexico (Sinervo et al. 2010), we can expect that *G. berlandieri*, *G. flavomarginatus*, and *G. morafkai*, in Sinaloa, at least, will be adversely affected by rising temperatures. Detailed predictions are not available. Given predictions for effects of climate change on salamanders in the southeastern U.S. (Milanovich et al. 2010), we expect that *G. polyphemus* also will be subjected to increasing periods of drought and rising temperatures. Because their physiological tolerances are similar to those of desert tortoises, we would expect that *G. polyphemus* will be able to adapt to those changes. There are differences in response to rainfall in these species, however (McCoy et al. 2011). Increasing fragmentation of the habitat of *G. polyphemus* will create additional stress. We do not know how climate change will impact the ability of *G. polyphemus* to move to more suitable habitats, but it is unlikely to be helpful. Therefore, we are left with serious concerns about the impact of changing climate on all species of tortoises in North America in the 21st century.

Little doubt exists that substantial changes will occur in climate before substantial action is taken to slow down, halt, and reverse global warming. There is also little doubt that we do not have enough information to predict the detailed response of different populations of different species to those changes. Detailed studies are needed on the physiological ecology of all North American tortoises. Studies of the biophysical ecology, field metabolic rates, water balance, and thermoregulatory behavior are needed, especially for *G. polyphemus*, *G. flavomarginatus*, *G. berlandieri*, and *G. morafkai* throughout their ranges. Much research awaits another generation of scientists; the only question is whether there is enough time to do that research before the consequences of global warming outstrip our ability to generate the information needed for adaptive management of our tortoise species.

CONCLUSIONS

All tortoises face similar physical constraints, avoiding heat during the middle of the day in the hottest portion of the year, and retreating to burrows or surface refuges to survive both heat and cold. Tortoises control body temperature by balancing heat absorbed from the environment in the form of solar and thermal radiation, with heat loss and gain coming from conduction with the substrate and convection (wind) from the atmosphere, and heat loss from evaporation. The concept of operative temperature (T_e) is a useful predictor of daily and seasonal population activity of juvenile and adult tortoises and can be a good measure to compare thermoregulation of the different species of North American tortoises living in distinctly different microhabitats. All *Gopherus* species use retreats to avoid extreme environments and cannot survive without them. Tortoises adjust their physiologies to changing environmental conditions and the use of doubly labeled water, radio telemetry, and field observations provide insights into the energetics of these animals. Despite the adaptability of *Gopherus* species, all are threatened by global warming.

Acknowledgments

The authors were supported by a grant from NASA and by the Betz Chair of Environmental Science at Drexel University.

5 Reproductive Physiology of North American Tortoises

DAVID C. ROSTAL

The reproductive biology of each of North America's tortoises has been studied at least to some degree in all five species. *Gopherus agassizii* and *G. polyphemus* have received the most attention from researchers, and these two species are known to vary from each other in important aspects of their reproductive biologies. The complete seasonal reproductive pattern is known for *G. agassizii* from the eastern Mojave Desert (Rostal et al. 1994b, Lance and Rostal 2002), *G. flavomarginatus* from the Chihuahuan Desert of Mexico (Gonzalez 1995, Gonzalez et al. 2000), and *G. polyphemus* from southwestern Georgia, U.S. (Ott et al. 2000), but it is only partially known for the other two species of *Gopherus*. Although some aspects of the reproductive biologies of *G. morafkai* and *G. berlandieri* can be included in this chapter, no endocrine data are available. Thus, some of the generalizations presented in this chapter may need to be revised as more information on these species is gathered. In this chapter, I define the basic reproductive traits of North American tortoises; and when possible, I make interspecific comparisons, particularly of male and female characteristics, the timing of reproduction, reproductive endocrinology, oviposition, and sexual dimorphism. The influences of upper respiratory tract diseases on reproduction also are discussed.

MALE CHARACTERISTICS

The male reproductive system of *Gopherus* spp. is similar to other turtles. All males possess a paired reproductive tract with testes located intraperitoneally (within the body cavity) and near the kidneys. The epididymides are coiled and are adjacent to the testes. Spermatogenesis (sperm formation) occurs in the seminiferous tubules of the testes, and sperm is stored in the epididymis until mating, which normally occurs in the late summer and fall. During mating, the male inserts its intromittent organ (or phallus; the equivalent of the penis) into the female cloaca and semen is transferred internally to the female's oviducts. The intromittent organ in tortoises tends to be round in shape when everted.

The testicular cycle is most well studied in *G. agassizii* (Rostal et al. 1994b, Lance and Rostal 2002; fig. 5.1). A distinct seasonality exists for the cycle of this species. When male tortoises inhabiting the eastern Mojave Desert emerge from hibernation in April, the testes are fully regressed and the lumen of the seminiferous tubules contains only primary spermatogonia, sertoli cells, and cellular debris (the latter seemingly from the prior cycle). The epididymides, on the other hand, are fully packed with spermatozoa. The spermatozoa are quickly put to use, as the tortoises undergo an abbreviated mating period soon after emergence from hibernation. By May, the seminiferous tubules contain only Sertoli cells and spermatogonia, and the cellular debris is gone. In early July, the seminiferous tubules show active cell division of the spermatogonia and abundant spermatids. By September and October, spermatogenesis is complete, seminiferous tubules are at the greatest diameter, and the lumens are full of mature spermatozoa.

Spermatogenesis is testosterone dependent and stimulation of gonadal recrudescence (recurrence) in the male appears to be influenced by temperature (Rostal et al., 1994b, Lance and Rostal, 2002). Seasonal testosterone levels are near their nadir (low point) in both male *G. agassizii* and *G. polyphemus* when they emerge from hibernation in late March and early April (fig. 5.2). Mating activity following emergence presumably does not generally fertilize the current year's eggs, as females in most populations ovulate and produce eggs shortly after emergence or in some cases actually emerge from hibernation with fully shelled eggs in the oviduct. Male testosterone begins to rise as the summer temperatures increase. Male testosterone and corticosterone levels reach their maximum levels in August and September when spermatogenesis is complete and fall mating begins. Male–male aggression is also at its peak at this time. It may be a general rule for *Gopherus*

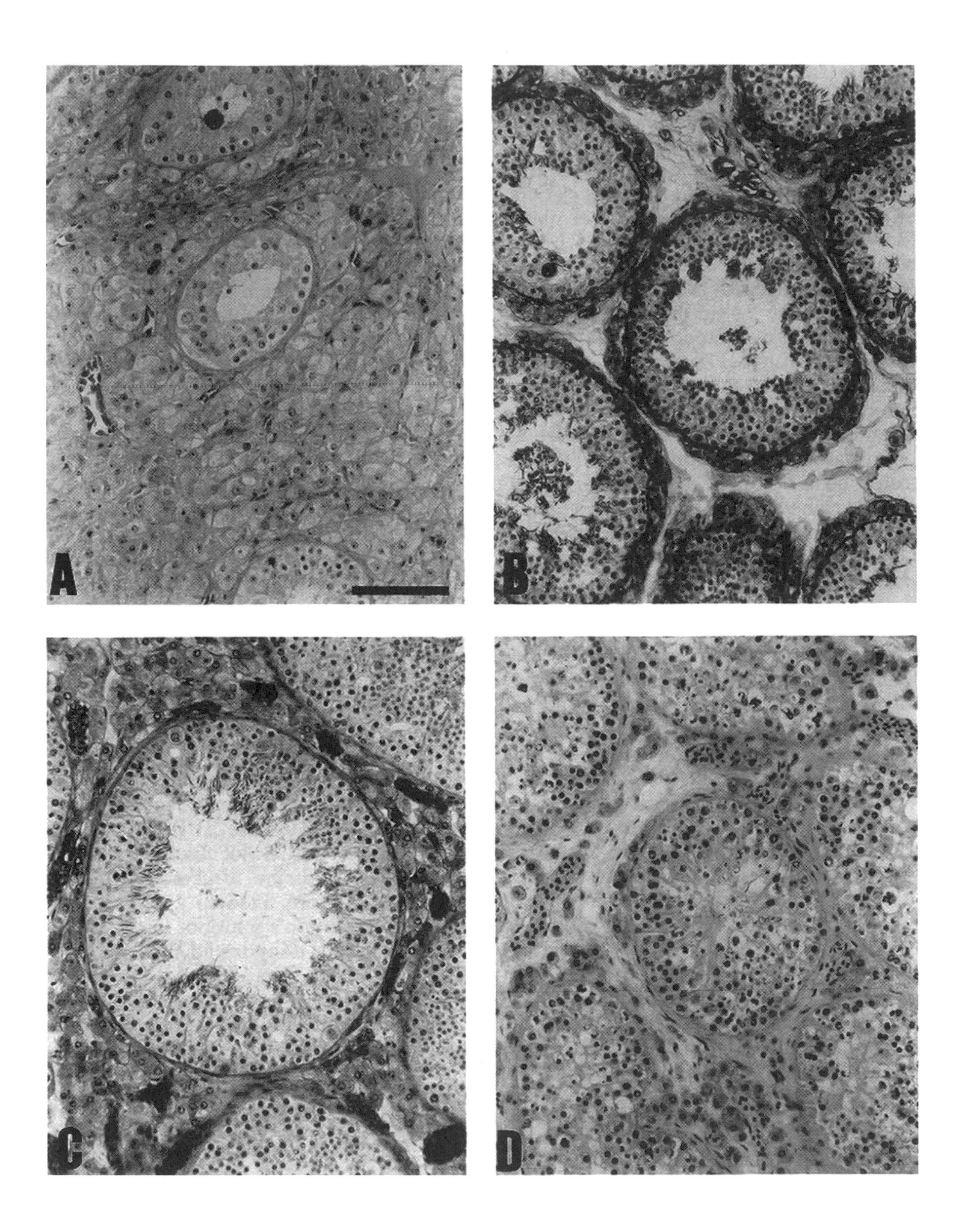
A
B
C
D

spp. that testosterone rises during the late summer and falls prior to hibernation, as this pattern has been observed in all populations studied to date.

FEMALE CHARACTERISTICS

The female reproductive system of tortoises has been studied more thoroughly than the male system, particularly for North American species. Basically, the female reproductive anatomy of tortoises is one of paired organs located abdominally. Adult females have a pair of ovaries with expanded stroma (a type of tissue) and paired oviducts at least 1.5 cm in diameter, suspended in the body cavity. Both ovaries function synchronously during reproduction (that is, they both produce eggs at the same time) and contain previtellogenic follicles, vitellogenic follicles, atretic follicles, and scars, depending on the season. Oviducts (which transport and shell the egg) are long and also function synchronously when undergoing changes in preparation for reproduction (e.g., thickening of the oviductal wall, development of the albumin gland and shell gland regions). In most species, ovulation occurs in the spring following emergence from hibernation; however, *Gopherus morafkai* displays a summer ovulation that corresponds to summer rains. The relationship between mating and ovulation—if any—is poorly known, and remains something to be worked out in detail. Sperm storage is prevalent in *Gopherus* spp., and it is entirely possible that there is no link between mating and ovulation: females may simply store the sperm they need and use it when the time is right for ovulation.

No specialized sperm storage structures are present in tortoises. Instead, sperm is stored in the ducts of the albumin glands in the upper region of the oviduct (Gist 1989, Rostal et al. 1994b). Following ovulation, it appears that sperm are passively moved from the gland ducts. Fertilization occurs

Fig. 5.1. (opposite) Histology of *Gopherus agassizii* testis during a spermatogenic cycle. Bar in 1A = 100μm. All four micrographs are at the same magnification. *A,* section from a testis from an animal collected in May. Seminiferous tubules are completely regressed and contain only spermatogonia and sertoli cells. The Leydig cells appear greatly hypertrophied and fill the interstitial area. Leydig cells are the cells that produce testosterone in the testis. *B,* section from a testis from an animal collected in July. Spermatogenesis has progressed to stage 4 to 5 of McPherson et al. (1982). Spermatocytes and spermatids are abundant and a few mature spermatozoa are present. The Leydig cells are also smaller in May. *C,* section from a testis from an animal collected in October. Seminiferous tubules are at their greatest diameter and spermatogenesis at a maximum. The Leydig cells again appeared hypertrophied. *D,* section from a testis from an animal collected in April. The lumen of the seminiferous tubules are filled with debris from the previous cycle. Large numbers of spermatogonia and spermatocytes are still present. The Leydig cells are not particularly abundant and are only moderately developed. From Rostal et al. (1994b).

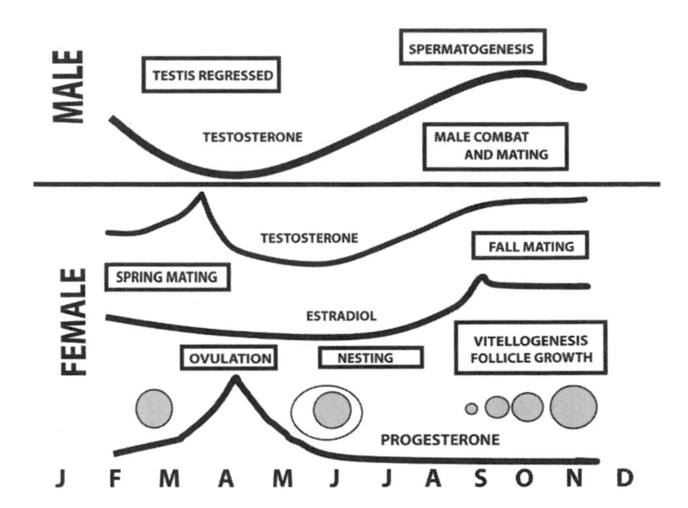

Fig. 5.2. Generalized reproductive cycle for *Gopherus agassizii, G. polyphemus,* and *G. flavomarginatus.* Modified from Rostal et al. (1994b).

in the upper oviduct where the sperm is stored. The details of fertilization are not known. Following fertilization, membranes and albumin layers are laid down around the ovum. Once a developing ovum passes the isthmus of the oviduct, it enters the shell gland region where the calcified shell is laid down on the slowly rotating ovum. The completed eggs are then held in this region of the oviduct until nesting. In females that lay multiple clutches of eggs, large preovulatory follicles are present in the ovary, in addition to shelled eggs.

Female tortoises display distinct seasonal cycles not only for estradiol, but also for circulating testosterone, progesterone, total calcium, and vitellogenin (fig. 5.2). All of these levels are affected by or affect the ovaries. As the ovarian follicles mature and increase in size from previtellogenic to vitellogenic (see below), the synthesis of testosterone increases. Circulating testosterone increases in association with ovarian recrudescence (revival after inactivity) prior to the mating period. At mating, the ovary probably is fully developed in tortoises, and the entire complement of follicles for the nesting season likely is present, although this part of the reproductive cycle is poorly studied in some species. Coincident to the onset of mating, testosterone and estradiol are at or near their maximum levels. As the nesting season progresses, subsequent clutches are ovulated and serum testosterone and estradiol levels are observed to decline to their nadir in June and July, the period of late nesting. During internesting periods, females—those that lay multiple clutches at least—may contain both fully shelled eggs in their oviducts and multiple preovulatory follicles in their ovaries. These eggs and follicles can fill the majority of the body cavity.

REPRODUCTIVE CYCLE

Vitellogenesis

The first stage of reproduction in the female is vitellogenesis, the process of ovarian and follicular growth. During vitellogenesis, estradiol (E2) is secreted into circulation by the granulosa cells of the previtellogenic ovarian follicles, in response to gonadotropin secretion by the pituitary gland. Circulating estradiol subsequently stimulates the liver to secret the vitellogenin protein. The vitellogenin protein then is taken up by the maturing follicles in the ovary. Vitellogenesis results in the incorporation of yolk proteins into the developing oocyte within the follicle. Vitellogenesis has been demonstrated to be estradiol dependent in a variety of reptiles (Ho 1987). Elevated levels of estradiol in captive *Gopherus agassizii* and wild *G. polyphemus* have been recorded during the fall months prior to hibernation, concurrent with follicular growth (Rostal et al. 1994b, Ott et al. 2000, Lance and Rostal 2002). In some populations, elevated levels of estradiol also can occur following emergence from hibernation, when the final increase in follicle size occurs one to two months prior to nesting.

Vitellogenesis occurs mostly in the fall, prior to hibernation (Rostal et al. 1994b, Ott et al. 2000, Lance and Rostal 2002). At this time, follicles are nearly fully developed and consistent in size (Rostal et al. 1994b). Vitellogenesis is completed in the spring shortly before or after emergence from hibernation. In *G. agassizii* that lay two or three clutches at approximately 30-day intervals, there appears to be a second or third vitellogenic surge if food is available.

Ovarian Cycle

Ovarian follicle development and size have been monitored in semicaptive and wild tortoises from several locations (Rostal et al. 1994b; Lance and Rostal 2002; Ott et al. 2000; R. Averill-Murray, personal communication; unpublished data; fig. 5.3). Female tortoises display distinct seasonal cycles in circulating testosterone, estradiol, progesterone, and total calcium. The female ovary undergoes an associated or prenuptial recrudescence prior to the mating period. Follicular maturation and ovarian recrudescence have been confirmed using ultrasonography in both captive *G. agassizii* and wild *G. polyphemus* (Rostal et al. 1994b, unpublished data). As the ovarian follicles mature and increase in size from previtellogenic to vitellogenic, the synthesis of testosterone increases. Circulating testosterone increases in association with ovarian recrudescence prior to the fall mating period. Mating in both species is observed prior to hibernation and following emergence from hibernation in the spring (Rostal et al. 1994b, personal observations). In *G. polyphemus,* only one clutch of eggs is produced following emergence from hibernation (Ott et al. 2000). In *G. agassizii,* however, two clutches are common, sometimes three if conditions are good (Rostal et al. 1994b, Lance and Rostal 2002).

Follicle size (diameter) is related to ovum size and the amount of energy invested by the female in reproduction. The energy is stored as yolk platelets in the ovum of the follicle; and these platelets plus the other substances essential for embryogenesis will produce both the hatchling and the yolk reserve that the hatchling will need for the first few months of its life. Variation in follicle size within and between females ultimately is expressed as variation in hatchling size between populations and species. Ovarian follicle size has been measured in wild populations of *G. agassizii, G. polyphemus,* and *G. morafkai* (Rostal et al. 1994b; R. Averill-Murray, personal communication; unpublished data).

Courtship and Mating

Males begin to establish territories in the fall, engaging in combat by using their gulars in an attempt to flip one another over (Ruby and Niblick 1994; fig. 5.3). Courtship and mating occur either in the fall or spring in most species of *Gopherus.* Once nesting begins, a significant decline in mating activity is observed. At this time, females have fully developed ovaries with large preovulatory follicles (Rostal et al. 1994b). Males have elaborate courtship behaviors, which include head bobbing and biting of the female forelegs (Ruby and Niblick 1994). It is not unusual to find more than one male *G. polyphemus* at the entrance to a single female's burrow during the fall and spring (personal observation).

Seasonal reproduction in tortoises is controlled by hormonal function. Testosterone appears to function in regulating seasonal reproduction in both male and female tortoises. The long-term elevation of testosterone in males appears primarily to have a physiological role, but also may have a behavioral role in priming specific regions of the brain. Male testosterosterone levels rise in July and August in *G. agassizii, G. flavomarginatus,* and *G. polyphemus* (Gonzalez 1995, Rostal et al. 1994b, Ott et al. 2000, Lance and Rostal 2002). This rise in male testosterone is correlated with spermatogenesis and male mating behavior during the fall. Within the preoptic region of males, the brain nuclei associated with reproduction are sensitive to elevated testosterone levels. In female *G. agassizii, G. flavomarginatus,* and *G. polyphemus,* however, testosterone appears to be involved directly in ovulation in the spring when the females emerge from fall hibernation and shell up eggs for oviposition (Gonzalez 1995, Rostal et al. 1994b, Ott et al. 2000, Lance and Rostal 2002). As circulating levels of testosterone decline following ovulation, females no longer are receptive and will avoid interactions with males (Rostal et al. 1994b, Ott et al. 2000, Lance and Rostal 2002). In some cases, females even have been observed to be aggressive toward males (personal observation).

Chin glands (also called mental glands) are unique to the genus *Gopherus.* Large-chain fatty acids are released from the chin gland, and probably serve to provide information related to species identity, territory, and rank (i.e., identification of conspecifics; Alberts et al. 1994) during periods of courtship and mating. The chin glands are more developed in males than in females. In *G. agassizii,* chin gland size varies seasonally and is largest during the fall months, corresponding with increased testosterone levels in the male. This period is also when there is an increased observation of mating and male-male combat in the species. Although chin glands do enlarge in male *G. polyphemus* in the fall, the enlargement is not as great as in *G. agassizii,* which may be related to smaller home ranges and higher densities of *G. polyphemus* resulting from limited upland habitat in the southeast United States (chapter 9).

Fig. 5.3. *A,* the enlarged gular scutes (arrow), as seen in this male *Gopherus polyphemus*, are made of keratin and begin to enlarge when circulating levels of testosterone increase during puberty. *B,* enlarged gular scutes are a male secondary sex characteristic and used in male–male combat during the mating season, when testosterone levels are elevated.

Ovulation

The first ovulation of the nesting season occurs after courtship and mating are completed. Subsequent ovulations following the first nesting event occur within 48 hours after the completion of nesting, and specific hormonal events occur during this period. Levels of progesterone, which is primarily associated with ovulation in turtles, increase sharply 24 to 48 hours following nesting. A significant increase in progesterone is first observed in March in both *G. agassizii* (Lance and Rostal 2002) and *G. polyphemus* (Ott et al. 2000). *In G. flavomarginatus,* progesterone was not observed to increase as sharply; however, it did elevate above levels observed during the nonnesting season (Gonzalez 1995). Comparable data are not available for *G. morafkai* or *G. berlandieri.* In many populations that hibernate, females have been observed to emerge from hibernation having ovulated an early clutch (Rostal et al. 1994b, Ott et al. 2000, personal observation). These females most likely used sperm stored from fall matings (Rostal et al. 1994b), and were actively thermoregulating within their burrows in March, prior to complete spring emergence (Ott et al. 2000). As the nesting season *of G. agassizii* progresses, and multiple clutches are laid, a steady decline in progesterone levels is observed (Gonzalez et al. 2000, Lance and Rostal 2002). Progesterone levels monitored at the time of nesting also were observed to decline with each subsequent clutch a female laid (fig. 5.2). In *G. polyphemus,* which lays only one clutch of eggs in a given year, progesterone levels decline rapidly following ovulation and nesting such that progesterone levels are near their nadir in July (Ott et al. 2000, unpublished data).

Luteinizing hormone (LH) is suspected to increase sharply following nesting, concurrent with the observed surge in progesterone and ovulation of the ovum in species that produce multiple clutches, such as *G. agassizii* and *G. flavomarginatus.* LH has not been measured in tortoises to date; but, based on sea turtle data, LH is thought to stimulate ovulation in concert with progesterone (Licht et al. 1979, 1982; Wibbels et al. 1992). If spring conditions are good in the Mojave Desert, female *G. agassizii* will undergo a second, and sometimes a third, activation of follicles, to be used for subsequent nesting (Rostal et al., 1994b). If fully mature follicles are not ovulated, they remain in the ovary and undergo atresia (Rostal et al. 1994b, Lance and Rostal 2002). Atresia is the process by which energy originally stored as yolk platelets in the oocyte during vitellogenesis and follicular growth is reabsorbed from the oocyte and potentially used by the fasting turtle. The follicle reduces in size over time and eventually becomes a corpora albicans, leaving behind a small scar.

Oviductal Egg Development

Ultrasound has been used successfully to track oviductal egg development in both captive and wild tortoises (Rostal et al. 1994b, Lance and Rostal 2002, unpublished data; fig. 5.4). Ovarian follicles (fig. 5.4, *top left*) are ovulated in response to luteinizing hormone (LH) and progesterone surges. Once ovulated, the ovum is first fertilized in the upper albumin gland region of the oviduct. As the fertilized ovum travels through the albumin gland region of the oviduct, it is surrounded by a nonechoic albumin layer (fluid which does not reflect sound waves because of the lack of calcium). Once the albumin layer is completed, the ovum plus albumin enters into the shell gland region of the oviduct where the shell membrane and the eggshell forms (Fig. 5.4, *top right*). In tortoises, fertilized eggs continue to develop further after being shelled, up to 30 days. During this period, the shell thickens and the ability to monitor development using ultrasound is reduced because of the heavy reflectance of ultrasound waves by the heavy impregnation of the shell by calcium crystals (fig. 5.4). Following the completion of the nesting season, the remaining follicles that have not been ovulated undergo atresia.

Oviposition

The process of egg laying (oviposition) is thought to be the most dangerous process for an adult turtle. During this process, a female must leave the safety of her burrow and be exposed to predators and other risks for as long as an hour, as she locates suitable sites for egg incubation. Most of our understanding of the physiology of egg laying in tortoises is based on data from *G. agassizii* and *G. polyphemus* (Rostal et al.1994b, Lance and Rostal 2002, Ott et al. 2001). Data for *G. flavomarginatus, G. morafkai,* and *G. berlandieri* are not available, but oviposition is presumed to be similar to other tortoise species, as the process appears fairly conserved in turtles (Blanvillain et al. 2010). Often, egg laying occurs on the apron of a burrow but may be outside the area of the burrow or may even be deep within the burrow (Rostal et al. 1994b, personal observation; chapter 6). Soil suitability and rainfall play key roles in the egg-laying process. If the soil is not structurally suitable or is too dry, the female will not be able to dig a proper egg chamber without the soil giving way and continuously filling the egg chamber. During egg laying, a series of specific behavioral and endocrine events occur in turtles, and levels of specific hormones change over the course of the nesting season. LH and progesterone surge at ovulation and levels of progesterone are observed to decline at nesting. Levels of testosterone and estradiol, which were high prior to ovulation, are at their lowest levels at nesting. Arginine vasotocin (AVT) is thought to increase in tortoises during oviposition, peaking at the time that the first egg is laid and subsequently declining to near prenesting levels within a 30- to 50-minute period as eggs are laid. The results of Figler et al. (1989) from sea turtles, as well as the successful use of AVT to induce ovulation in *G. polyphemus* (Demuth 2001), support the hypothesis that AVT has a physiological role in oviducal smooth muscle contraction and oviposition in tortoises.

Corticosterone, Thyroid Hormone, and Phospholipids

Corticosterone levels have been measured in *G. agassizii, G. flavomarginatus,* and *G. polyphemus* (Gonzalez 1995, Ott et al. 2000, Lance et al. 2001, Lance and Rostal 2002). Lance et al. (2001) showed that corticosterone levels were strongly correlated with testosterone in male *G. agassizii* and that levels

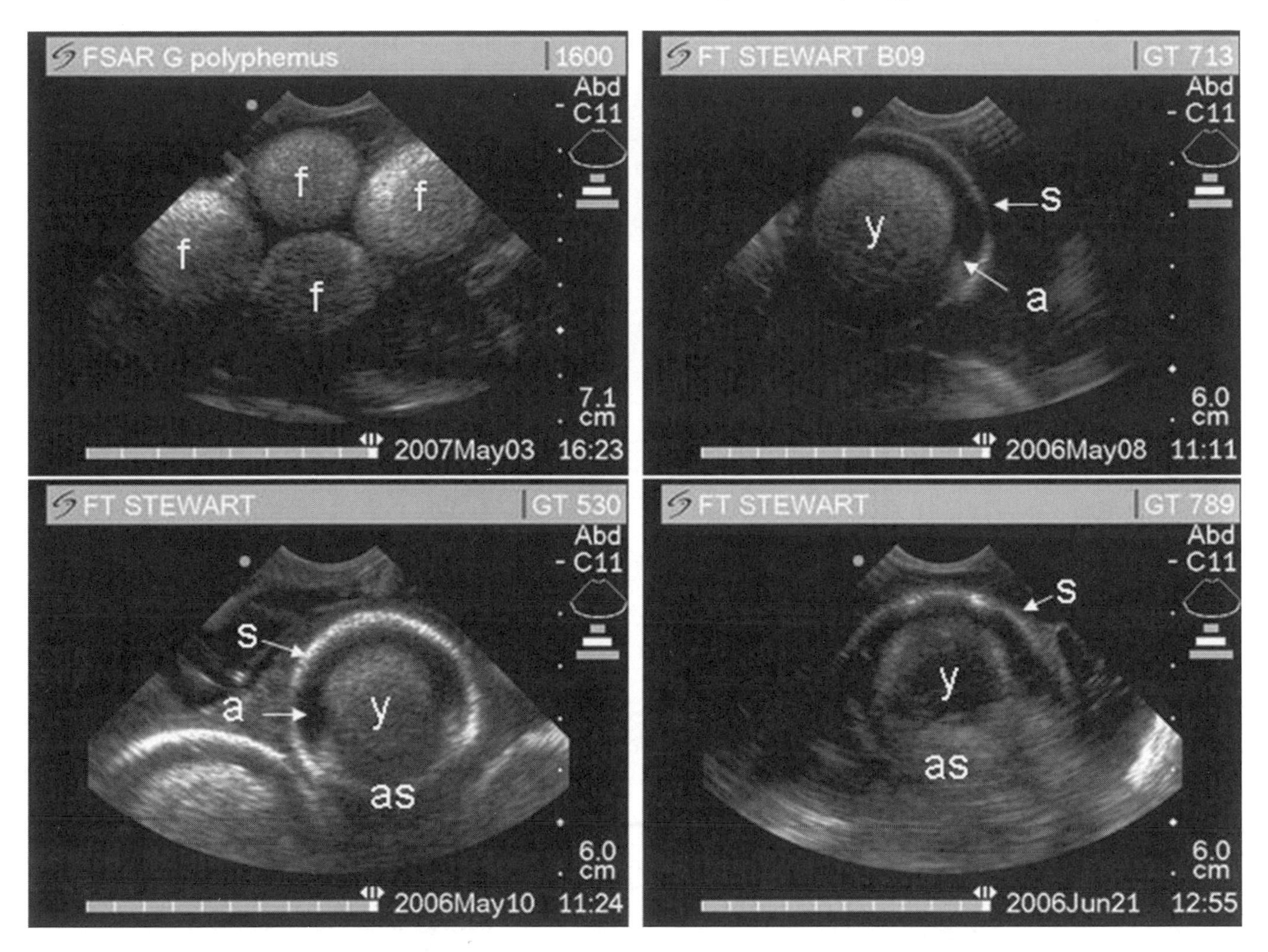

Fig. 5.4. Ultrasonography of *Gopherus polyphemus* ovaries and oviductal egg development. *Top left,* ultrasound image of a large vitellogenic follicle (*f*; 2.1 cm diameter) nearing ovulatory size prior to hibernation (October) in the ovary. *Top right,* ultrasound image of a recently ovulated oviductal egg (less than ten days postovulation) showing a well defined yolk (*y*; 2.3 cm diameter) and a thinly calcified shell (*s*). *Bottom left,* ultrasound image of a fully developed oviductal egg (between 20 and 30 days post-ovulation) showing a less defined yolk (*y*; 2.3 cm diameter) and a well calcified shell (*s*). Resolution is poorer at this stage due to the heavily calcified shell. Oviductal eggs were normally observed on both sides of the tortoise and could be manually palpated at this stage. *Bottom right,* fully developed oviductal egg (30+ days of development) showing full calcification of the shell (*s*) and limited visibility of the yolk (*y*). The acoustic shadow (*as*) covers half or more of the egg. This egg is lower in the oviduct than previous stages and is ready for ovipositing (nesting).

were much higher in males than in females. Female *G. agassizii* did not show as strong a seasonal pattern, but levels did spike in April and May, when spring mating and nesting is observed (Lance et al. 2001, Lance and Rostal 2002). In male *G. agassizii,* the highest levels of corticosterone observed were associated with a peak in spermatogenesis and male–male combat associated with fall mating (Ruby and Niblick 1994). Lance et al. (2001) note that these elevated levels are well below the range that would be induced by stress (20 ng/ml), and represent a natural seasonal pattern that appears associated with the natural reproductive cycle. In *G. polyphemus,* Ott et al. (2000) observed much higher corticosterone levels, which may represent stress from sitting in traps up to several hours before sampling (Lance et al. 2001, Lance and Rostal 2002). Corticosterone levels reported for *G. flavomarginatus* are also high, ranging from 5.14 to 31.67 ng/ml in males and 8.81 to 20.15 ng/ml in females. Average monthly levels above 10 ng/ml suggest that these animals also were stressed before sampling (Gonzalez 1995). No data on corticosterone levels are available for *G. berlandieri* or *G. morafkai.*

Thyroid hormone (T4) has been measured only in *G. agassizii* (Kohel et al. 2001, Lance and Rostal 2002). Seasonal patterns have been observed in both male and female *G. agassizii,* with females displaying a single peak in the spring and males displaying two peaks, one in spring and one in late summer. In both males and females, the highest levels were observed following emergence from hibernation in April. Males show a second elevation in T4 in late summer, as they begin to combat other males and undergo gonadal recrudescence and spermatogenesis in preparation for fall mating. This elevation in T4 corresponds with the beginning of elevation of testosterone and corticosterone in late summer (Lance and Rostal 2002).

Total lipids, triacylglycerol, phospholipids, and cholesterol associated with vitellogenesis are elevated in female *G. agassizii* during the fall, when ovaries are being prepared for the following spring nesting season (Lance et al. 2002, Rostal and Lance 2002). In contrast, lipid levels are greatest in the spring prior to testicular recrudescence and spermatogenesis in male *G. agassizii* (Lance et al. 2002, Lance and Rostal 2002). No data are available for the other *Gopherus* spp.

INFLUENCES OF UPPER RESPIRATORY TRACT DISEASE ON REPRODUCTION

Declines in wild populations of *Gopherus agassizzi* led to its federal listing (chapter 18). Early population declines were attributed largely to habitat loss and degradation, but more recent declines have been attributed to disease, in particular upper respiratory tract disease (URTD). The disease also is known in *G. polyphemus* (chapters 8, 18). Many studies have investigated how URTD is transmitted, and its impact on population ecology. Among other consequences, the disease affects a tortoise's ability to smell, which may interfere with the many functions based on odor, such as feeding and conspecific recognition (Alberts et al. 1994). Vertical transmission of the disease from mother to offspring also has been a concern (Schumacher et al. 1999), but whether the influence of URTD on reproduction is enough to cause population declines is unknown.

The acute effects of upper respiratory tract disease on reproduction in *G. agassizii* were monitored from 1991 to 1993 at the Desert Tortoise Conservation Center (DTCC) (unpublished data). The disease broke out in several research pens that had been set up to study reproduction following the listing of *G. agassizii* in 1991 (Rostal et al. 1994b). The outbreak appeared to result from the introduction of undiagnosed diseased individuals into several of these pens (Jacobson et al. 1995). The disease was allowed to run its course, because no effective treatment was available. Initial results showed that tortoises that had low titers of antibodies to the causative agent of URTD, *Mycoplasma agassizii,* and mild signs of URTD, such as serous exudate from the nares, were able to reproduce, but that tortoises that had high titers of antibodies to *M. agassizii* and severe signs of URTD, such as purulent exudate from the nares and dirt encrusting the front legs, failed to reproduce. The latter group of individuals also displayed relatively low hormone levels, remained in their burrows for extended periods, and did not come out to feed as often as the other tortoises.

In 2000–2001, 28 of the original 30 female tortoises and 16 of the original 20 male tortoises at the DTCC were still alive in their original research pens and still had their identification markings. Of these 28 female tortoises, 18 that originally tested positive for antibodies to *M. agassizii* (ELISA positive) and showed signs of the disease in 1992 were still alive and ten that originally tested negative for antibodies to *M. agassizii* (ELISA negative) and never showed signs of the disease were still alive. "ELISA" is an acronym for Enzyme-Linked ImmunoSorbent Assay, which is used to detect particular proteins—in this case antibodies. Equal numbers of ELISA positive and negative tortoises died or disappeared during the eight years between studies. During the summers of 2000 and 2001, 16 of the 18 ELISA-positive tortoises and nine of the ten ELISA-negative tortoises produced eggs. Several females that had stopped reproducing during the outbreak of the disease recovered and were reproducing again. Hatching success and hatchling size were similar for eggs produced by ELISA-positive and -negative females, during both years. Seventy percent of animals that tested positive for antibodies in 1992 never showed signs of URTD or exhibited signs only upon initial emergence from hibernation in 2000 and 2001. Many tortoises that previously tested positive for antibodies could not be distinguished from those that tested negative, based on signs of URTD during late summer and fall of 2000. Furthermore, *M. agassizii* could only be identified using PCR from those animals showing signs of URTD, and not necessarily from animals that had previously tested positive for antibodies.

These data support the conclusions that (1) tortoises with persistent titers to *M. agassizii* are capable of reproduction over extended time periods, (2) that ELISA-positive tortoises maintained with adequate nutrition can produce similar size clutches to free-ranging tortoises and to captive tortoises that have not been exposed to the *Mycoplasma,* and (3) that ELISA-positive tortoises produce hatchlings that are free of signs of the disease and are indistinguishable from hatchlings from URTD-negative animals (Schumacher et al. 1999). Given these results from captive tortoises, understanding the effect of URTD on reproduction in wild populations is critical. Although URTD also has been identified in *G. polyphemus,* studies of the disease have focused on transmission and exposure levels, and studies on the effects of the disease on reproduction do not exist. It is plausible that populations on good habitat with nonclinical or subclinical levels of mycoplasmal infection display normal reproduction and viable offspring (unpublished data).

Stress is a key factor in understanding the onset of URTD. Distinguishing acute stress from chronic stress is important. Animals under chronic stress should lack the normal stress response observed in a variety of vertebrates. Injection of adrenocorticotropic hormone (ACTH) should result in a rapid increase in circulating corticosterone released from the adrenal gland (Lance and Rostal 2002). This response was demonstrated in previous studies at the DTCC. Whether ELISA-positive tortoises have a compromised immune response is unknown. Whether animals with chronic URTD display normal stress responses also is unknown.

CONCLUSIONS

The reproductive physiology of North American tortoises is best understood for *Gopherus agassizii* and *G. polyphemus.* Males and females display distinct seasonal reproductive cycles. These cycles are adapted to the environmental conditions encountered throughout their ranges. For the three species for which data are available (*G. agassizii, G. flavomarginatus,* and *G. polyphemus*), both testicular and ovarian maturation are prenuptial, occurring predominately in the fall (Rostal et al. 1994b, Gonzalez et al. 2000, Ott et al. 2000, Lance and Rostal 2002). The seasonal patterns of testosterone, estrogen, and progesterone are similar to those reported for other turtle species (Blanvillain et al. 2010); however, further research is needed to clarify their physiological functions. Corticosterone, thyroid hormone, and lipid levels associated with seasonal reproduction have received the least attention. They have been studied mostly in *G. agassizii* and to a lesser extent *G. polyphemus* and *G. flavomarginatus* (Gonzalez et al. 2000, Ott et al. 2000, Lance et al. 2001, Lance and Rostal 2002, Lance et al. 2002).

Embryonic Development, Hatching Success, and Temperature-Dependent Sex Determination in North American Tortoises

David C. Rostal
Thane Wibbels

A variety of reptiles possess temperature-dependent sex determination (TSD) (reviewed by Janzen and Paukstis 1991, Janzen and Krenz 2004), including many endangered and threatened species. TSD is of significance in the conservation of endangered reptiles because it can produce highly skewed sex ratios (Bull and Charnov 1989; Mrosovsky and Provancha 1992; Mrosovsky 1994; Hanson et al. 1998; Wibbels et al. 1991b, 1998; Wibbels 2003, 2007), which can affect the reproductive ecology and survival of species (Mrosovsky and Yntema 1980; Morreale et al. 1982; Wibbels 2003, 2007; Girondot et al. 2004). Management of an endangered or threatened reptile exhibiting TSD should include monitoring of the sex ratios of hatchlings produced within populations (Vogt 1994; Lovich 1996; Wibbels 2003, 2007). Monitoring of hatchling sex ratios often is avoided because it requires sacrificing hatchlings. If certain characteristics of TSD are known, however, such as the thermosensitive period, transitional range of temperatures, and pivotal temperature, then hatchling sex ratios can be predicted by monitoring incubation temperatures (Hanson 1998; Wibbels 2003, 2007; LeBlanc et al. 2012). Further, such information can also facilitate the manipulation of sex ratios in hatchery programs in an effort to enhance their recovery (Vogt 1994; Mrosovsky and Godfrey 1995; Wibbels 2003, 2007).

In addition to its conservation significance, understanding the characteristics of TSD also can provide insight into the physiology and evolution of TSD. For example, comparison of pivotal temperatures among and between species may suggest if and how environmental temperature selects for specific transitional range of temperatures and pivotal temperatures (Bull et al. 1982; Janzen and Paukstis 1991; Wibbels et al. 1991b, 1998; Cevalier et al. 1999; Wibbels 2003). The period of temperature sensitivity has been examined in some reptiles with TSD (reviewed by Wibbels et al. 1991a, 1994; Rostal et al. 2002), and this information is essential for investigating the physiological and molecular basis of TSD in reptiles (Place and Lance 2004, Rhen et al. 2007, Rhen and Schroeder 2010). Four species of tortoises have been shown definitively to possess TSD: *Testudo graeca* (Pieau 1972), *T. hermanii boettgeri* (Eendebak 1995), *Gopherus agassizii* (Spotila et al. 1994, Lewis-Winokur and Winokur 1995) and *G. polyphemus* (Burke et al. 1996). An unpublished dissertation also suggests thermal effects on sex determination in *Geochelone carbonaria* (Mallman 1994). In species of turtles with TSD, nest temperature influences the sex ratio of hatchling turtles, while in species with genetic sex determination, the sex ratio of hatchlings is independent of incubation temperature. In species of turtles with TSD, low nest temperatures commonly produce more male hatchlings and high nest temperatures produce more female hatchlings.

This chapter provides a general characterization of TSD in North American tortoises. Two species have been studied extensively: *Gopherus agassizii* (Spotila et al. 1994, Rostal et al. 2002, Baxter et al. 2010) and *G. polyphemus* (Demuth 2001). These two species are found at the extreme geographic ranges for tortoises in North America (Ernst and Lovich 2009). Limited field data are available for *G. morafkai* but no data at all for *G. berlandieri* or *G. flavomarginatus*. That all members of the genus *Gopherus* appear to be closely related suggests that TSD may function similarly among them. The proper conservation and management of all five species likely requires an understanding of their reproductive biologies, inclusive of TSD.

TEMPERATURE, INCUBATION DURATION AND SUCCESS, AND SEX RATIO AT HATCHING

Several studies have investigated both the effects of constant temperature and naturally fluctuating temperatures on incubation times, hatching success, and hatchling sex ratios in North American tortoises (summarized in table 6.1). Under constant temperature regimes, both *Gopherus agassizii* and *G. polyphemus* show relatively long incubation durations (time period from laying until hatching): from 70 days to more than

Table 6.1. Incubation temperature, incubation duration, hatching success, and sex ratio of hatchlings for studies of *Gopherus agassizii, G. berlandieri,* and *G. polyphemus*

Species	Location	Incubation Type	Temp (°C)	Days	% Hatch Success	% Male	References
G. agassizii	E Mojave, Nevada	Lab (constant temp)	25			100	Lewis-Winokur; Winokur 1996
			27			100	
			28			100	
			29			100	
			29.4			100	
			31			41	
G. agassizii	E Mojave, Nevada	Lab (constant temp)	26.0	125	50	100	Spotila et al. 1994; Rostal et al. 2002
			28.1	89	96	100	
			29.0	87	100	100	
			30.6[a]	72	93	96	
			31.3	78	100	50	
			32.8	68	93	0	
			33.0	73	90	0	
			34.0	76	93	0	
			35.3	85	29	0	
G. agassizii	C Mojave, California	Experimental nests		90	79	40	Baxter et al. 2008
G. berlanderi	S Texas			88–118	60.6		Judd and McQueen 1980
G. polyphemus	SW Georgia	Field			86.0		Landers et al. 1980
G. polyphemus	C Florida	Field			67–97		Smith 1995
G. polyphemus	N Florida	Field		105	82.3[a]		Butler and Hull 1996
G. polyphemus	South Carolina	Lab (constant temp)	26	114.6	63	100	Burke et al. 1996
			29	97	89	75	
			32	86.3	78	0	
G. polyphemus	C Florida	Lab (constant temp)	26	112.57	34.78	100	Demuth 2001
			28	94.93	63.63	78.6	
			29	88.64	58.33	64.3	
			30	83.07	62.50	33.3	
			31	79.86	60.87	7.2	
			32	77.22	40.90	0	
		Exp *in-situ* Nest 1	34	85[b]	4.15	0	
		Exp *in-situ* Nest 2	na	87.75	83.33	40	
			na	90.5	66.66	50	
G. polyphemus	SE Georgia	Lab (constant temp)	27.9	97.2	84.45[b]	100[c]	Rostal and Jones 2002
			31.4	82.2		0[c]	

[a] Protected nests.

[b] Combined for both temperatures.

[c] Sex ratio predicted based on Demuth 2001.

125 days, depending upon temperature. For the two species, both low temperatures (26°C) and high temperatures (35°C in *G. agassizii* and 34°C in *G. polyphemus*) result in lower hatching success and hatchling growth rates (Spotila et al. 1994, Demuth 2001, Rostal et al. 2002). Incubation duration for natural nests of the two species range between 80 and 100 days under fluctuating temperatures, but show highly variable mean hatching success rates, ranging from 28% to 86% (table 6.1). This observed incubation duration appears to be the maximum possible within the northern part of the ranges of *G. agassizii* and *G. polyphemus,* where nesting occurs in May and June and hatching occurs in August and September, prior to temperatures decreasing to levels below those promoting normal activity in the two species (chapter 9). In *G. morafkai,* eggs are laid from June to August, and incubate until September or October in Arizona (Averill-Murray et al. 2002a). Nest sites in *G. morafkai* have been documented under large boulders, and temperature recording from these locations support a microhabitat temperature similar to those of *G. agassizii* nests in the Mojave Desert (R. Averill-Murray, personal communication; unpublished data). No evidence for overwintering of hatchlings of either species exists, as has been documented for some *Emydid* species that range into the northern part of North America (Ultsch 1989; Costanzo et al. 1995, 2004, 2008).

TEMPERATURE EFFECTS ON SEX DETERMINATION

A general characterization of TSD in *Gopherus agassizii* and *G. polyphemus,* including estimates of the effects of specific temperatures on sex determination, indicates that both species possess a "Male:Female" pattern of TSD (Type 1a; Ewert et al. 1994), in which relatively cool incubation temperatures produce males and relatively warm temperatures produce females. The transitional range of temperatures (range in which the sex ratio shifts from producing 100% males to 100% females; Mrosovsky and Pieau, 1991) extends from approximately 30.5°C to 32.5°C in *G. agassizii* (Spotila et al. 1994, Rostal et al. 2002) and 28.0° C to 31.5° C in *G. polyphemus* (Burke et al. 1996, Demuth 2001). The pivotal temperature of 31.3°C (producing an approximate 1:1 sex ratio; Rostal et al. 2002) is slightly lower than the 31.8°C predicted by Spotila et al. (1994). For *G. polyphemus,* the pivotal temperature is approximately 29.5°C (Demuth 2001).

The pivotal temperature for *G. agassizii* is one of the highest reported for any turtle (Alho et al. 1985, Ewert et al. 1994, Mrosovsky 1994, Wibbels et al. 1998, Wibbels 2003). The relatively high pivotal temperature estimated for *G. agassizii* may represent a prime example of how environmental temperature may select for pivotal temperature; in this case, a high pivotal temperature for a tortoise that inhabits a warmer xeric environment than other tortoise species studied (chapter 9). Because of the shallow nest depths produced by *G. agassizii,* little buffering of nest temperatures occurs, and, as a result, mean nest temperatures are higher (Baxter et al. 2010; Rostal, unpublished data). In contrast, *G. polyphemus* in the southeastern U.S. displays a pivotal temperature similar to those of other temperate turtle species (Bull et al. 1982, Janzen and Paukstis 1991, Wibbels et al. 1991a, Burke et al. 1996, Demuth 2001).

PERIOD OF SEX DETERMINATION

The thermo-sensitive period has been elucidated only for *Gopherus agassizii* (Rostal et al. 2002; fig. 6.1). A temperature-shift experiment, from a female-producing temperature (34°C) to a male-producing temperature (29°C), showed that embryos became committed to becoming females between stages 15 to 18 (embryonic staging from Yntema 1968). Embryonic staging is based on a series of developmental characteristics from stage 1 to 26 (fully formed hatchling). The most critical stages, from a sex determination perspective, are 14 to 21. During this period, in addition to gonadal formation, development of organs, such as eyes, limbs, and carapace, is observed. These stages of development can be correlated with temperature and number of days of incubation. Using key developmental structures, it is possible to stage the embryo. During temperature shifts from the male-producing temperature to the female-producing temperature, the sex of embryos remained labile as late as stage 21. These findings suggest that temperature sensitivity can begin before stage 15, and, depending on the specific temperature-shift regimen, some embryos remain sensitive to a temperature shift until at least stage 21 and have similar thermo-sensitive periods to several other species of turtles (Yntema 1979, Bull and Vogt 1981, Pieau and Dorizzi 1981, Yntema and Mrosovsky 1982, Wibbels et al. 1991a). Precise comparisons between studies are confounded by the fact that the specific temperatures and shift protocols used in each study can significantly affect the results, however (Wibbels et al. 1991b). For example, the length of the thermo-sensitive period of sex determination can vary slightly depending upon the specific male and female temperatures used in the experimental design (Wibbels et al. 1991b). Therefore, the results should be considered a general estimate of the thermo-sensitive period. The temperature-shift experiment with *G. agassizii* also showed that the sexual fate of embryos incubated at the female-producing temperature becomes irreversible earlier than embryos incubated at the male-producing temperature, a finding that has been reported for several reptile species exhibiting TSD (reviewed by Wibbels et al. 1991a, Merchant-Larios et al. 1997). The opposite finding has been reported for species of turtles, however (Bull and Vogt 1981, Pieau and Dorizzi 1981). Such variation could relate to interspecific differences in TSD, or to differences in the temperatures and protocols used in the various studies (Wibbels et al. 1991b).

The thermo-sensitive period, from embryonic stage 15 to stage 21, encompasses day 15 to 23 of incubation through day 28 to 42 of incubation, depending upon incubation temperature (table 6.2). This period represents the second quarter of the 70- to 125-day incubation period of *Gopherus* spp. In many species of turtles, the thermo-sensitive period of sex determination approximates the middle third of the incubation period, however (reviewed in Wibbels et al. 1991b, Wibbels 2003). This difference may be a result of the relatively long incubation duration of *G. agassizii,* which spends a relatively large amount of time in later embryonic stages, as they grow and assimilate yolk. Considering the incubation durations reported for the various North American tortoises (table 6.1), it is plausible that the second quarter of the incubation period may be the best approximation of the thermo-sensitive period

Table 6.2 Staging data for two species of *Gopherus*

	G. agassizii		*G. polyphemus*	
Stage of Development	Days at 29 C	Days at 34 C	Days at 27.5 C	Days at 31.5 C
12	na	na	na	14
15	23	14	25	18
16	25[a]	17	na	na
17	27	20	28	22
18	31	23	na	na
19	34	25	43	28
20	37	26	na	na
21	42	28	46	40
22	44	32	na	na
23	na	36	na	na
Hatching	86.5	75.5	103	92

Sources: Data for *G. agassizii* from Rostal et al. 2002. Data for *G. polyphemus* from Rostal, unpublished data.

Note: Embryonic staging based on Yntema 1968.

[a]Estimated.

of sex determination. Based on incubation studies including embryonic staging (table 6.2), sex is determined by day 43 of incubation at the latest, leaving half or more of the incubation period to develop from stage 21 to stage 26 (hatchling). At stage 21, embryos have limbs with scales and toes, a carapace with the beginning of osteoderms and scutes, and a head with eyes and beak. The remainder of incubation results in a larger and fully formed embryo. Future studies documenting the thermo-sensitive period of sex determination are obviously warranted in the other *Gopherus* species.

Gonadal differentiation in *G. agassizii* (Rostal et al. 2002) is similar in both chronology and morphology to that reported for other turtles with TSD (Yntema and Mrosovsky 1980, Pieau and Dorizzi 1981, Wibbels et al. 1991a, Merchant-Larios et al. 1997). Gonads begin to show sexual differentiation between stages 18 to 21, and exhibit distinct sexual dimorphism by stage 23. The onset of temperature sensitivity precedes any obvious sexually dimorphic differentiation of the gonads at the temperatures examined. Temperature-shift experiments (fig. 6.1) indicate that, at the female-producing temperature, the sexual fate of the embryo becomes irreversible before any distinct sexually dimorphic differentiation of the gonads; but, at the male-producing temperature, the sexual fate of the embryos becomes irreversible at a time when the gonads are already beginning to undergo sexual differentiation. In both *G. agassizii* and *G. polyphemus,* the gonads exhibit distinct sexual differentiation (i.e., stage 23) by approximately day 32 or 46 of incubation, depending upon incubation temperature (i.e., 34°C or 29°C in *G. aggassizii*, and 31.5°C or 27.5°C in *G. polyphemus,* respectively; Rostal et al. 2002; Rostal, unpublished data). Sexual differentiation in *G. agassizii* and *G. polyphemus* is similar to that described in other turtles, with cortical proliferation occurring at female-producing temperatures, and sexual cord development occurring in the medullary region of the gonad at male-producing temperature (fig. 6.2).

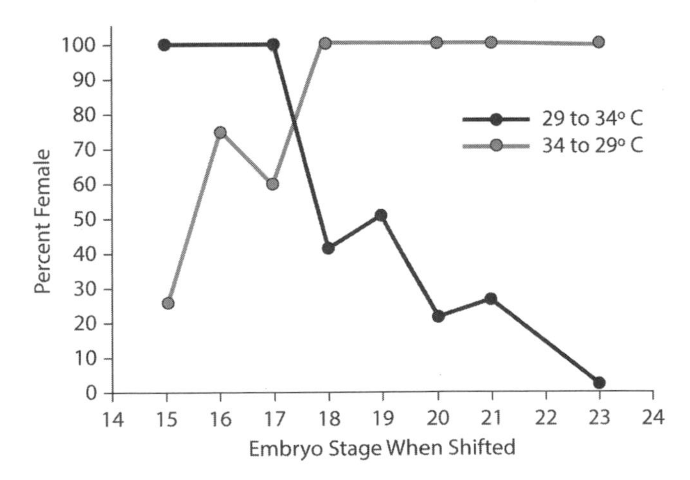

Fig. 6.1. Results of temperature-shift experiments, in which embryos of *Gopherus agassizii* were shifted from a male-producing temperature (29°C) to a female-producing temperature (34°C), or vice versa, once during incubation. From Rostal et al. (2002), Allen Press Publishing Services.

INCUBATION TEMPERATURE, MORTALITY, AND EMBRYONIC DEVELOPMENT

Hatching success rates for *Gopherus agassizii* and *G. polyphemus* were high at incubation temperatures ranging from 28°C to 34°C (table 6.1). Hatching success rates were low just outside this range, however, at incubation temperatures of 26°C and 35.2°C. Percent survival at 34°C was 95%, while percent survival at 35.2°C was only 30%. These findings are consistent with those reported previously for *G. agassizii* (Spotila et al. 1994). At temperatures that produced high hatching success rates, the total time of incubation ranged from 68 (32.8°C) to 89 (28.1°C) days. Monitoring of development at 34° C (76 days) and 29° C (87 days) indicated that embryos developed from stage 1 through 23 over a period of approximately 32 or 44 days, respectively, and then required from 44 to 43 days, respectively, to develop from stage 23 through 26 (i.e., hatchling). Interestingly, previous studies of other turtles suggested that lower incubation temperatures produced better hatching success (Packard et al. 1987, 1991). Once again, it is plausible that *G. agassizii* has adapted to the relatively high habitat temperatures not only in regards to sex determination, but also in regards to development and hatching success.

Baxter et al. (2008) created pseudonests for *G. agassizii* and recorded a seasonal trend in sex ratio production: early nests produced mostly males and later nests produced all females. The percentage of incubation time above and below pivotal temperature in those nests correlated with the observed sex ratios (i.e., warmer nests produced females and cooler nests produced males). Temperatures within natural nests fluctuate, however (fig. 6.3), and depending upon the magnitude of the fluctuations and the degree to which the temperatures fluctuate between the male and female temperatures (i.e., above and below the pivotal temperature for a specific species), interpretation of temperature data becomes more difficult (fig. 6.4). Models have been proposed to account for the effects of temperature fluctuations on TSD (Georges et al. 2004), but the accuracy of those models needs to be tested (Les et al. 2007, Eich 2009, LeBlanc et al. 2012). Thus, further studies are needed to address situations in which temperatures fluctuate widely around the pivotal temperature; however, in some nests, temperatures remain above or below the pivotal temperature and clearly indicate the production of either males or females (Hanson et al. 1998, Eich 2009).

TEMPERATURE DEPENDENCE AND CLIMATE CHANGE

Global climate change may have especially severe effects on species with TSD (Janzen 1994, Hays et al. 2003, Hawkes et al. 2007, Schwanz and Janzen 2008, Chaloupka et al. 2008, Fuentes et al. 2009, Hays et al. 2010, Mitchell and Janzen 2010,

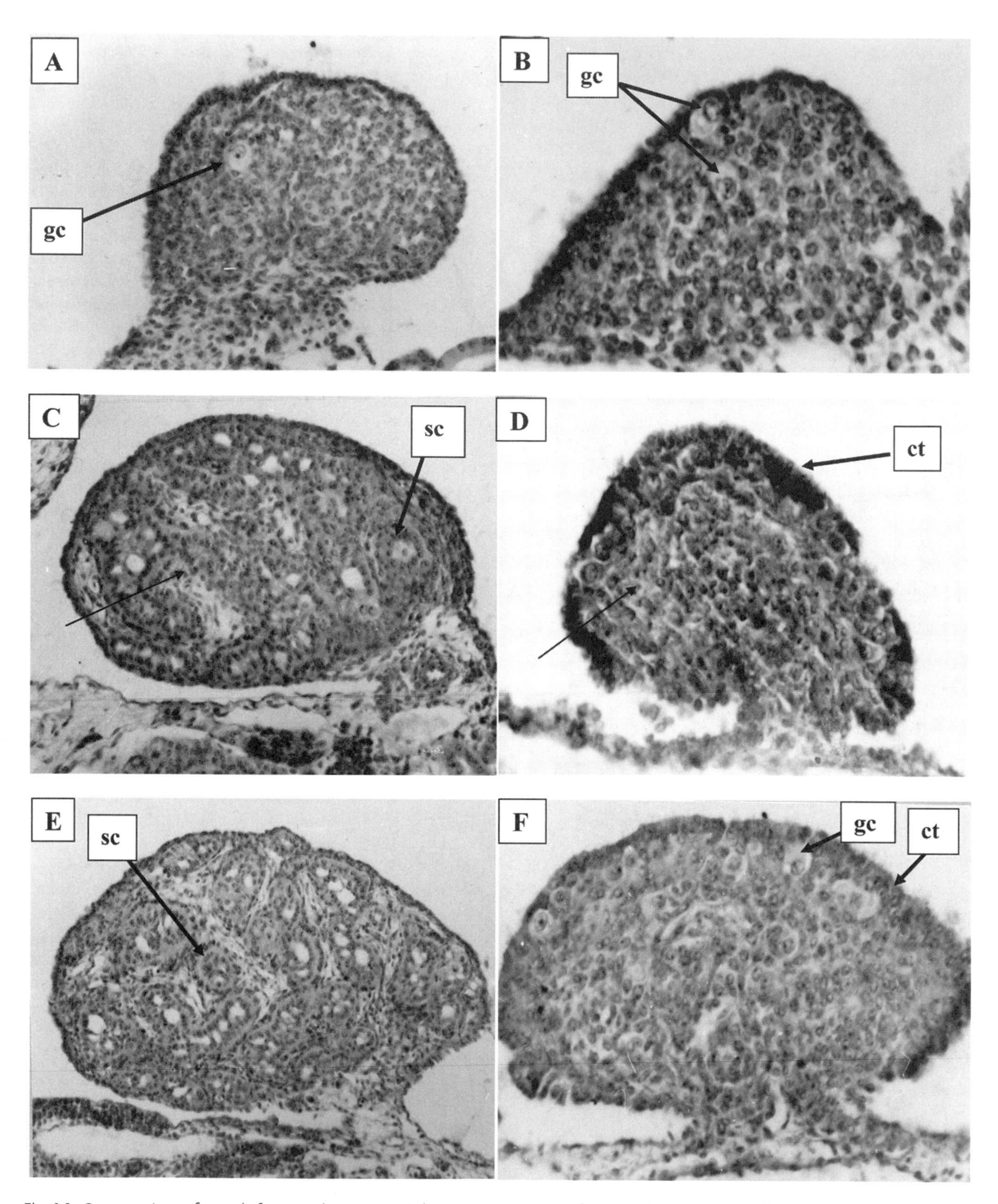

Fig. 6.2. Cross sections of gonads from *Gopherus agassizii* during various stages of embryonic development. Gonads from a male-producing temperature (29°C) are shown in *A* (embryonic stage 17/18), *C* (stage 21) and *E* (stage 22/23). Gonads from a female-producing temperature (34°C) are shown in *B* (stage 17/18), *D* (stage 21), and *F* (stage 22/23). By stage 22/23, sexually dimorphic changes are apparent, with medullary sex chords (*sc*) becoming prominent at the male-producing temperature. By stage 22/23 at the female-producing temperature, the cortical region (*ct*) has begun to thicken and germ cells (*gc*) often are associated with the cortex. Developmental staging based on Yntema (1968). From Rostal et al. (2002), Allen Press Publishing Services.

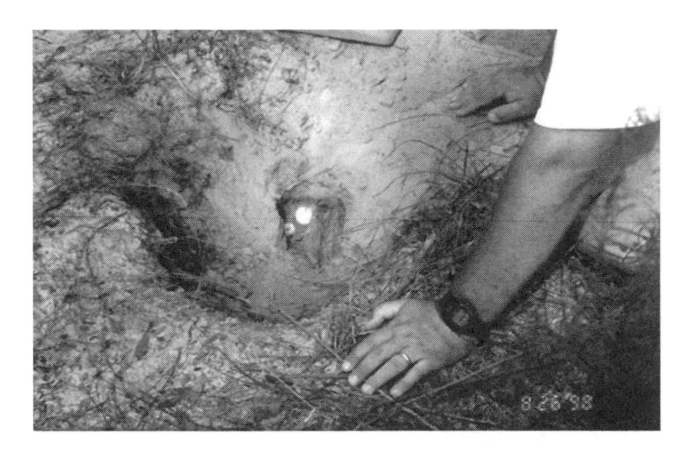

Fig. 6.3. *Gopherus polyphemus* nest located in the burrow apron in Georgia (*top*), and nest site location for *G. morafkai* under a large boulder in Arizona (*bottom*).

Patino-Martinez et al. 2012). An average global temperature increase of 1 to 4°C over the next century (IPPC 2007) creates a dilemma for reptiles with TSD because they typically have a transitional range of temperatures that only spans a few degrees C. Some species with TSD in fact already produce female-biased sex ratios, and in some cases, extremely so (reviewed by Shine 1999, Wibbels 2003). Studies of several populations of *G. agassizii* and *G. polyphemus* indicate that adult sex ratios tend to be roughly 1:1 (table 6.3), suggesting that these species may have some leeway and could shift to producing female-biased sex ratios in response to increased climatic temperatures. In laboratory studies, pivotal temperatures are based on eggs from multiple females; therefore, the reported pivotal temperature is an average for the population studied. In large populations in the wild, however, natural variation occurs, and individual females may pass on different pivotal temperatures to their progeny. Variation in pivotal temperature among females in a population is called "clutch effect." Such variation also may occur among populations. At least in the case of *G. agassizii*, however, an increase of a few degrees C could push temperatures into ranges resulting in embryonic mortality (e.g., 34° to 35°C; Rostal et al. 2002).

As temperatures increase, tortoises may shift their timing of nesting. A previous study of *Chrysemys picta* indicates that turtles may begin nesting earlier in a season in response to increased environmental temperature (Schwanz and Janzen 2008). It is also possible that tortoises may alter nest site location. Nesting location plasticity appears to be common in *Gopherus*. Female *G. agassizii* and *G. polyphemus* are known to nest under vegetation, on the burrow apron, and at different distances down their burrows. Soil temperature profiles taken from Florida to South Carolina show that temperatures in the southern portion of the range of *G. polyphemus* tend to be lower than those recorded in the northern portion of the range, but that the 80+ days required for incubation cannot be met in most of South Carolina. Although temperature data-loggers placed in sandhill habitat in northern South Carolina displayed suitable temperatures, the duration of these suitable temperatures was roughly 60 to 70 days before they decreased below viable levels (Rostal, unpublished data). Most other turtle species nesting in South Carolina, including sea turtles, require only 45 to 60 days for incubation. Tortoises appear constrained by their longer incubation duration to lower latitudes. A similar pattern was confirmed between Arizona and northern Nevada for *G. agassizii* (Rostal, unpublished data). Regardless, if environmental temperatures continue to increase as predicted over the next century, climate change has the potential to produce extreme sex ratios and detrimentally affect the hatching success of North American tortoises.

CONCLUSIONS

Specific details of TSD for *Gopherus morakai, G. flavomarginata,* and *G. berlanderi* are completely missing. The thermosensitive period of sex determination has been studied only in *G. agassizii* and, therefore, studies in other *Gopherus* species are warranted. Further research on natural nest temperatures and hatchling sex ratios from those nests would provide useful data. Such work can provide a resolute understanding of the transitional range of temperatures under natural incubation regimes and a better understanding of the effects of natural fluctuating temperatures on sex determination. This sort of information can provide considerable insight into the ecology and evolution of *Gopherus*.

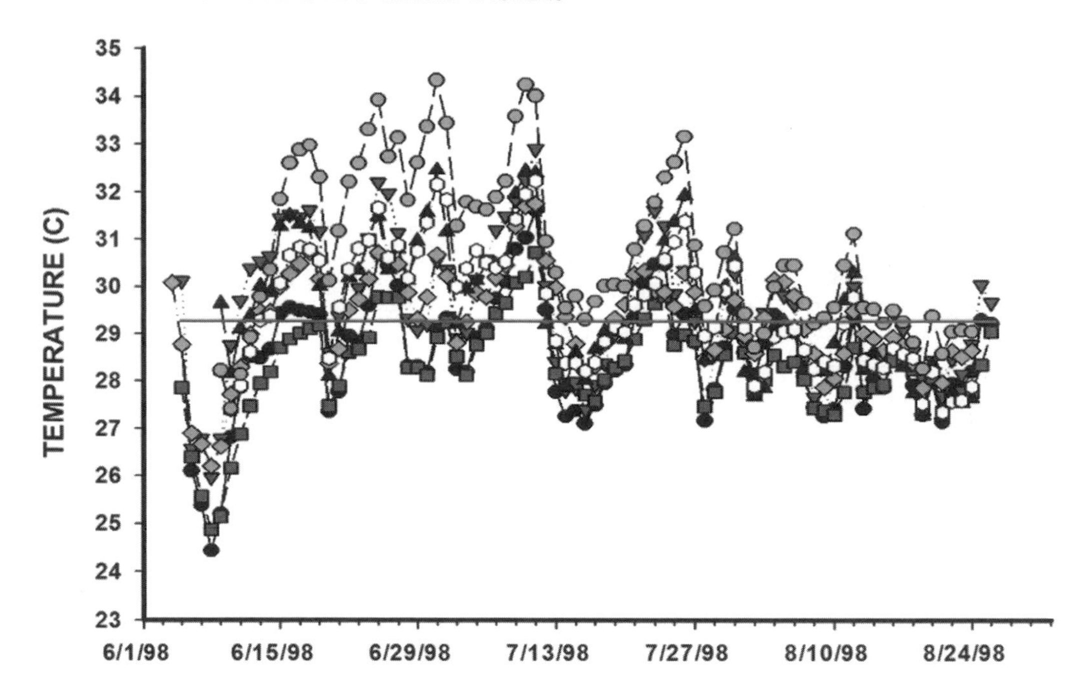

Fig. 6.4. Natural nest temperatures for *Gopherus polyphemus* from Fort Stewart, Georgia, in 1998. Each line represents an individual nest and each symbol represents the mean temperature per day for each nest. The horizontal line represents the pivotal temperature estimated for *G. polyphemus*.

Table 6.3 Adult sex ratios observed in natural populations

Species	Location	# of Males	# of Females	Sex Ratio (M/F)	Reference
G. agassizii	Beaver Dam, Utah			1:1.51	Woodbury and Hardy 1948
G. agassizii	Beaver Dam, Utah			2.33:1	Coombs 1977
G. agassizii	China Lake, California			1:1.78	Luckenbach 1982
G. berlandieri	S Texas	36	32	1:0.88	Rose and Judd 1982
G. berlandieri	S Texas	261	243	1:093	Hellgren et al. 2000
G. agassizii (morafkai)	Tiburŏn Island, Sonora			1:0.82	Osorio and Bury 1982
G. polyphemus	SE Georgia			1:1.1	Jones 1996
G. polyphemus	Mississippi and Louisiana	46	41	1:0.89	Smith et al. 1997
G. polyphemus	Florida			2:1	Diemer 1992b
G. polyphemus	Merritt Island NWR, Florida	227	213	1.07:1	Demuth 2001
G. polyphemus	George L. Smith SP, Georgia	30	38	1:1.27	Rostal and Jones 2002
G. polyphemus	Fort Stewart, Georgia	34	41	1:1.21	Rostal and Jones 2002

Studies of the potential impact of global climate change on North American tortoises should be strongly encouraged. The studies should focus on how tortoises can adapt to changes in environmental temperature. For example, research on the seasonal timing of reproduction could provide insight how sex ratio production may be affected by global climate change. It is plausible that tortoises may begin nesting earlier in response to increased environmental temperature. Temperature has been reported to be a cue in the timing of seasonal reproduction in many reptiles, and increases in global temperature could alter the timing of nesting. Because the precise molecular and physiological pathways underlying TSD are not clear, it is unknown if the pivotal temperatures and transitional range of temperatures in TSD could evolve in such a short time period to adapt to changes in environmental temperature. Considering their sensitivity to temperature, tortoises could prove to be an excellent model for examining the effects of global climate on the reproductive ecology of reptiles.

7 Growth Patterns of North American Tortoises

HENRY R. MUSHINSKY

The rate of growth of an individual tortoise affects the time it takes to achieve the minimum size for sexual maturity, and therefore has a significant influence on the number of offspring an individual can produce in its lifetime. Rapid growth of hatchling and juvenile tortoises also makes them less vulnerable to predation. Hence, understanding the factors that influence tortoise growth rates can provide insights into several life history traits. Growth rates of tortoises are variable: in general, individuals grow very rapidly until they begin to reproduce, at which time growth slows appreciably. As ectotherms, they depend upon their thermal environment to elevate body temperature and have the capacity to decrease their rates of growth to persist during times and at places of low resource availability (Andrews 1982). Turtles are thought to have functionally indeterminate growth: the epiphyses of the long bones remain cartilaginous throughout life. The growth zones of their bones narrow as they age, but turtles retain the capacity to increase in size until death (Haines 1969). As pointed out by Andrews (1982) several decades ago, the difficulty in evaluating growth of long-lived reptiles is that few data exist for individuals observed over a long time period; that difficulty remains today.

The outcome of indeterminate growth becomes biologically significant if the largest males gain the most copulations and/or the largest females produce the most eggs in a population. Evidence exists that larger *Gopherus polyphemus* males have a reproductive advantage over smaller males (Moon et al. 2006); therefore, larger size may translate into increased offspring production. A benefit of large size for female *G. polyphemus* is that as they grow, they add one mm in average egg diameter for every 32 mm increase in carapace length (Small and MacDonald 2001). If a large egg produces a relatively large hatchling, then the large size may translate into increased survival. As well, because larger females produce larger clutches of eggs than smaller females (Iverson 1980), large size may facilitate fitness of the female through increased survival of her young. In an exceptional case study, female *G. polyphemus* that established burrows on reclaimed phosphate-mined land in central Florida continued relatively rapid growth as mature individuals and produced unusually large clutches for the species (Small and Macdonald 2001).

GROWTH OF EMBRYOS, NEONATES, AND JUVENILES

Growth begins in the egg and is fueled by energy and nutrients derived from the female (chapter 5). The yolk sac that provides the essentials for growth is absorbed at the time of hatching or very shortly thereafter. Yolk material that remains at the time of hatching is called reserve yolk; it provides the initial source of energy for the neonate (Ewert 1979). Moisture available to the developing embryo influences the size of the hatchling and the amount of reserve yolk remaining at the time of hatching. Inadequate moisture for normal development can decrease the amount of yolk absorbed for growth of the embryo, and if not fatal, can result in hatching of relatively small individuals with a bolus of reserved yolk that remains outside the body cavity for a day or two before it is absorbed into the gut (Arata 1958, Linley and Mushinsky 1994). Postnatal lecithotrophy was defined by Lance and Morafka (2001) as the time of continued growth and development after hatching that is fueled by nutrients and energy provided by the yolk. As much as 50% of the caloric value of yolk lipids may be conserved as posthatching reserve for growth and mobility of North American tortoises. Lance and Morafka (2001) proffered that this nutritional stage could continue for the first six months of life. Accordingly, a postnatal lecithotroph *Gopherus agassizii* could hatch in September, disperse to its first shelter, inoculate itself with fermenting anaerobes via coprophagy, grow, burrow, and hibernate without ever ingesting conventional nutritional forage. Coprophagy is the ingestion of tortoise fecal matter that is done to inoculate the digestive tract of a hatchling tortoise with bacteria that facilitate the digestion of plant material.

Table 7.1. Growth rates of captive *Gopherus* tortoises, mm per year

Size class	N	Original	Increase	%	N	Original	Increase	%
	Gopherus polyphemus				*Gopherus agassizii*			
50–74	1	56.0	7.0	12.5	3	58.9	7.1	12
75–99	1	94.0	13.0	13.8	3	83.8	11.6	13.9
100–124	1	118.0	17.0	14.4	3	115.7	13.1	11.3
124–149	6	139.3	17.5	12.6	2	137.4	11.8	8.6
150–174	7	164.7	7.5	4.6	3	160.0	9.0	5.5
175–199	11	185.1	9.0	4.9	2	184.6	8.3	4.5
200–224	3	203.0	8.0	3.9	2	212.5	14.3	6.71
225–249	3	232.0	7.0	3.0	3	237.5	7.2	3.0
250–274					4	260.5	6.3	2.4

Sources: G. polyphemus data adapted from Goin and Goff 1941. *G. agassizii* data adapted from Patterson and Brattstrom 1972.

Hatchling size of *G. agassizii* is dependent on egg size, water availability during development, and incubation temperature (Spotila et al. 1994). Relatively large eggs produced larger hatchlings that retained their size advantage over smaller hatchlings. Small hatchlings had not caught up to larger members of cohorts nearly four months after hatching. All individuals grew at about the same rate; so, large eggs produced large hatchlings that were larger than siblings at 120 days of age. Eggs and hatchlings reared at 30.6°C grew significantly more rapidly than those reared at 28.1° or 32.8°C, and eggs reared at 35.3°C lost weight during incubation. Substrate moisture content affects hatching success; high moisture content (4% soil moisture) was lethal at 26.0°C and also at 35.3°C. At this high temperature, the eggs absorbed sufficient moisture to crack the shells, but the cracked eggs were repaired by the researchers and the amount of egg buried in the sand was reduced to facilitate the development of the embryos. At high soil moisture levels, the shell of the *G. agassizii* takes up moisture from the sand faster than it evaporates water vapor through the portion of the shell exposed to air. Dry sand (0.4% soil moisture) permits good growth and egg survivorship if the temperature is above 26.0°C and below 35.3°C.

The phrase "neonatal Eureptilian amniotes" (excludes birds) was coined by Morafka and colleagues (2000) to describe the first 10% of ontogeny of a young reptile prior to sexual maturity. The phrase recognizes the complex interaction of the environment (including the oviduct, nest, and habitat surrounding the nest) with the egg during the embryonic development and the earliest stage of life. If a tortoise requires more than ten years to become sexually mature, then the first year of life is its neonatal year.

Normal growth and development require a complete and nutritious diet for all ages of tortoises in a population. The diet of the neonate or juvenile tortoise can be quite different from that of conspecific adults. For example, juvenile *G. polyphemus* consumed 26 genera of plants, 16 of which were selected positively; and the most common plant, the grass genus *Aristida,* was not eaten during most of the year (Mushinsky et al. 2003). Although forbs such as *Liatris* and *Dyschoriste* are included in the diets of adult tortoises, they are the most important components in the diets of juveniles as they contain relatively high concentrations of nitrogen or other elements that are essential for proper growth and development (Garner and Landers 1981). Numerous researchers have reported on the growth of captive tortoises. Early reports implied that growth was highly predictable and nearly invariable, while more recent reports recognize a high degree of plasticity and variation that is dependent on nutrition. Table 7.1 is a summary comparison of the reported growth rates for *G. agassizii* and *G. polyphemus,* indicating that for a given size of tortoise, the relative increase in carapace length was quite similar. Based on a small collection of wild-caught individuals, Legler and Webb (1961) reported that the growth of *G. berlandieri* during the first six years increased by 15, 36, 21, 15, 10, and 9%, respectively, followed by 3–9% increases for seven years, except for a growth surge at age 12 of 16%. Maximum annual growth rate of the *G. flavomarginatus* measured in wild individuals at the Mapimi Biosphere Reserve near Durango, Mexico, occurs in subadults near maturation: 44 mm in females at 288 mm carapace length (CL), and 26.4 mm in males at 248 mm CL (Adest et al. 1989). The carapace length of *G. berlandieri* increased by 50% during year one of life, followed by 6% for years three to five; then, about 5% per year until age 18 (Auffenberg and Weaver 1969). Other researchers have reported that *G. berlandieri* increased in CL by 56% during the first year, from 42.2 to 65.9 mm (n = 8; Judd and McQueen 1980). *Gopherus berlandieri* is the smallest and *G. flavomarginatus* is the largest of the five taxa. Growth data on these two species are limited.

A stunning example of just how plastic the growth pattern of juvenile tortoises can be was presented by Jackson et al. (1976). High-quality nutrition and lack of dormancy during the first few years of life produced extraordinary growth in *G. agassizii*. Hatchlings were obtained from a clutch laid in captivity. The yolk sacs were absorbed in two days, and all began to eat and drink water immediately. Four individuals, all males, were reared on a substrate of ground alfalfa pellets, which was consumed both intentionally and unintentionally while ingesting a broad array of chopped fruits and vegetables offered twice per day. Individuals were allowed to graze on grass, clover, and flowers outdoors for up to two hours each afternoon. Extraordinary growth was observed during the first three years: increases averaged 176%, 43%, and 11%, respectively. Relative to typical growth rates of wild *G. agassizii,* these four individuals grew to the equivalent size of 11-, 17-, and 18-year-old tortoises during the three years of captive growth. As a consequence of the rapid growth, these individuals expressed distinct abnormal shell morphology, caused by the rapid accumulation of scutes, which created a pyramidal shape of the vertebral and lateral scutes. In a second paper on the growth of the same desert tortoises, Jackson et al. (1978) reported that they grew by 11.7% and 4.2% in their fourth and fifth years. These captive raised four-year-olds were equivalent to the size of 20-year-old wild individuals. The four males showed courtship behavior during the fourth year when a female was placed in their pen; and one individual examined by necropsy had mature, motile sperm in the epididymides.

GROWTH AND ENERGETICS

An interesting series of papers has addressed the energetic costs of growth in neonate reptiles. Nagy (2000) compared the costs for neonate reptiles, young and adult birds, and mammals. Mammals and birds have much higher baseline metabolic costs because of endothermy; energy for growth is in addition to these basal metabolic costs. Adult reptiles expend small amounts of energy on growth; essentially they nearly stop growing once maturity is achieved, but young reptiles must grow rapidly to survive and require large amounts of energy to fuel that growth. Nagy concluded that there was little evidence indicating that high energetic costs of growth in rapidly growing neonate reptiles increased their metabolic rates above those expected for adult reptiles of the same body mass. Resource needs, including food supply and space requirements, of neonates may not be unusually high and they may not be more vulnerable to resource shortages than adults. Compared to birds and mammals, neonate reptiles compensate for energetic costs of growth by having different patterns of behavior than adults, so as to reduce energy expenditures for other processes and maintain metabolic rates that are typical of similar-sized adults. Because the metabolic costs of growth in reptiles have been difficult to detect, a set of experiments was designed to quantify any added costs of growth by comparing energy expenditures of young growing *Gopherus agassizii* with similar-sized, but adult, Parrot-Beaked Tortoises (*Homopus areolatus*) (Brown et al. 2005). The study was conducted both indoors under highly controlled conditions and outdoors under more variable conditions, but within enclosures. The two species were matched for body size to minimize the body size effects. In outdoor enclosures, mass-adjusted field metabolic rates of the juvenile *G. agassizii* were not higher, and, in fact, were significantly (29%) lower than those of the adult Parrot-Beaked Tortoises. In controlled indoor environments, standard metabolic rates did not differ between species at any of the temperatures tested (20°, 25°, and 30°C), thus failing to provide any support for the idea that energetic costs in juveniles should be high because of the demands of growth. The effects of early life stage and intensity of growth on juvenile growth costs are intriguing issues for further research; the authors proposed long-term studies of tortoises from hatchling to adulthood to determine the energetic costs of growth early in life.

Two growth models have been used to examine the growth patterns of the *Gopherus polyphemus*; the details of these models are presented in Frazer and Ehrhart (1985). The general von Bertalanffy and logistic equations, with CL as the measure of tortoise size, were used by Mushinsky et al. (1994) to model the growth of tortoises in central Florida. Because the researchers did not have known-age tortoises older than 20 years, they used Fabens' (1965) method for fitting modified (growth interval) equations to data obtained from recaptures of individuals. As judged by residual mean square values, the logistic growth-interval equation described the growth pattern of the *G. polyphemus* better than the von Bertalanffy equation. The same two growth models were applied to capture-recapture data collected by Aresco and Guyer (1999b); and, in contrast, the von Bertalanffy model provided the better fit to their data. The later study was done in south-central Alabama, where tortoises grow more slowly than in central Florida, taking about 20 years to achieve sexual maturity, rather than nine to eleven years. Difference in growth patterns of tortoises in these two populations is quite startling. For example, ten-year-old tortoises in central Florida have a mean carapace length of 208 mm, while the ten-year-old tortoises in south-central Alabama have a carapace length of only 153 mm. Aresco and Guyer (1999b) suggest that the forage plants available at their study sites, pine plantations, are not adequate to fuel tortoise growth in a manner adequate to fit the logistic growth model. In summary, these two studies present strong evidence for the value of nutritional quality for the well-being of tortoises.

ALLOMETRY AND SEXUAL DIMORPHISM

Proportional changes in body dimensions contribute to the development and shape of the adult body form. The most detailed study of allometric growth was done by Landers et al. (1982): they measured body width and thickness and related those measures to changes in CL in *Gopherus polyphemus.* From hatching through 17 years of age—which is the approximate age at maturity in the study population—the major axis of growth was antero-posterior: width and thickness increased at 73% and 41% of CL, respectively. The increase in CL was greatest during the first four to eleven years of life; and from age five to seventeen, growth was at the same relative rate in all three dimensions. In this southern Georgia population, as individuals approached maturity, annual increments of all body dimensions decreased markedly, but females grew slightly faster and became larger than males. Allometric relations of mature tortoises were different between sexes. In mature males, incremental increases in body width were nearly equal to increases in CL, while body thickness increase was more rapid, relative to CL increase. In mature females, CL and body width continued to increase at about 1/3 the rates of 12–17-year-old tortoises, while body thickness increased at a slower rate. These allometric changes produced females that were longer and wider relative to their body thickness than they were at ages 12–17.

Sexual size dimorphism (SSD) likely is a product of which sex achieves sexual maturity first, rather than the product of sexual selection (Gibbons and Lovich 1990). *Gopherus polyphemus* females were larger than males in the southern Georgia population studied by Landers et al. (1982); but, as pointed out by Mushinsky et al. (1994), the degree of SSD reflects the number of years one sex requires to become mature, relative to the other sex. In southern Georgia, females continue relatively rapid growth for three to five years longer than males, producing considerable sexual size dimorphism, while in central Florida, both sexes mature at ages nine to eleven years, and males and females are much less dimorphic in body size at the same age (Mushinsky et al. 1994). Additional dimorphic characteristics are the gular projection (larger in males, but subject to wear and breakage), and several characters associated with egg-laying by females, including the thickness of the anal shield, anal notch dimensions, and the anal width (McRae et al. 1981). The depth of the plastral concavity was the best character for determining the sex of the *G. polyphemus* in Florida (Mushinsky et al. 1994). A detailed phylogenetic analysis of SSD in 82 taxa of turtles concluded that both male and female size are significantly correlated with fecundity traits, meaning larger turtles produce larger clutches, but the degree of SSD does not correlate with fecundity traits, indicating that the sexual dimorphism, per se, does not influence traits associated with reproduction (Gosnell et al. 2009).

Within the genus *Gopherus,* species exhibit two patterns of sexual size dimorphism. Adult females of the *G. flavomarginatus* (Morafka et al. 1989) and *G. polyphemus* (Landers et al. 1982) tend to be larger than males of the same age. Alternatively, adult males of the *G. berlandieri* (Rose and Judd 1982) and *G. agassizii* (Berry 1989) tend to be larger than females of the same age. This pattern reflects the evolutionary history of the genus, with the closely related *G. polyphemus* and *G. flavomarginatus* being the largest members of the genus.

Fig. 7.1. Plastron of a hatchling *Gopherus polyphemus* (*top*), with a visible umbilical scar (*arrow*). Plastron of a *G. polyphemus* in its third year of life (*bottom*), with two complete growth rings (*arrows*). This individual is experiencing rapid growth during the summer of its third year of life, as evidenced by the width of the new scutes (that appear lighter in color than the older scutes) being formed along the midline of the body. Photos courtesy of W. Hentges and A. Hathaway.

COUNTING GROWTH RINGS AND AGING TORTOISES

As a tortoise grows, it forms "growth rings" on its shell that have been used to assess the age of tortoises, much like counts of tree rings can be used to age a tree (fig. 7.1; chapter 16). Wilson and colleagues (2003) have provided a detailed analysis of the use and misuse of estimating the age of turtles by counting scute growth rings or annuli. As they point out, no detailed investigations of the development of the keratinized covering of turtle shells have been done. Not all species of turtles retain their scutes, as do the North American tortoises. In tortoises, the margins of the old scute layers form a shallow pyramid of successive scute layers, each new layer slightly larger than the previous one. A critical assumption of assessing age by counting growth rings is that the rings are annular (i.e., one growth ring is formed per year, in which case a growth ring reflects growth during a single complete year of development). Under some circumstances, this assumption certainly holds, but one cannot assume that one scute ring equals one year of growth in all cases. Growth of a tortoise need not be uniform and consistent. During periods of slow growth, a relatively thin layer of the epidermis is keratinized to form a depression in the scute that can appear as an incomplete layer. Incomplete growth rings have been considered to be "false rings" (Germano 1988). To be confident that one can age tortoises in a population, one must recapture marked individuals over a long time period and recapture at least some of them at less than yearly intervals to calibrate growth rings with years of age (Wilson et al. 2003). Even then, the data can only be interpolated, not extrapolated, within the population.

Growth of a tortoise is strongly correlated with resource abundance and food intake. Some researchers (e.g., Germano 1988, 1992, 1994) support the notion that scute growth rings form after growth spurts, with most growth occurring during a well-defined portion of the year when resources are abundant. This type of growth produces one distinct growth ring per year. Other researchers (e.g., Tracy and Tracy 1995) argue that scute growth rings form as structural adaptations to strengthen the shell as body size increases, and, therefore, that scute growth rings are added as a function of an increase in body size, not age. Tracy and Tracy (1995) conducted a controlled laboratory growth experiment with *Gopherus agassizii* neonates of known age that they raised for several years. One- and two-year-old individuals in their study had anywhere from 0 to 14 rings; few had the one to two rings expected if rings were annular. Tracy and Tracy (1995) detailed how their results had potentially devastating ramifications for estimates of the ages of free-ranging tortoises based upon counts of scute growth rings. For example, Germano (1992), using counts of scute rings, estimated that most adult *G. agassizii* don't live beyond 30 years; yet, Brussard et al. (1994) provided evidence that they may live 50 years or more. More recently Curtin et al. (2009) used skeletochronology to determine that members of this species do live beyond 50 years. Because population viability analysis of rare species depends upon accurate estimates of age at maturity and generation length, overestimates of age because individuals accumulate more than one growth ring per year would lead one to believe that generation time is shorter than it actually is (Berry 2003b).

The literature on the use of scute ring counts to age tortoises, or turtles in general, has been controversial. After reviewing 145 published papers that used counts of scute rings to age turtles, Wilson et al. (2003) concluded that aging turtles from such counts is feasible in some types of studies, and at some locations, but only after carefully calibrating relationships between scute ring counts and time. Of the 145 papers, 101 did not undertake any calibration. Two studies of *G. polyphemus* and one of *G. berlandieri* illustrate how counts of scute rings can be calibrated, as well as the inherent limitations of the method.

During a ten-year study in central Florida, Mushinsky and colleagues (1994) captured and recaptured 284 *G. polyphemus* individuals 571 times, and 40 adult females were radiographed to determine the minimum size at first reproduction. At about 11–12 years of age, the growth rate of the tortoises declined markedly, and the researchers could not make accurate counts of scute rings. The slow growth rate was not sufficient to produce individually discernible rings. They were able to test the validity of using scute counts to age individuals less than 12–13 years of age because many of the marked individuals were recaptured several times during the ten years of the study. No estimated age, based on counts of scute rings, varied by more than one year for known-age individuals. The researchers also established that the rapid growth detected in juveniles (when scute rings are most distinct) slowed greatly when sexual maturity was reached. Some individuals in this population achieved sexual reproduction in nine to ten years. From age one to twelve years, individuals grew at the average rate of 18.9 mm CL per year, and the rate declined to about 3% per year as individuals approached 20 years. Because accurate determination of the age of tortoises at this site was verified by recaptured individuals, the researchers were able to determine growth patterns of some for up to 20 years of age.

The second detailed study of tortoise growth was done in south-central Alabama (Aresco and Guyer 1998), a location with a shorter growing season than the previous study in central Florida. *G. polyphemus* ages calculated from measurements of scute rings were compared to ages determined from mark-recapture data. The number of scute rings produced during capture intervals matched the actual number of years of the interval in all tortoises (n = 39) less than 15 years old. Slow growth prevented accurate scute measurements on most mature tortoises older than 15 years. The researchers found false scute rings on 27% of the juveniles and subadults captured in one year, 1996, and postulated that these false scute rings were formed in 1995 because the study site experienced low rainfall that year and growth was very limited. False rings are those that are incomplete and/or very shallow (Germano 1988).

A substantial publication on the poorly studied *G. berlandieri* summarized more than four years of data obtained by a capture-mark-recapture study (Hellgren et al. 2000). While growth of individuals was not the focus of the research, the researchers presented some pertinent data. They were able to age the tortoises by counts of growth rings, and verified their counts with 1020 captures of 835 individuals. They calculated mean annual growth rates of juvenile tortoises (n = 52) to be 15.9 +/– 0.7 mm per year, from a carapace length of 40 mm at hatching. This study provided strong evidence for rapid growth of juveniles until sexual maturity was achieved, at an age of five years and a CL of 120 mm.

TORTOISE GROWTH IN RELATION TO CLIMATE AND HABITAT QUALITY

The literature on the relationship between North American tortoise growth patterns and various measures of climate is ambiguous. By far, the most comprehensive analysis of the growth patterns of North American tortoises was completed by Germano (1994). He accumulated 20 years of monthly data on precipitation, mean maximum, and mean minimum temperatures from numerous weather stations within the geographic ranges of the four species. Annual precipitation levels were highest for *Gopherus polyphemus* (1162–1593 mm), followed by *G. berlandieri* (472–982) and *G. flavomarginatus* (310–376). Within the ranges of the two western species, he reported precipitation levels of 278–664 mm per year in Sinaloa, 140–324 mm per year in the Sonoran Desert, 101–223 mm per year in the eastern Mojave Desert, and 102–169 mm per year in the western Mojave Desert. Maximum and minimum temperatures did not vary greatly across the ranges of the species, suggesting that all North American species are exposed to nearly the same thermal extremes, although the highest summer temperatures occurred within the Sonoran Desert and eastern Mojave Desert, and averaged 6–7° C higher than temperatures within the ranges of *G. polyphemus* and *G. flavomarginatus*. Given that the five species are herbivores, and plants respond favorably to precipitation, Germano (1994) anticipated a positive relationship between annual precipitation and growth rates and patterns. He used four growth measurements to investigate their relationship to the environmental data. All growth data came from measures of scute rings or other morphological measurements of living and preserved tortoises that were fitted to growth curves. The four measures used were asymptotic size, weighted mean annual growth rate, percentage of asymptotic size achieved at curve inflection, and the time period in years required to grow from 10 to 90% of the asymptotic size. None of four growth measurements were correlated with precipitation among the species of *Gopherus,* although mean annual growth was negatively correlated with mean annual precipitation for *Gopherus agassizii*. The underlying assumption was that each of his four metrics should be related to the precipitation data he had collected over a 20-year period. He concluded from his analysis that the environment is not the main determinant of growth rates of North American tortoises.

A recent study compared longevity and growth strategies of the western species in two American deserts (Curtin et al. 2009). These researchers used skeletochronology to assess age and growth patterns. In the Mojave Desert rainfall is irregular and resources are more limited than in the Sonoran Desert. Individuals from the western Mojave Desert grew faster than those from the Sonoran and achieved sexual maturity in 17–19 years compared to 22–26 years to sexual maturity for the Sonoran females. The oldest individuals from the Sonoran Desert were 54-year-old males and 43-year-old females. The oldest individuals from the western Mojave Desert were 56-year-old males and 27-year-old females. Male *G. agassizii* are larger than the females. This difference in longevity and body size of males and females possibly results from females channeling all or most of their nonmaintenance energy into reproductive effort. Because of the faster growth of *G. agassizii*, *G. morafkai* of similar size are older in age and older at death. Tortoises at both locations have a significant relationship between size and age for all adults. The results of the Curtin et al. (2009) study illustrate the significant influence growth rate can have on essential life history parameters, such as age at first reproduction and longevity. Frequent droughts and low annual rainfall in the western Mojave Desert can cause chronic physiological stress, which was proposed as the major driving force behind the two contrasting life histories and growth strategies of the two species. Interestingly, Wallis et al. (1999) reported that tortoises residing in the eastern Mojave Desert produce eggs at a smaller size, produce smaller eggs, and lay more eggs than those residing in the western Mojave Desert.

The influence of rainfall on growth of the *G. agassizii* was documented by a long-term study in southern Nevada. Starting with 15 individuals one to four years old between 1963 and 1965, Turner et al. (1987b) tracked their growth for more than two decades. The tortoises were in three nine ha enclosures but otherwise exposed to natural variations in climate and food availability. In 1987, these tortoises were as old as 26 years; no difference in the growth of males and females was evident. Tortoises grew to plastron lengths of 100 mm in six to seven years, 130 mm in 10–11 years, 150 mm in 13–14 years, and were 215 mm at 24 years of age. In a follow-up paper (Medica et al. 2012) based on these same tortoises, the authors report that males continued to grow more rapidly than females, and at age 30, body size differences became significant, and annual tortoise growth was strongly correlated with winter rainfall. Because of their long-term data set, the authors were able to address statements made in the literature that *G. agassizii* expresses rapid growth between the ages of 15 and 20 years (Patterson and Brattstrom 1972). A similar idea was put forward by Landers et al. (1982) that *G. polyphemus* has an apparent "surge" in growth at carapace lengths of 100–120 mm. The data obtained from this long-term study in southern Nevada refute the notion of rapid growth of *G. agasizii* between 15 and 20 years old. Rather, the authors tracked rainfall and plant productivity during various years and concluded that rapid growth of tortoises in their enclosures reflected unusually good environmental conditions and not some intrinsic process. Evidence to refute the idea that the *G. polyphemus* has a growth surge was published by Mushinsky et al. (1994) and Aresco and Guyer (1998, 1999b). Young juveniles of all North American tortoise species grow rapidly; annual growth seems to reflect local climatic conditions and availability of nutritious plants much more so than any intrinsic pattern of growth. Body condition is a measure of the relationship between body mass and size and was evaluated recently by McCoy et al. (2011). A comparison of body condition indices indicated that mean body condition was lower and seasonal fluctuations were of lesser amplitude in *G. polyphemus* than *G. agassizii*. Body condition of *G. agassizii* was correlated with rainfall, but body condition of *G. polyphemus* was not. Changing rainfall patterns caused by global warming may put both species at risk.

Many natural habitats within the ranges of North American tortoises are now being invaded by nonnative plants. One of the first studies to address the possible influence of nonnative plants on *G. agassizii* biology was conducted by Hansen et al. (1976), who studied diets in three locations by examination of fecal pellets. Although this study did not address the influence of nonnative plants on the growth of the individuals directly, it did document substantial changes in tortoise diets at a location heavily colonized by nonnative plants. For example, foxtail brome (*Bromus rubens*) and redstem filaree (*Erodium cicutarium*), both nonnatives, made up 20% of the diet at the Grand Canyon site and 85% of the diet at the Beaver Dam Wash site, indicating that tortoises are forced to eat what is available for them to consume. The potential detriment of being a general herbivore was addressed by a more recent study on the nutritional quality of foods of juvenile

G. agassizii (Hazard et al. 2009). These researchers compared the growth (weight gain or loss) of young tortoises fed four different foods: *Achnatherum hymenoides,* a native grass; *Schismus barbatus,* a nonnative grass; *Malacothrix glabrata,* a native forb; and *Erodium cicutarium,* a nonnative forb. The greatest nutritional difference among the four diets was between the forbs and grasses, not between the native and nonnative plants. The grasses were higher in fiber content and contained less digestible energy than the forbs. The nonnative forb yielded more energy and nitrogen per unit dry mass than the native forb. Young tortoises gained weight rapidly when eating either of the forbs but lost weight when eating either of the grasses. The researchers noted that tortoises ingested less volume of the grasses than of the forbs, and may not have ingested sufficient quantities of grasses to maintain body condition. Although there appeared to be few nutritional differences within the food types (grasses or forbs) the authors offered a broader perspective that is cause for concern for the future of the *G. agassizii.* As the dominant nonnative, nutritionally poor, invasive grasses increase in biomass by spreading across the range of the species, they are causing a reduction in native forb diversity. If tortoises are forced to switch from native forbs to less nutritious grasses and/or spend more time searching for the more nutritious forbs, there may be ecological and nutritional consequences for the species.

CONCLUSIONS

Tortoises have anatomical adaptations that likely enable them to grow until death, yet the first years of rapid growth have the greatest biological significance. Soil moisture and temperature strongly influence embryonic development and the amount of reserve yolk available to fuel posthatching growth. All five species of *Gopherus* exhibit sexual size dimorphism and exhibit sexually distinguishing characteristics. Age of individuals is difficult to determine. The use of scute rings to age tortoises must be done with caution and verified by independent data. When suitable data are available, the general von Bertalanffy and logistic equations have been used successfully to model the growth of tortoises. Growth of juvenile tortoises illustrates the extreme plasticity of growth patterns. Habitat quality strongly influences growth rate and the number of years needed to achieve the minimum size for sexual maturity. Forbs are highly nutritious plants that are especially important to neonatal Eureptilian amniotes for rapid growth and development. Well-planned and executed experiments have failed to provide support for the idea that energetic costs in juveniles should be high because of the demands of growth.

Health Issues of North American Tortoises

Elliott R. Jacobson

In many areas within their ranges, North American tortoises (*Gopherus polyphemus, G. berlandieri, G. agassizii, G. morafkai,* and *G. flavomarginatus*) are in decline. Causes for decline include increased predation by domestic and nondomestic animals, invasive plant introductions resulting in diminished forage quality, dramatic automobile-related mortality as a consequence of highway construction, fires, protracted droughts, disease-associated mortality, and probably the most important, land-use practices resulting in loss of habitat and fragmentation of populations. As populations become more constricted and density increases, epizootics (outbreaks of disease in animal populations) have a greater probability of occurring. Continued development of prime habitat into homes and shopping centers has displaced large numbers of tortoises. Although relocation has been used to mitigate displacement, there is little data to show consistent long-term success.

Health assessment is being used more and more to try to determine the "relative health" of key populations of animals that are being used for monitoring overall health of remaining populations. Most of this work involves *G. agassizii* and *G. polyphemus.* Health assessment is tricky business, however, and evaluating a tortoise is no easy affair. In this chapter, guidelines for health assessment of individuals and populations of tortoises are reviewed. The most important diseases of North American tortoises also are reviewed and discussed, because understanding the range of diseases that have been described from these tortoises is an important component of a health assessment program.

HEALTH ASSESSMENT

Health assessment of wild tortoises, as with other animals, includes assessing the abiotic and biotic factors that affect their geographic distribution. Quality of forage and water availability need to be assessed when determining the status of tortoises. Little to nothing is known about the effect of toxicants that are applied to the environment in which these animals live, nor the effects of those potential toxicants naturally found in both native and introduced plants that are eaten by tortoises. In most cases, it is not cost effective to conduct a thorough health assessment of every individual tortoise in a population. Consequently, an epidemiologist and statistician should be consulted to determine the minimum number of individuals that need to be sampled. Available resources will influence the number of tortoises sampled, frequency of sampling, number and type of assays performed, and duration of sampling.

Guidelines for field evaluation of *Gopherus agassizii* (Berry and Christopher 2001) and *G. polyphemus* (Wendland et al. 2009) have been published. No such information is available for the other species. Published field sheets (fig. 8.1) should be used for collecting high quality qualitative and quantitative data on desert tortoise health and disease, and serve as a guide for systematic clinical evaluation of other tortoises. Guidelines and health assessment forms developed by Berry and Christopher (2001; fig. 8.1) were an outcome of collaborations between wildlife biologists, veterinarians (clinicians, clinical pathologists, pathologists, epidemiologists), and toxicologists. Notes, which many biologists take in the field, should be included. As part of assessment, digital images of the entire tortoise and surrounding habitat should always be taken. Digital images of a tortoise include dorsal, ventral, all four limbs, left and right lateral of the neck and head, and frontal, showing the nares. These images provide a permanent record of the appearance of the tortoise. An individual identification of the tortoise and a date should be added to each image.

Prior to physically handling the tortoise, it is important to observe the animal in the environment. Recognizing the normal behavioral repertoire for tortoises at different times of the year (and day) is very important to note. A tortoise may manifest behavioral changes prior to signs of clinical disease. For example, if a tortoise is above ground at inappropriate times of the year, when temperatures are suboptimal, this behavior may be an indication of illness.

FIELD WORKER(S):

Handler(s) ______________________

Recorder(s) ______________________

Observer(s) ______________________

STUDY SITE ______________________

UTM (WGS 84) Easting ____________

Northing ____________

County ________________

State ________________

Date (ddmmmyyyy) ______________________

Tortoise ID _____ Sex _____ Captype ____

MCL (mm) _____ Gular (mm) _____

PLN (mm) _____ Weight (g) _____

Time (PST) spent handling start _____

end _____

On a plot? yes/no

New growth? yes/no SWC _____

Photos ______________

TRANSMITTER:

Transmitter frequency? ______________________

Transmitter number? ______________________

Who attached transmitter? ______________________

Notes

LOCATION:

At cover site? Cover site type?

Tag # _____ burrow __

entering __ pallet __

exiting __ shrub __

on mound __ cave __

inside* __ rock shelter __

*specifically where ______________________

TEMPERATURES (ºC):

At 1.5m ____ 1cm ____ Soil ____ FWS 2″ start ____

Activity? Not at cover site? FWS 2″ end ____

resting __ in open __

basking __ other __

walking __ Interacting with another tortoise? yes/no

feeding __ sex ____ size ____ number ____

Type of interaction(s): ______________

POSTURE/BEHAVIOR:

Behavior appropriate for the time of day? yes/no If no, describe: ______________

Behavior appropriate for the season? yes/no If no, describe: ______________

Can withdraw tightly into shell? yes/no If no, describe: ______________

Alert and responsive? yes/no If no, describe: ______________

Limbs, head hanging limp or loose? yes/no If yes, describe: ______________

Lethargic? yes/no If yes, describe: ______________

FORELIMBS

Right normal/abnormal If abnormal, describe: ______________

Left normal/abnormal If abnormal, describe: ______________

HINDLIMBS

Right normal/abnormal If abnormal, describe: ______________

Left normal/abnormal If abnormal, describe: ______________

OTHER

Tail normal/abnormal If abnormal, describe: ______________

FORELEGS (adjacent to face)

Dried dirt on forelegs? yes/no/unk

Moisture on forelegs? yes/no/unk

Dried exudate on scales (glossy with dried exudate)? yes/no/unk

Scales cracking (from exudate)? yes/no/unk

Fig. 8.1. Health profile form for *Gopherus agassizii.* From Berry and Christopher (2001).

Field workers(s) ______________ Date (ddmmmyyyy) ______________

Study site name ______________ Tortoise ID ________ Sex ________

CHIN GLANDS

Site	Size	Drainage	Severity (Rate 1–4)		Color of Drainage
R Gland	normal/swollen	present/absent	____	____	clear/cloudy/white/yellow/green
L Gland	normal/swollen	present/absent	____	____	clear/cloudy/white/yellow/green

INTEGUMENT

Integument dull?	yes/no	If yes, describe location:	____________
Integument glossy?	yes/no	If no, describe:	____________
Normal elasticity?	yes/no	If no, describe:	____________
Abnormal skin peeling?	yes/no	If yes, describe:	____________

Supplemental system for grading the beak, nares, eyes, and chin glands of desert tortoises. Instructions: depending on subject, circle one or more options. Rating system: 1=normal, 2=mild, 3=moderate, 4=severe

BEAK & NARES

Site	State	Severity (Rate 1-4)	Notes
Moisture (e.g. If beak is damp or wet, describe cause)			
Beak	dry/damp/wet	____	____________
R Nare	dry/damp/wet	____	____________
L Nare	dry/damp/wet	____	____________
Exudate			
Beak	none/dried/wet	____	color: clear/cloudy/white/yellow/green
R Nare	none/dried/wet	____	color: clear/cloudy/white/yellow/green
L Nare	none/dried/wet	____	color: clear/cloudy/white/yellow/green
Bubbles			
R Nare	yes/no	____	____________
L Nare	yes/no	____	____________
Occlusion (wet mucus, dry mucus, dirt, plant material)			
R Nare	none/partial/complete	____	____________
L Nare	none/partial/complete	____	____________
Normal Foraging Stains?			
Beak	yes/no/no evidence	____	If no, describe: ____________
Dirt on/in			
Beak	yes/no	____	____________
R Nare	yes/no	____	____________
L Nare	yes/no	____	____________

ORAL CAVITY

		Notes
Observed?	yes/no	____________
Discharge present?	yes/no/unk	____________
Plaques or ulcers present?	yes/no/unk	____________
Smells?	yes/no/unk	____________
Color of membranes?	unk/white, pink, yellow, other	____________

Fig. 8.1. continued

Field workers(s) ______________________ Date (ddmmmyyyy) ______________

Study site name ______________________ Tortoise ID __________ Sex __________

EYES: Palpebrae (Palp), Periocular (Perioc), Globe

Variable		State	Upper Palp	Lower Palp	Upper Perioc	Lower Perioc
			Severity (Rate 1–4)			
Discoloration						
	R Eye	yes/no/unk	______	______	______	______
	L Eye	yes/no/unk	______	______	______	______
Edema						
	R Eye	yes/no/unk	______	______	______	______
	L Eye	yes/no/unk	______	______	______	______
Crusts (includes dry mucus)						
	R Eye	yes/no/unk	______	______	______	______
	L Eye	yes/no/unk	______	______	______	______

Discharge

R Eye yes/no/unk If yes, is it wet/dried ______

L Eye yes/no/unk If yes, is it wet/dried ______

Other Lesions of Palpebra and Periocular Area

R Eye yes/no/unk ______ trauma/necrosis/peeling scales/other: ________

L Eye yes/no/unk ______ trauma/necrosis/peeling scales/other: ________

Sunken/Recessed Eyes

R Eye yes/no/unk ______

L Eyc ycs/no/unk ______

Eye Swollen or Bulging in Appearance

R Eye yes/no/unk ______ dorsal ______ lateral

L Eye yes/no/unk ______ dorsal ______ lateral

Degree of Openness of Palpebra

R Eye normal (100% open)/partially closed If partially closed, ______ % is closed.

L Eye normal (100% open)/partially closed If partially closed, ______ % is closed.

Condition of Globe (circle all that apply)

R Eye clear/bright/mucus present/dull/cloudy/inflamed sclera

L Eye clear/bright/mucus present/dull/cloudy/inflamed sclera

Conjunctiva

R Eye yes/no/unk If yes, ______ % is exposed.

L Eye yes/no/unk If yes, ______ % is exposed.

Other obvious lesions

R Eye none/corneal ulcers/corneal abrasions Other: ______________________

L Eye none/corneal ulcers/corneal abrasions Other: ______________________

Is plant material in eye? yes/no/unk R eye/L eye ______________________

Is dirt/sand in eye? yes/no/unk R eye/L eye ______________________

Fig. 8.1. continued

Field workers(s) ______________________ Date (ddmmmyyyy) ______________

Study site name ______________________ Tortoise ID __________ Sex __________

RESPIRATION/BREATHING

		Notes
Smooth?	yes/no/unk	
Wheezing?	yes/no/unk	
Rasping or clicking?	yes/no/unk	

BLOOD SAMPLE*

Name of sticker(s) ______________________

Total No. of needle sticks	_____	Total No. of tubes	_____
Tube 1 % that is blood	_____	% that is lymph	_____
Location ____________		Sample size (ml)	_____
Tube 2 % that is blood	_____	% that is lymph	_____
Location ____________		Sample size (ml)	_____
Tube 3 % that is blood	_____	% that is lymph	_____
Location ____________		Sample size (ml)	_____
Tube 4 % that is blood	_____	% that is lymph	_____
Location ____________		Sample size (ml)	_____

NASAL LAVAGE SAMPLE

Nasal lavage sample taken: yes/no

SP4 added to sample: yes/no

Sample size (ml): _____

*** Please make sure each tube is properly labeled. For, example:**
☒ 2121, G. agassizii
Fort Irwin Translocation
San Bernardino Co., CA
21 September 2006
Blood/Plasma Mix Tube 1** of 2
****Where Tube 1 matches the information given on the left.**

URINE/URATES

Did tortoise urinate? yes/no

Urine color: ____________

Urine volume (ml): _____

Viscosity: ____________

Particulates? yes/no

Particulate color: ____________

Particulate volume (ml): _____

REHYDRATION

Was tortoise rehydrated? yes/no

Amount rehydrated: _____

Amount of time in water: _____

Table 1. System for grading signs of trauma in desert tortoises. The carapace, plastron, gular, and integument on limbs and head should be rated separately.

I. Distribution: specify by plastron, carapace, gular, limbs, or head
   - 1 = not present; no signs of trauma
   - 2 = mild; covers ≤ 10% of plastron, carapace, gular, head, or limbs
   - 3 = moderate; covers 11–40%
   - 4 = severe; covers > 40%

II. Severity of lesions, both historic and present
   - 1 = no trauma
   - 2 = mild; minor chews or chips, no bone exposed; 1-3 damaged scales or 1 damaged toenail
   - 3 = moderate; small areas of bone exposed (healed or fresh injuries); several scales damaged on one or more limbs; > 1 toenail damaged
   - 4 = severe; missing large areas of scute or bone; bone exposed; soft tissue damage to limbs

III. Chronicity of trauma
   - 1 = no trauma
   - 2 = healed or healing injuries (state whether trauma is healed or healing)
   - 3 = fresh injuries

Fig. 8.1. continued

Field workers(s) ______________________ Date (ddmmmyyyy) ______________

Study site name ______________________ Tortoise ID __________ Sex __________

PARASITES and SPINES: *show location on the diagram below

Parasite present? yes/no Type: __________ Number: _____ Size(s): __________

Spine present? yes/no Type: __________ Number: _____ Size(s): __________

SIGNS OF TRAUMA: Severity 1 = no signs, 2 = mild, 3 = moderate, 4 = severe
Use Table 1 for definitions of mild, moderate and severe, for distribution, severity and chronicity of damage from trauma.

Bone/scute replacement? yes/no/unk Describe: ______________________

		Severity (Rate 1-4)	Location	Notes
HEAD	Distribution	_____	_______	______________
	Severity	_____	_______	______________
	Chronicity	_____	_______	______________
LIMBS	Distribution	_____	_______	______________
	Severity	_____	_______	______________
	Chronicity	_____	_______	______________
GULAR	Distribution	_____	_______	______________
	Severity	_____	_______	______________
	Chronicity	_____	_______	______________
CARAPACE	Distribution	_____	_______	______________
	Severity	_____	_______	______________
	Chronicity	_____	_______	______________
PLASTRON	Distribution	_____	_______	______________
	Severity	_____	_______	______________
	Chronicity	_____	_______	______________

Label all historic and recent trauma, show type in writing on form, e.g., injuries from tearing, gnawing, gnashing. Show areas of bone exposure and label as "bone exposed." Note scales removed from limbs, injuries or scars on limbs. Draw on toenails for each limb. Label each trauma as fresh, healed, or healing injuries.

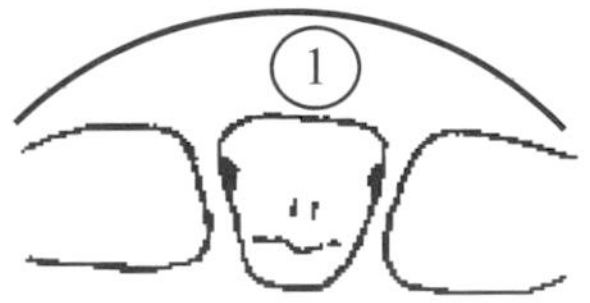

1

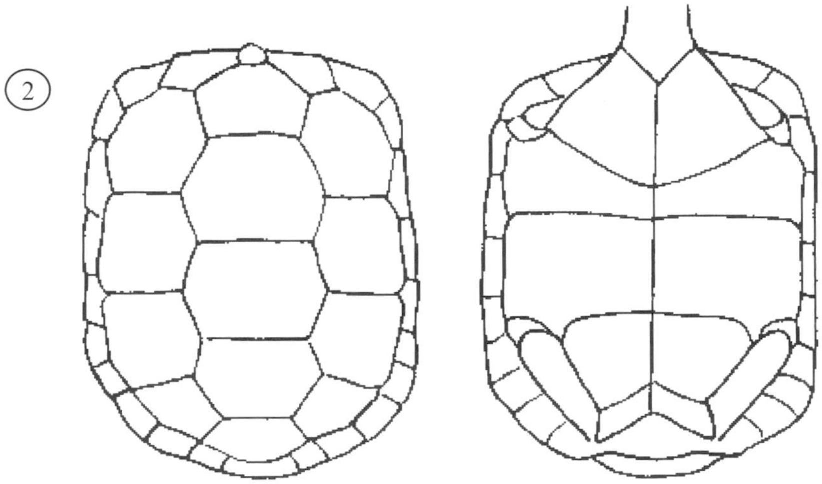

2

1. Damage to skin? yes/no
   If yes, describe:

2. Toes normal? yes/no
   If no, describe:

Fig. 8.1. continued

Table 2. System for grading shell lesions caused by disease, such as cutaneous dyskeratosis in desert tortoises. The carapace, plastron, and integument on limbs and head should be rated separately.

I. Distribution of disease signs: specify by plastron, carapace, limbs, or head
- 1 = not present; no signs of lesions
- 2 = mild; lesions manifested primarily at seams, covers less than 10% of plastron (or carapace or limbs, etc.)
- 3 = moderate; covers 11–40%
- 4 = severe; covers > 40%

II. Severity of lesions (from disease, e.g., cutaneous dyskeratosis)
- 1 = no lesions
- 2 = mild; discoloration follows edges of lifting laminae, lightly discolored, flaking
- 3 = moderate; discoloration extends over several layers of laminae, edges of laminae flaking, scutes may be thin in small areas, and potential exists for small holes and openings exposing bone
- 4 = severe; some scutes or parts of scutes eroded away or missing and bone exposed, eroded, or damaged

III. Chronicity of trauma
- 1 = no lesions
- 2 = old lesions; no apparent recent activity, signs of regression or recovery; development of healthy, normal laminae is apparent at seams of scutes
- 3 = active, current lesions

Fig. 8.1. continued

Field workers(s) ______________________ Date (ddmmmyyyy) ______________

Study site name ______________________ Tortoise ID __________ Sex __________

EVIDENCE OF SHELL/BONE DISEASE **Describe**

		Describe
Scutes laminae peeling/missing?	yes/no/unk	______________
Scutes depressed/concave?	yes/no/unk	______________
Pitting?	yes/no/unk	______________
Fungal areas? (draw onto diagram and label)	yes/no/unk	______________

SIGNS OF LESIONS FROM DISEASE: Severity 1 = no signs, 2 = mild, 3 = moderate, 4 = severe
Use Table 2 for definitions of mild, moderate and severe for distribution, severity and chronicity.

Bone/scute replacement? yes/no/unk Describe: ______________

		Severity (Rate 1–4)	Location	Notes
HEAD	Distribution	____	______	______________
	Severity	____	______	______________
	Chronicity	____	______	______________
LIMBS	Distribution	____	______	______________
	Severity	____	______	______________
	Chronicity	____	______	______________
GULAR	Distribution	____	______	______________
	Severity	____	______	______________
	Chronicity	____	______	______________
CARAPACE	Distribution	____	______	______________
	Severity	____	______	______________
	Chronicity	____	______	______________
PLASTRON	Distribution	____	______	______________
	Severity	____	______	______________
	Chronicity	____	______	______________

Legend: Lesions from cutaneous dyskeratosis (specifically), other lesions (enter symbols), describe in writing. Note lesions and extent on limbs.

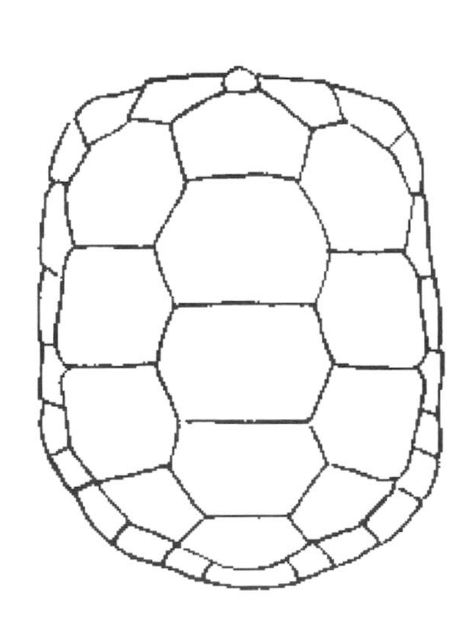

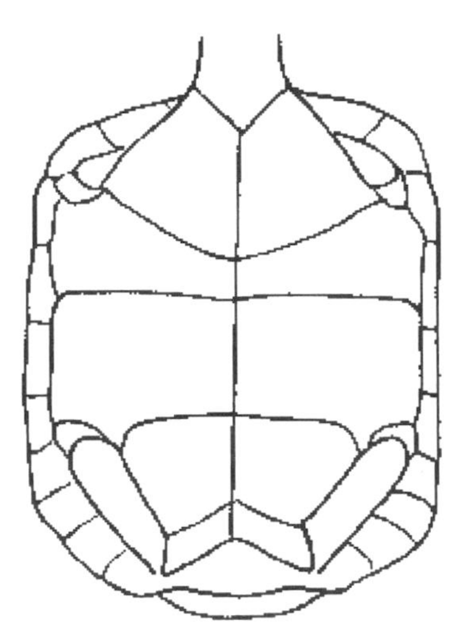

Fig. 8.1. continued

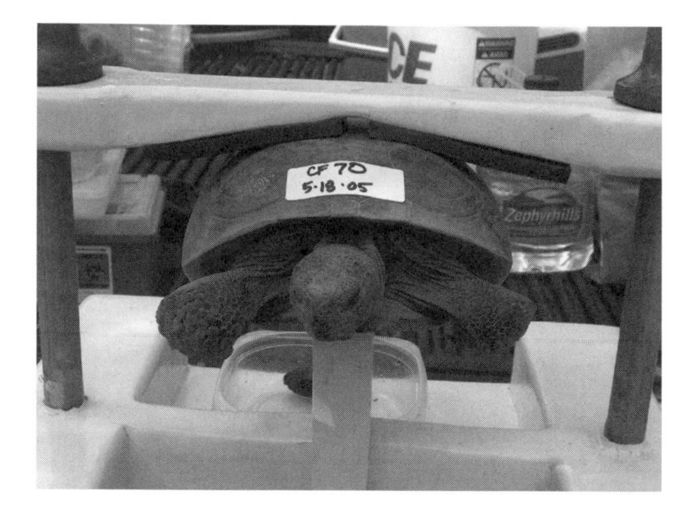

Fig. 8.2. Restraint device used to simplify the handling and collection of a blood sample from *Gopherus polyphemus*. Photo courtesy of L. Wendland.

Tortoises can be difficult animals to examine physically. Many species are capable of withdrawing their limbs and head into the margins of their shell when threatened, becoming almost impossible to extricate manually. Depending upon what is assessed, a minimum of two persons may be necessary to evaluate a tortoise and collect data. Restraint devices have been devised to simplify the handling and collection of blood samples (fig. 8.2). Above all, remember that examination of a frightened tortoise requires a good deal of patience. Also remember that myopathy (muscle damage) may occur if excess force is needed to extend their limbs or head. If the tortoise urinates when handled, it subsequently needs to be soaked in water prior to release, because the urine in the bladder may serve as a source of water in times of drought. Thus, losing water when handled can compromise the survival ability of a tortoise.

If blood is to be collected for hematologic and biochemical analysis, the quicker it can be collected after the tortoise has been disturbed, the more reliable the data. Ideally, collection should be done within 10 minutes of first handling. While a blood sample can be obtained from many tortoises using manual restraint, at times, immobilization or sedation is required. Injectable drugs such as ketamine (2–30 mg/kg) and telazol (2–4 mg/kg), given intramuscularly, have been used for immobilization of tortoises and other chelonians (Schumacher 2012). Medetomidine has been used either alone (150 µg/kg) or in combination (100 µg/kg) with ketamine (5 mg/kg) for sedation. One benefit of using medetomdine is that it can be reversed using atimpemazole (5 times the dose of medetomidine) (Sleeman and Gaynor 2000). The addition of either midazolam (0.1 mg/kg), buprenorphine (0.02 mg/kg), or morphine (0.5 mg/kg) to the above combination provides more reliable immobilization (D. Heard, personal communication). Because it may take up to 24 hours for a tortoise to recover fully from the effects of these drugs, if it is to be released on site, it may need to be housed (placed in a box or container) and released the next day. For a detailed review of sedation and anesthesia of chelonians, see Schumacher (2012).

Various collection sites, each having advantages and disadvantages, can be used for obtaining blood from chelonians (Jacobson et al. 1992, Jacobson 1993, Hernandez-Divers et al. 2002). The only peripheral vessels that can be visualized are the jugular vein and carotid artery that often bulge the overlying skin in the lateral cervical region of the neck. A 23-gauge butterfly catheter is commonly used. The catheter can be flushed with sodium or lithium heparin in saline (500 U/10 ml saline) to reduce clot formation. If serum (rather than plasma) is needed for a particular test, then blood should be collected with unheparinized catheters and placed in tubes lacking an anticoagulant. Applying direct pressure over the venipuncture site can prevent hematoma formation and additional loss of blood. Other commonly used sites for sampling are the brachial and subcarapacial veins (Hernandez-Divers et al. 2002). The subcarapacial venipuncture site is located underneath the carapace, posterior to the nuchal scute and beneath the anterior portion of the 1st vertebral scute, cranial or anterior to the 8th cervical vertebra. Inserting the needle a few millimeters posterior to the skin–carapace junction and then angling the needle dorsally toward the 8th cervical vertebra provides access to the site.

Some investigators have used toenail trimming to collect small quantities of blood. There are two conflicting opinions concerning this method of blood sampling. One view is that the procedure does not cause perceivable pain (as evidenced by the lack of a withdrawal reflex in most tortoises), it minimizes stress in comparison to restraint for venipuncture, and it requires minimal technical training. The opposing view is that there is potentially more pain induced by trimming a toenail than venipuncture, that risk of infection is greater than with venipuncture, and that samples may not be of good value for serologic testing. Both biologists and veterinarians have taken positions on both sides of this debate. The author's personal view is to avoid toenail clipping. The author believes that the need for minimal technical training is not sufficient to justify toenail clipping as a preferred method of blood collection.

Once blood has been collected, it should be aliquoted into several tubes containing lithium heparin as the anticoagulant for performing a complete blood count (CBC) and plasma biochemical analyses. A blood tube for a CBC should be kept on ice until it is processed. Immediately after collection, tubes for plasma biochemical determinations should be centrifuged, the plasma removed, and placed in a cryotube and either submitted on ice for analyte (biochemical) determinations or placed in liquid nitrogen or an ultrafreezer until the analyses are performed. Certain blood values have been reported for healthy (Christopher et al. 1999) and ill (Christopher et al. 2003) *G. agassizii* and healthy *G. polyphemus* (Taylor and Jacobson 1981). Studies of *G. agassizii* are the best to date, and for now, can be used as reference values in assessing blood analyte values of other North American tortoises.

Complete blood counts (CBC) include: packed cell volume (PCV), total white blood cell (WBC) count, estimated WBC count, differential WBC count, icterus index, plasma protein concentration, fibrinogen concentration, evidence of cellular toxicity, platelet morphology, and red blood cell (RBC) morphology. For PCV determination, a microhematocrit tube containing ammonium-heparin can be centrifuged for 2 min in a microhematocrit centrifuge. Methods of determining the total WBC count (WBC/μl) include use of a hemocytometer and Natt-Herrick's or phloxine B solutions or estimation from a blood film stained with Wright-Giemsa stain (for details, see Campbell 1996.) A 100-cell differential WBC count (heterophils, lymphocytes, basophils, eosinophils, monocytes, and azurophils) is determined by light microscopic examination of Wright-Giemsa stained blood films. Fibrinogen is measured using a heat precipitation method and a refractometer.

Plasma can be obtained using a microcentrifuge for microtainer tubes and a universal centrifuge for larger tubes. A bench-top biochemical autoanalyzer (such as a Hitachi; Roche Diagnostics, Indianapolis, Indiana) can be used to measure biochemical analytes: albumin, alkaline phosphatase (ALP), albumin/globulin (A/G) ratio, aspartate aminotransferase (AST), blood urea nitrogen (BUN), total calcium, chloride, cholesterol, creatine kinase (CK), gamma glutamyl transpeptidase (GGT), globulin, glucose, magnesium, phosphorus, potassium, sodium, total protein, triglycerides, and uric acid. A portable biochemical analyzer such as an i-STAT Portable Clinical Analyzer (Heska Corporation, Loveland, Colorado) or Abaxis VetScan (Abaxis, Union City, California) can be used in the field to measure some of the most clinically important analytes in blood. Because different analyte values have been obtained using different analyzers for sea turtles (Wolf et al. 2008, Atkins et al. 2010), it is probably the same for other chelonians. Because no single analyzer is ideal for all situations, it is important to realize that values may vary when using different machines, and that environmental conditions also may affect the accuracy and precision of a machine.

The relationship between weight (Wt) and carapace length in the midline (MCL) has been used to assess condition of tortoises (Jackson 1980). To determine if the relationship between Wt and MCL could be used to discriminate between healthy *G. agassizii* and those with clinical signs of upper respiratory tract disease (URTD), the logarithm of MCL was regressed on the logarithm of W for both groups (Jacobson et al. 1993). Although a significant difference was found between the regression lines for the two groups—tortoises with clinical signs weighing about 7% less than those without signs—several tortoises with clinical signs weighed the same as tortoises of the same size without clinical signs. The presence of uroliths, a large volume of coelomic exudate, and intestinal gravel accumulations can elevate body weight of ill tortoises. For evaluating body condition of wild tortoises, some authors (Nagy et al. 2002) have recommended that a condition index (CI) simulating a physical density value should be used. In this study, the CI was based on the ratio of Body Mass (Wt) to the approximate volume (MCL × width × height). The highest CI values for all individual tortoises recaptured in all seasons were defined as their "prime CIs." The authors found that the prime CI for *G. agassizii* in their study averaged 0.64 g/cm^3 (range = 0.6–0.7 g/cm^3).

For evaluating tortoises in the wild, location can be very important. Finding a tortoise outside its burrow on a cold day often is an indication of a health issue. Observing ambulation, movements and position of the head, and response to approach should be noted (fig. 8.1). Disposable gloves often are used when picking up a tortoise for evaluation. The shell should be examined for the quality of its scutes. Seams between adjacent scutes should be examined for evidence of new growth. Abnormal patterns of keratinization, scute desquamation, flaking, fissures, discoloration, ulceration, erythema, and necrosis should be recorded. The soft integument should also be examined for lesions and presence of ticks. Legs should be extended gently to look for any lesions in the areas around both proximal and distal ends of the limbs. Asymmetric patterns of toenail wear may be associated with lameness. Other abnormalities include missing, deformed, or discolored toes. Limbs should be examined for soft tissue swelling, muscle atrophy, and painful responses to manipulation. The joints of opposite limbs should be compared to identify swelling and to evaluate range of motion. The large scales on the forelimbs should be examined for abnormal patterns of keratinization such as dyskeratosis (Jacobson et al. 1994).

Use of an ophthalmoscope is essential in performing a thorough eye examination. This is a relatively small tool that has great value. The eyes and surrounding tissues should be symmetrical. The relative position of each globe in the orbit should be noted. The cornea and lens should be clear and transparent. The anterior chamber of the eye should be free of blood or exudate. Recognizing changes in palpebral and conjunctival thickness is an important component of the physical examination, but often very difficult to categorize accurately. Examples of various stages of change can be found in Berry and Christopher (2001) and Wendland et al. (2009).

The beak should be examined for fractures and malocclusion, which may predispose tortoises to dehydration. The presence of moisture or dried mucus on the beak may indicate oral infection, ulceration, or irritation from a foreign body. In order to examine and sample the oral cavity, tortoises will require extension of the neck and manual restraint of the head; care should be taken to hold the head from the base of the skull, avoiding pressure on tympanic membranes and mandibular joints. A dull curved dental scaler or paper clip may be placed over the tip of the lower beak and the jaws gently pried open. Large tortoises may require chemical restraint. Once the mouth is open, the investigator should note any abnormalities such as plaques, ulcers, accumulation of caseous (resembling cheese or curd) material in choanae (internal nares) (see appendix), and changes in mucosal color (fig. 8.1).

The nares should be patent, symmetrical, and free of discharges or dried mucus. Cutaneous erosion or depigmentation around or below the nares may indicate a history of chronic nasal discharge. Even if the nares are dry, nasal exudate may be present in the nasal cavity. Presence of hidden nasal discharge can be evaluated by pressing on the intermandibular tissue, which pushes the tongue dorsally into the choanae and expels any free exudate in the nasal cavity out through the external nares.

The tympanum should be observed for any noticeable bulging. Middle and inner ear abscesses are occasionally seen in tortoises. Mental glands should be checked for abnormal discharges. The cloaca should also be checked for lesions or prolapse.

Specific diagnostic procedures are useful in determining the etiology of some diseases that have epizootic potential. Tortoises exhibiting stomatitis (appendix), with or without a caseous exudate, may be lightly anesthetized and swabs or scrapings of oral epithelium collected. Inflammatory exudates and ulcerative lesions should be submitted for cytological examination. Inflammatory exudates and ulcerative lesions should be cultured for bacteria (aerobic and anaerobic) and fungi, and a smear of the material on a glass slide should be submitted for cytological examination. Intranuclear herpesvirus inclusions may be observed, but polymerase chain reaction (PCR; Rabinow 1996) is the definitive diagnostic method used in determining the presence of genomic sequences of herpesvirus and other pathogens, such as *Mycoplasma*. Coelomic palpation (appendix) can be performed in animals of adequate size. Ascites (appendix), bladder stones, shelled eggs in the oviduct, and other conditions can be detected. Wet mounts of fecal smears, as well as fecal flotation, should be employed in examinations for enteric parasites. Most chelonians are colonized by a variety of nematodes and ciliate protozoans that are, for the most part, nonpathogenic. Treatment of wild tortoises for parasites should be avoided unless well justified. Presence of ticks should be noted and locations recorded. Representative ticks should be collected and placed in 70% alcohol.

COLLECTING BIOPSIES

Techniques for collecting biopsies from chelonians and other reptiles can be found elsewhere (Beaupre et al. 2004), and are only briefly reviewed here. Of all the reptiles, chelonians present the greatest challenge for biopsy, especially when lesions involve the shell. The turtle shell is a very hard biological structure that makes biopsy somewhat difficult. Some form of general anesthesia is generally needed, whether it is an injectable agent or inhalant. All sites should be scrubbed with a disinfectant (such as chlorhexidine) prior to biopsy. While an individual is under anesthesia, a rotary power saw (Dremel Mototool; Dremel Mfg. Co., Racine, Wisconsin) can be used to cut a wedge out of the shell. Ideally, the biopsy should include normal tissue along with the diseased component. The size of the tissue sampled and its thickness will vary with the size of the tortoise and the extent of the lesion. In an adult tortoise, for soft tissue, a 5 mm biopsy punch is commonly used. A portion should be fixed in neutral buffered 10% formalin (NBF) for histopathological evaluation and a portion (with the most superficial contaminated portion removed) submitted for microbial culture and molecular diagnostics. The portion for microbial culture can be placed in an ultrafreezer until histological findings have been reported. Histopathological findings often will direct the clinician on how to proceed with disease diagnostics, saving on cost.

When the biopsy is completed, the biopsy site should be cleansed with chlorhexidine. If the shell is biopsied, the site should be covered with a sterile plastic drape that is slightly larger than the wound. Using a cyanoacrylate surgical glue (Vetbond; 3M, St. Paul, Minnesota), the drape is bound to the intact shell that is adjacent to the biopsy site. The author generally does not see significant bleeding as a consequence of shell biopsy. If significant bleeding does occur, the wound can be packed for a few minutes with a small portion of sterile gauze, until blood loss ceases. Whatever is placed in the shell biopsy site will impede healing; so the author avoids filling the defect with any material, such as bone wax.

For biopsy of soft tissue, the skin should first be cleansed with chlorhexidine or ethanol. Next, 2% xylocaine (volume used will vary with size of animal and size of site) is infiltrated around the biopsy site. If the sample is to be cultured, sterile saline is used instead of chlorhexidine or ethanol. If there is epidermal involvement, a biopsy punch can be used for collecting the sample. Following biopsy, the edges of the tissue can be brought together with a forcep and surgical glue can be applied. The tissue is held together until the glue has hardened. If the skin requires suturing, an absorbable monofilament material is routinely used.

If a subcutaneous mass is present, fine-needle aspiration can be performed. After several passes through the mass, the plunger is released and the needle removed from the mass. A biopsy of the mass also may be needed. In such situations, the skin overlying the mass is incised with a sterile scalpel blade and a biopsy of the mass is collected. The specimen should be cut into several portions for use in cytological preparations, histopathology culture, and molecular diagnostics.

Additional diagnostics, such as laparoscopy, ultrasound, and radiology often are used for diagnosing health problems of tortoises and other turtles in captivity (Penninck et al. 1991, Rostal et al. 1994a, Hernandez-Divers et al. 2009). These diagnostics are commonly used to evaluate the reproductive status of tortoises in the field.

NECROPSY

If an animal dies or is euthanized, a necropsy should be performed. For details, see Homer et al. (1998). A standardized necropsy form should be used to record information. If available, a summary of the clinical course of each animal should be added to the necropsy form. For wild animals found dead in the field, the field health assessment form should be attached to the necropsy report. Photographs should be taken of the entire carcass in the field, both dorsally and ventrally, and of any lesions recognized. An individual identification of the tortoise and a date should be added to each image.

Necropsies start on the outside and move internally. The exterior of the animal should be thoroughly examined, describing all gross abnormalities. Lesions can be added to drawings of the tortoise. During a necropsy, the shell is severed at the bridge, using an electrical saw. The plastron is removed, and internal organs examined. All liquid and solid contents of the bladder are removed, measured, and a sample of the liquid portion is submitted for urinalysis. Samples of all significant lesions should be placed in neutral buffered 10% formalin (NBF) for histopathology. Since NBF will only penetrate 6 mm in 24 hours, make sure tissues are <6 mm thick, to allow adequate fixation. The NBF to tissue volume ratio should be 10:1. If hard tissue, such as long bone, is collected, it should be fixed in a container separate from the soft tissues, to allow adequate penetration and fixation. Digital images should be taken of all internal organs.

The overall appearance of the animal will dictate whether to continue with a full necropsy. If the animal is in an advanced state of postmortem change, collection of tissues for histopathological evaluation will be unrewarding. Although less than ideal, fresh carcasses that are quickly frozen can be submitted at a later date for full necropsy. While freezing artifact is commonly encountered in such cases, the tissue generally is in a state that allows identification of major lesions.

Total weights are obtained for heart, liver, and kidneys. Samples of organs are collected for histopathology: tongue; left and right salivary glands; left and right chin glands; left and right nasal cavities; glottis; trachea; tracheal bifurcation; left and right lungs; left and right lobes of the liver and gall bladder; left and right thymus/parathyroid; thyroid; heart; esophagus; mid-stomach; pylorus; cranial-, mid-, and caudal small intestine; cranial-, mid- and caudal colon; spleen; pancreas; left and right kidneys and adrenals; left and right gonads; brain; cervical spinal cord; left femur; pectoralis muscle and associated fat; skin; left and right ear canals; vertebral scute 3; peritoneum; left forelimb scales; and shell. The nasal cavity (NC) is removed from other structures of the head and cut in the midline using a rotary electric device fitted with a circular stainless steel blade. Next, swabs from each cavity and lungs are obtained for microbial culture and PCR testing for pathogens such as *Mycoplasma*. Swabs of the tongue and submandibular glands should be collected for herpesvirus PCR. Small portions of each NC are often collected in 2.5% glutaraldehyde for transmission electron microscopy (TEM). Once all needed samples are collected, each half of the head containing the NC is placed in NBF for 48 hours, followed by placement in a decalcification solution (Cal-Ex decalcifier; Fisher Scientific, Fair Lawn, New Jersey) until tissues are soft enough (generally 48 hours) to be sectioned in a microtome for light microscopy. The author typically collects 30–40 6 μm sections from each NC, with sectioning performed in the dorsal to ventral plane. The first section and then every 10th section are stained for light microscopy with hematoxylin and eosin. For evaluation of the NC for upper respiratory tract disease (URTD), previously published criteria are used to classify the lesions as normal, mild, moderate, and severe (Jacobson et al. 1995).

Pathology Services

Two private pathology laboratories that specialize in exotic animal pathology, including tortoises, are: Northwest ZooPath, Monroe, Washington, and Zoo/Exotic Pathology Service, West Sacramento, California. The Veterinary Pathology Service, College of Veterinary Medicine, University of Florida, Gainesville, has had many years of experience working up biopsies of live tortoises and performing complete necropsies and histopathological evaluations on those that are submitted dead. If a tortoise is found recently dead, ideally it should be refrigerated (not frozen) and delivered (on ice) to a pathology service as soon as possible. Freezing will result in artifactual changes in tissues that may interfere with the histopathological assessment.

INFECTIOUS DISEASES

Captive tortoises are susceptible to a variety of pathogens. As more tortoises are submitted for necropsy, new diseases and infectious agents are found. Several diseases have emerged as major health problems for wild North American tortoises. More diseases have been reported in wild *Gopherus agassizii* than in other North American tortoises. This difference has resulted from the listing of *G. agassizii* as threatened, which committed federal agencies to determine causes of mortality. Starting with viruses, the major diseases of North American tortoises will be reviewed.

Herpesvirus

Herpesviruses are DNA viruses that replicate within the nucleus of cells. Certain herpesviruses can remain latent for prolonged periods. Of viral diseases, herpesviruses have surfaced as important pathogens in tortoises. The first report of a herpesvirus-like agent in a tortoise was a 6-year-old captive *G. agassizii* (Harper et al. 1982). Other papers of herpesvirus in *G. agassizii* followed (Pettan-Brewer et al. 1996, Martinez-Silvestre et al. 1999). In a recent report (Johnson et al. 2005), a captive *G. agassizii* from California with a severe pharyngitis was found to have a herpesvirus infection. This herpesvirus was determined to be distinct from a previously described herpesvirus in imported tortoises in Japan (Une et al. 1999). Recently, gene sequences for this novel herpesvirus were identified in two *G. agassizii* using PCR (Jacobson et al. 2012). The importance of this virus in wild tortoises is yet to be determined.

Numerous species of imported and captive tortoises in the pet trade have been reported to have herpesvirus infections, including Argentine tortoises (*Chelonoidis chilensis*) and spur-thighed tortoises (*Testudo graeca*) from private colonies in the United Kingdom and Greece and Hermann's tortoises (*T. hermanni*), Russian tortoises (*Agrionemys horsfieldii*), and marginated tortoises (*T. marginata*), also in the United Kingdom (Jacobson 2007a). More than one strain of tortoise herpesvirus has been identified (Marschang et al.1999, Marschang et al. 2001). Using PCR amplification of a conserved region of two genes (Origgi et al. 2004, Johnson et al. 2005), at least four genomic types of tortoise herpesviruses have been identified to date (Jacobson et al. 2012). While one genomic type may be relatively benign in its natural host, host switching can occur when exotics come in contact with native tortoises. The nonnative host often is more susceptible to an active (lethal) infection (Picco et al. 2010).

Ranavirus

This genus of DNA viruses is a member of the family Iridoviridae. *Ranavirus* (also referred to as *Iridovirus*) has a broad host range, having been shown to be capable of infecting fish, amphibians, and reptiles. The pathogenicity of *Ranavirus* in amphibians varies with the life stage of the susceptible amphibian. It is most pathogenic in embryos and tadpoles, and least pathogenic in adults. Diagnostically, PCR is the preferred method to determine presence of this virus.

The first report of a *Ranavirus* infection in a wild tortoise involved a *G. polyphemus* in Florida that had signs of URTD (Westhouse et al. 1996). Beginning in August 2003, *Ranavirus* infection was identified in additional *G. polyphemus* in Florida, a dead captive Burmese star tortoise (*Geochelone platynota*) in a zoological collection in Georgia, multiple eastern box turtles (*Terrapene carolina carolina*) at a nature sanctuary in Pennsylvania, and wild eastern box turtles in Georgia and Florida (Johnson et al. 2008). All tortoises and box turtles had overlapping signs, including cervical edema, palpebral edema, rhinitis, and stomatitis-glossitis (appendix).

Although much evidence points to *Ranavirus* as a major pathogen of eastern box turtles, it appears to be sporadic in *G. polyphemus*. A serological survey performed on 658 plasma samples obtained from wild *G. polyphemus* across Florida revealed that only eight (1.2%) were seropositive (appendix; Johnson et al. 2009).

Chlamydia and *Chlamydophila*

Chlamydia and *Chlamydophila* are obligate intracellular bacteria that were formerly (prior to 1999) grouped together in the genus *Chlamydia*. Several species of *Chlamydophila* have been identified in reptiles, including chelonians (Jacobson 2007b). Bacteria consistent with *Chlamydia* were identified in a group of juvenile desert tortoises being kept in outdoor pens as part of a research project (Johnson 2012). Clinical findings included nasal discharge, dried feces around the vent from diarrhea, soft shells, and poor body condition. Histological examination of several tortoises identified colitis, conjunctivitis, tracheobronchitis and pneumonia associated with intracellular bacteria in epithelial cells. The bacteria had morphologic, staining, and ultrastructural features consistent with *Chlamydia*.

Mycoplasmosis

Rhinitis and chronic URTD have been reported in *G. agassizii* from multiple sites in the Mojave Desert of the southwestern U.S. (Jacobson et al. 1991) and in *G. polyphemus* in the southeastern U.S. (McLaughlin et al. 2000). Although primarily seen in adult tortoises in the wild, under experimental conditions, all age groups are susceptible. A novel mycoplasma, *Mycoplasma agassizii*, was isolated from affected tortoises, and transmission studies confirmed a causal relationship between *M. agassizii* and URTD in both *G. agassizii* (Brown et al. 1994) and *G. polyphemus* (Brown et al. 1999b). A second distinct species, *M. testudineum*, was isolated from both species (Brown et al. 2004), and a recent report indicates an association with URTD (Jacobson and Berry 2012). Transmission studies are needed to understand the role of this microbe in URTD.

Although supporting data are lacking, various extrinsic and predisposing factors probably are involved in outbreaks of URTD. Clinically silent infections may become exacerbated by environmental stress (Brown et al. 2002). Environmental perturbations, such as prolonged drought, may influence the periodicity of outbreaks in populations of *G. agassizii* known to have mycoplasmosis. For *G. polyphemus*, severe environmental stressors may include hurricanes, with flooding of burrows, or very cold winters. Capturing and transporting of tortoises during relocation, restocking, and repatriation efforts also may be significant sources of stress that result in overt disease. The release of ill captive tortoises may be a significant factor accounting for the presence of URTD in certain populations.

To perform seroepidemiological and other disease-related studies, an enzyme linked immunosorbent assay (ELISA) was developed to determine the presence of antimycoplasma antibodies in the plasma of *G. agassizii* (Schumacher et al. 1993). The test was validated using experimental transmission studies with both *G. agassizii* (Brown et al. 1994) and *G. polyphemus* (Brown et al. 1999b). The ELISA has been refined, drawing on the accumulation of an immense database of ELISA results from more than 20,000 plasma / serum samples (Wendland et al. 2007).

Using matrix population models and Markov chain models for temporally autocorrelated environments, Perez-Heydrich et al. (2011) studied the influence of chronic recurring disease epizootics on host population dynamics and persistence. They used URTD within natural populations of *G. polyphemus* as a model system. The outcome of this study revealed that the impact of disease on host population dynamics depended primarily on how often an epizootic event occurred within the affected population, rather than how long the epizootic persisted within the infected population.

In the past, ELISA testing was used as a convenient method for determining the ultimate disposition of *G. polyphemus* and *G. agassizii* in certain parts of their ranges. The policy of euthanasia of *G. agassizii* in Nevada can be found in the 1996 "Protocol for the Prevention of the Transmission of Disease among *Gopherus agassizii* at the Desert Tortoise Conservation Center and Transfer and Holding Facility," which was developed by Southern Nevada Environmental Incorporated and the Bureau of Land Management. This policy was only adopted in Nevada and not in any other state where the species is found. Brown et al. (2002, 505) stated: "There are inadequate scientific data to provide definitive guidelines for the disposition of seropositive tortoises." This statement was partially responsible for termination of the ELISA-based euthanasia program in Nevada in June 2007 (R. Averill-Murray, personal communication). Continued research over the past decade indicates that this statement still holds true. Euthanasia of seropositive (appendix) tortoises eliminates animals that might otherwise have provided valuable reproductive and genetic contributions to wild populations. Relocation of seropositive tortoises may result in spread of mycoplasmosis to susceptible animals, however, and could have detrimental impacts on recipient populations. Thus, when making management decisions on the basis of the *M. agassizii* ELISA, it is critical to establish clear goals for the tortoise population of interest, to determine a necessary sample size to meet the goals for detection, and finally, upon receipt of results, to consider the predictive values of the test before implementing any policy.

Fungal Shell Disease and Pneumonia

Necrotizing scute disease (NSD) is a fungal condition affecting captive and free-ranging *G. berlandieri*, and caused by *Fusarium incarnatum* (formerly *F. semitectum*) (Rose et al. 2001). It is a slowly progressing shell disease in which degradation and necrosis of the epidermal lamellae translate into whitish or discolored scutes. The disease is disfiguring, but appears to remain confined to the carapacial scutes, does not disseminate, and is not fatal. The disease was reproduced experimentally in *G. berlandieri*, but could not be induced in ornate box turtles (*Terrapene ornata*). A similar-appearing shell disease has been seen in *G. polyphemus* in Florida presented to the Zoological Medicine Service, University of Florida. Suboptimal environmental conditions, such as increased humidity in the burrows of the tortoises and above-normal rain, may predispose these animals to infection.

Fungal shell and skin disease has been seen in several salvaged *G. agassizii* (Homer et al. 1998). Some cases probably are secondary to trauma. Additionally, two *G. agassizii* with clinical signs of respiratory disease had scattered septate and branching fungal hyphae in the lung. Based on the morphology of the fungal hyphae and ovoid yeasts, there appeared to be a dual infection with *Aspergillus* sp. and *Candida* sp. The author has also seen several cases of fungal pneumonia in wild *G. polyphemus* showing signs of respiratory disease. Fungi often cause infection secondary to some predisposing factor, such as poor sanitation, high humidity, malnutrition, or overcrowding.

Cryptosporidium

Cryptosporidium is a coccidial protozoan that is an important parasite of wild and captive reptiles (Upton et al. 1989). At least 57 different reptile species have been reported to be infected, including snakes, lizards, and tortoises (Donoghue 1995). Under light microscopy, *Cryptosporodium* was observed lining the gastric mucosa of a chronically ill *G. berlandieri* (Jacobson 2007c). This tortoise had severe diffuse atrophic gastritis.

Significant Pathogens in Exotic Tortoises

Large numbers of tortoises and other reptiles are imported annually into the United States. Carried along with them is an array of infectious agents, several of which have the potential to cause mortality in *Gopherus*. A novel *Siadenovirus* (member of the virus family Adenoviridae) was identified in a group of 105 Sulawesi tortoises (*Indotestudo forsteni*) that were illegally imported into the United States (Rivera et al. 2009). Upon arrival, they were confiscated by the U.S. Fish and Wildlife Service and distributed to several zoological institutions and private individuals. Most of these tortoises were anorexic (loss of appetite) and lethargic, and had ulcerative lesions in the oral cavity, nasal and ocular discharge, and diarrhea. Histopathological examination revealed a variety of lesions, including rhinitis, pneumonia, hepatitis/hepatic necrosis, and encephalitis, often associated with intranuclear inclusions consistent with a viral infection. Polymerase chain reaction testing of multiple tissues and plasma from 41 out of 42 tortoises were positive for an adenovirus, which was characterized by gene sequence analysis as a novel *Siadenovirus*.

An unusual unclassified intranuclear coccidia was reported in two severely ill radiated tortoises (*Geochelone radiata*) that were captive bred and reared in the United States (Jacobson et al. 1994). Light microscopic examination of multiple tissues revealed nephritis and pancreatitis, with an intranuclear protozoan in renal epithelial cells, hepatocytes, pancreatic acinar cells, and enterocytes. Electron microscopic examination identified the organism as an intranuclear coccidian. Additional cases of intranuclear coccidiosis in tortoises were subsequently reported, including several radiated tortoises with proliferative pneumonia and a leopard tortoise (*Geochelone pardalis*) that had an inner ear infection (Garner et al. 2006). This pathogen continues to be a problematic and insidious parasite in collections of tortoises across the United States.

Laboratories for Molecular Diagnostics of Pathogens

Samples for PCR detections of a variety of pathogens in tortoises can be sent to: April Childress (Childressa@ufl.edu), College of Veterinary Medicine, University of Florida, Gainesville, FL 32610. For mycoplasma diagnostics, contact Dr. Mary Brown (mbbrown@ufl.edu), College of Veterinary Medicine, University of Florida, Gainesville, Fl 32610.

NONINFECTIOUS DISEASES

Cutaneous Dyskeratosis

High mortality rates and a shell disease, originally described as shell necrosis, were observed in the population of *Gopherus agassizii* (in the Colorado Desert, on the Chuckwalla Bench Area of Critical Environmental Concern, Riverside County, California). This disease was subsequently named cutaneous dyskeratosis (Jacobson et al. 1994). It is a disease of the keratinaceous layer of shell that is disrupted by multiple crevices and fissures. The lesion commenced at seams between scutes and spread toward the middle of each scute in an irregular pattern.

In the most severe lesions, dermal bone showed osteoclastic resorption, remodeling, and osteopenia (reduced bone mass). Although the disease was present on the carapace, plastron, and thickened forelimb scutes, the plastron was more severely affected than other areas of the integument. Special staining indicated a loss of the normal integrity of the horny material covering affected scutes. For the most part, the epithelial cells that formed a pseudostratified layer under affected portions of each scute remained intact. Although the location and histologic appearance of the lesion were compatible with a dyskeratosis and were suggestive of either a deficiency disease or toxicosis, the exact cause of the disease could not be determined.

Urolithiasis

Urolithiasis refers to an accumulation of uric acid calculi anywhere in the urinary system. In tortoises, it is primarily seen in the urinary bladder (cystolithiasis). It has been seen sporadically in populations of both captive and wild juvenile and adult *G. agassizii*. It is uncommon in *G. polyphemus*. Of 42 ill or dead *G. agassizii* that were received between March 1992 and July 1995 for necropsies from the Mojave and Colorado deserts of California, three cases of urolithiasis were seen (Homer et al. 1998). Factors that predispose animals to urolithiasis include excretion of calculogenic material in the urine, urinary pH, dehydration, vitamin A deficiency, and supersaturation of urine by stone-forming salts (Maxie 2007).

Gout

Gout is an accumulation of uric acid crystals in visceral organs such as the kidney, liver, pericardial sac, and/or skeletal joint (articular) spaces. Affected tortoises typically have elevated levels of uric acid in the plasma. In the study involving 42 ill or dead *G. agassizii* salvaged from the Mojave Desert, only one necropsied tortoise had renal and articular gout. Predisposing factors for gout include dehydration, preexisting renal disease, exposure to a nephrotoxin, or excess animal protein in the diet (Maxie 2007). The exact cause in *G. agassizii* is unknown, however.

Oxalosis

Renal oxalosis, previously unreported in *G. agassizii*, was seen in an individual that had an obstruction of renal tubules with polarizing crystals, identified as calcium oxalate, throughout the kidney (Jacobson et al. 2009). The light microscopic changes seen in the kidney associated with crystals of calcium oxalate were considered sufficient to compromise renal function. In a retrospective study of necropsied *G. agassizii* where the kidney was evaluated, two additional cases were found having similar-appearing crystals in the kidney. The amount of crystallization in the kidney was categorized as minimal, however, and was considered an incidental finding and unlikely to affect renal function. *Gopherus agassizii* often experiences extended periods of dehydration and starvation during or after droughts (Berry et al. 2002), which may influence oxalate deposition in the kidney. In the last few decades, many alien plant species, such as the Saharan or Moroccan mustard (*Brassica tournefortii*) and fountain grass (*Pennisetum setaceum*), have invaded the Mojave and Colorado Deserts (Lovich 2000, Minnich and Sanders 2000). Although tortoises have not been observed to forage on these plants, no efforts have been made to determine if they will consume them under drought or experimental conditions, and no research has yet been conducted on the oxalate content of these plants. Possibly such invasive plants have higher oxalate content than the native plants that tortoises feed upon.

Drought

There are numerous anecdotal accounts of large die-offs of tortoises at the end of a long period of drought. Drought and subsequent dehydration and starvation are seen as contributors to poor condition and death in wild *G. agassizii* (Turner et al. 1984, Peterson 1996a). Drought also has been implicated in low growth rates of young individuals, reduced egg production and reproductive effort in females, reduced activity levels and movement, and low metabolic rates (Berry et al. 2002). During or following droughts between 1990 and 1995, 11 individuals of *G. agassizii* suffering from dehydration and starvation were salvaged from three field sites in the Mojave Desert (Berry et al. 2002). In the weeks and months preceding salvage, the tortoises behaved abnormally for the season and weather conditions (e.g., not entering burrows for hibernation in fall, remaining above ground overnight exposed to freezing temperatures). Rain sufficient to produce free-standing water fell in the vicinity of nine of the individuals, but only four showed evidence of drinking. Necropsy findings included atrophy or disappearance of the thymus, lack of subcutaneous fat adjacent to the proximal ends of the humeri, lack of coelomic fat, empty stomachs and upper intestines, and urolithiasis.

Trauma, Predation, Burns

In a study of salvaged tortoises in the Mojave Desert, traumatic injuries consisted of one tortoise entombed within its burrow, one tortoise burned in a brush fire, two tortoises struck by moving vehicles, and one tortoise attacked by a predator (Homer et al. 1998). At the University of Florida, approximately 25 injured tortoises are seen per year. The majority of these are either hit by cars or traumatized by dogs. Rehabilitation for these animals can take months.

CONCLUSIONS

Given the number of North American tortoises living and breeding on confined areas of land, there is hope that remaining populations can be managed for long-term persistence. Assessment of population health plays a critical role as a part of an overall management plan. Although it is impossible to assess health of all remaining populations of *Gopherus* spp., certain key populations need to be selected for long-term monitoring. Veterinarians have become more involved in such management programs and can provide help to conservation biologists who plan tortoise management programs.

Necropsy of ill or recently dead tortoises is an all important component of any health assessment program, to understand the causes of mortality and identify new emerging diseases.

APPENDIX 8.1

Glossary of Important Definitions Regarding Health and Disease

Acute inflammation. Initial response of the body to harmful stimuli achieved by the increased movement of plasma and leukocytes (especially granulocytes) from the blood into the injured tissues.

Alkaline phosphatase (ALP). A hydrolase enzyme responsible for removing phosphate groups from many types of molecules, including nucleotides, proteins, and alkaloids. In humans, alkaline phosphatase is present in all tissues throughout the entire body, but is particularly concentrated in the liver, bile duct, kidney, bone, and placenta.

Alanine transaminase (ALT). An enzyme that is also called serum glutamic pyruvic transaminase (SGPT) or alanine aminotransferase (ALAT). It is commonly measured clinically as a part of a diagnostic evaluation of hepatocellular injury, to determine liver health.

Analyte. A substance or chemical constituent that is undergoing analysis.

Ascites. Abnormal accumulation of fluid in the coelomic cavity.

Blood urea nitrogen (BUN). A BUN test measures the amount of nitrogen in the blood that comes from the waste product urea. Urea is made when protein is broken down in the body. Urea is made in the liver and passed out of the body in the urine.

Caseous. Resembling cheese or curd.

Chronic inflammation. Prolonged inflammation, which leads to a progressive shift in the type of cells present at the site of inflammation and is characterized by simultaneous destruction and healing of the tissue from the inflammatory process. Macrophages or histiocytes are generally seen at the site of the inflammation.

Clinical sign. Any indication of a medical condition that can be objectively observed.

Complete blood count (CBC). A CBC gives important information about the kinds and numbers of cells in the blood, especially red blood cells, white blood cells, and (for reptiles) thrombocytes. Hematocrit or packed cell volume is generally included.

Condition index. A scoring of the body condition using a noninvasive method for assessing health. Weight-to-volume ratios have been used for tortoises.

Creatine phosphokinase (CPK). Also known as creatine kinase or phospho-creatine kinase (and sometimes wrongly as creatinine kinase), an enzyme expressed by various tissues and cell types. Elevation is an indication of muscle damage.

Disease. "Any deviation from or interruption of the normal structure or function of any part, organ, or system (or combination thereof) of the body that is manifested by characteristic set of symptoms and signs and whose etiology, pathology, and prognosis may be known or unknown" (Dorland's Illustrated Medical Dictionary 1985).

Enzyme linked immunsorbent assay (ELISA). A biochemical technique used mainly in immunology and diagnostic medicine to detect the presence of an antibody or an antigen in a sample. When measuring antibody, a positive test indicates presence of a specific antibody (exposure). It does not provide direct information about an active disease process.

Epidemiology. Study of health-event patterns in a population. It is the cornerstone method of public health research and helps inform evidence-based medicine for identifying risk factors for disease and determining optimal treatment approaches to clinical practice and for preventive medicine.

Epizootic. A disease that appears as new cases in a given animal population, during a given period, at a rate that substantially exceeds what is "expected" based on recent experience. Epidemic is the analogous term applied to human populations.

Gamma-glutamyl transpeptidase. Also known as gamma-glutamyl transferase, it is an enzyme that transfers gamma-glutamyl functional groups. It is found in many tissues, the most notable one being the liver, and has significance in medicine as a diagnostic marker. It is unknown whether levels of this enzyme will rise in the presence of liver disease in tortoises.

Glossitis. Inflammatory change of the tongue.

Hematocrit or packed cell value (PCV). The percentage of red blood cell volume that is occupied by blood cells. It is normally about 20–30% for tortoises.

Infection. Invasion of the body with organisms that have the potential to cause disease.

Infectious disease. A disease resulting from the presence and activity of a microbial agent.

Mycoplasmosis. A collective term for infectious diseases caused by the microorganisms in the genus *Mycoplasma*.

Necropsy. Postmortem examination of an animal.

Negative predictive value. The proportion of subjects with a negative test result who are correctly diagnosed.

Palpation. Physical examination by pressure of the hand or fingers to the surface of the body, especially to determine the size or consistency of an underlying organ or mass.

Pathogen. An infectious agent (virus, bacteria, fungus, parasite) that has the potential to invade an animal's body and cause disease.

Pathogenicity. Ability of a pathogen to produce an infectious disease in an organism.

Pathology. Study and diagnosis of disease.

Plasma biochemicals. Inorganic and organic analytes in plasma (or serum) that are routinely used to evaluate the health of an animal.

Polymerase chain reaction (PCR). A technique in molecular biology to amplify a single or a few copies of a piece of DNA across several orders of magnitude, generating thousands to millions of copies of a particular DNA sequence.

Positive predictive value. In statistics and diagnostic testing, the proportion of subjects with positive test results that are correctly diagnosed.

Prevalence. In epidemiology, the total number of cases in the population, divided by the number of individuals in the population. It is used as an estimate of how common a disease is within a population over a certain period of time.

Sensitivity. The proportion of positives that are correctly identified (e.g., the percentage of sick animals correctly identified as having the condition).

Serology. The scientific study of blood serum and other bodily fluids. In practice, the term usually refers to the diagnostic identification of antibodies in the serum.

Seronegative. A serological test (using serum or plasma) with a negative result, indicating lack of specific antibodies against a pathogen such as *Mycoplasma agassizii*.

Seropositive. A serological test (using serum or plasma) with a positive result, indicating presence of specific antibodies against a pathogen such as *Mycoplasma agassizii*.

Serum neutralization. A method of determining presence of specific antibodies in serum based on the loss of viral infectivity through reaction of the virus with specific antibody. Virus and serum are mixed under appropriate conditions and then inoculated into cell culture, eggs, or animals. The presence of unneutralized virus may be detected by reactions such as cytopathic effects (CPE), haemadsorption/haemagglutination, plaque formation, and disease in animals.

Specificity. The proportion of negatives that are correctly identified (e.g., the percentage of healthy animals correctly identified as not having the condition).

Stomatitis. Inflammatory disease of the oral cavity.

Subacute inflammatory disease. A condition intermediate between chronic and acute inflammation, exhibiting some of the characteristics of each.

Subclinical. A condition in which an animal is infected with a pathogen but is not manifesting any clinical signs.

Symptom. Any manifestation of a condition that is apparent to the patient.

Upper respiratory tract disease (URTD). Disease affecting the upper respiratory tract, which includes the nares, choanae, nasal cavities, and orbital adnexal structures.

Zoonoses. Diseases of animals that are transmissible to humans.

9 Habitat Characteristics of North American Tortoises

Kenneth E. Nussear
Tracey D. Tuberville

North American tortoises are distributed in semiarid and temperate deserts and coastal regions of the southern United States and Mexico (Bury and Germano 1994). The five species currently recognized each have specific habitat requirements, which they fulfill through their selection of, and interaction with, unique habitat constituents. Below, we discuss the physiographic and geological associations, perennial and annual vegetation components, shelter sites, and climatic conditions associated with the species' habitats, as well as the potential threats to their habitat.

PHYSIOGRAPHY AND GEOLOGY

Gopherus agassizii is most commonly found in valley bottoms throughout the Mojave Desert and portions of the Sonoran Desert north and west of the Colorado River (fig. 9.1). It inhabits several distinct habitat types, from alkaline areas surrounding playas, through bajadas formed by alluvial fans at middle elevations, and into rockier slopes and intermittent dunes and valleys as its distribution encroaches on mountainous areas (Bury et al. 1994, Nussear et al. 2009). The species occurs on soils where burrow construction is possible (Burge 1978, Morafka and Berry 2002), ranging from sandy to sandy loamy to gravelly (Luckenbach 1982), and in rockier soils with caves that provide cover during summer aestivation and winter dormancy (Woodbury and Hardy 1948, Bulova 1994, Nussear et al. 2007). Individuals tend to avoid areas of exposed bedrock, which impedes digging shelter and nesting sites (Bramble 1982, Bury et al. 1994) and inhibits mobility in the steep terrain typical of this soil type (Bury et al. 1994, Nussear et al. 2009).

Gopherus morafkai lives in the Sonoran Desert south and east of the Colorado River in Arizona, extending southward into Sonora and Sinaloa, Mexico (Fritts and Jennings 1994, Van Devender 2002a, Murphy et al. 2011). The species also inhabits Mojave Desert habitats in Arizona, along the transition zone of the Mojave and Sonoran Deserts (Murphy et al. 2011), although some populations may be remnant populations of *G. agassizii* isolated when river channels shifted (McLuckie et al. 1999). The distribution of *G. morafkai* spans many subdivisions within the Sonoran Desert, including the Lower Colorado River Valley and Arizona Upland, margins of semidesert grasslands in the east, and Sinaloan thornscrub and deciduous forests in the south (Brown 1994, Turner and Brown 1994, Fritts and Jennings 1994, Bury et al. 2002, Van Devender 2002b).

Within its broad distribution of habitat types, *G. morafkai* occupies a wide range of soils. The Arizona Upland consists of rocky slopes and ridges, with underlying soils formed from volcanic, metamorphic, and sedimentary rocks (Van Devender 2002b). Individuals there live on rocky slopes and bajadas and in deeply incised washes with caliche caves (fig. 9.2). Individuals generally are absent from intermountain floors (Barrett 1990, Berry et al. 2002a, Averill-Murray and Averill-Murray 2005), but occupy the entire habitat above the bajada (Averill-Murray et al. 2002b). The rocky habitat of the Arizona Upland transitions in the south to the Plains of Sonora and there, much like in the Arizona Upland, individuals inhabit rocky slopes, bajadas, and arroyos (Fritts and Jennings 1994, Van Devender 2002a).

The two species of desert tortoise are the most closely related among the five species of tortoises in North America (chapter 3), and occupy a continuum of habitats throughout the Mojave and Sonoran Deserts, ranging from valley bottoms in the Mojave Desert to uplands in the hotter and wetter Sonoran Desert. It has been hypothesized that this upward shift in elevation in the Sonoran Desert is attributable to the temperature and precipitation differences between the Mojave and Sonoran Deserts: periodic flooding, high clay and silt content in soils, and corresponding low plant diversity are typical of the valley bottoms of the Sonoran Desert, especially in the lower Colorado River Valley (Turner and Brown

Fig. 9.1. Typical *Gopherus agassizii* Creosote scrub habitat, in the Mojave Desert, near Coyote Springs Valley, Nevada. Photo courtesy of K Nussear.

Fig. 9.2. Characteristic upland habitat of *Gopherus morafkai*, near Tuscon, Arizona. Photo courtesy of TC Esque.

Fig. 9.3. Dave Morafka surveying *Gopherus flavomarginatus* habitat. Photo courtesy of D Biggins and B Bury.

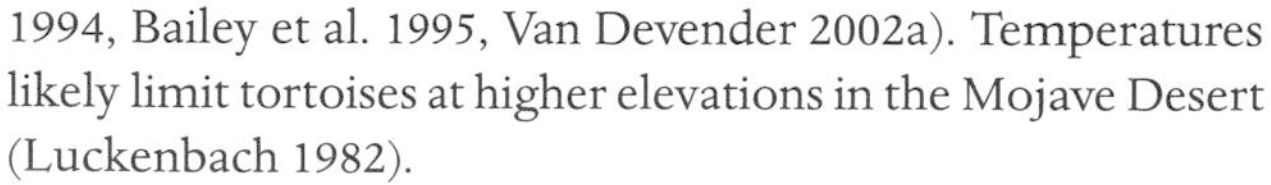

1994, Bailey et al. 1995, Van Devender 2002a). Temperatures likely limit tortoises at higher elevations in the Mojave Desert (Luckenbach 1982).

Gopherus flavomarginatus is precinctive to an area of approximately 6000 km² within the Mapimian subprovince of the Chihuahuan Desert (Morafka 1982). It occupies interconnected drainages in the Bolson de Mapimi, in portions of northeastern Durango, southwestern Chihuahua, and southeastern Coahuila, Mexico (Auffenberg 1969, Morafka 1982, Aguirre 1994). The species occurs exclusively in bolsons—rolling desert valleys typically containing a playa surrounded by hillsides and mountainous areas (Bury et al. 1988; fig. 9.3). Limestone ridges surround many of these bolson lowlands, which were once pluvial lakes in the late Pleistocene and early Holocene (Morafka 1982). Individuals live and construct their burrows in areas that are predominantly sand (50–72%; Lieberman and Morafka 1988) and fine gravel, with little clay or silt (7–25%; Bramble 1982, Morafka 1982), and with high alkalinity (Morafka 1982, Adest and Aguirre 1995). Individuals generally are absent from the flat lowland areas, which flood periodically during summer rains (Bury et al. 1988), but inhabit the gently sloping sandy areas below the rocky hilltops. Individuals also are rare in areas of higher slopes, as soils change from hard compacted sand/clay composition

Fig. 9.4. Thornscrub savannas with patches of brush and cactus and scattered open grassy areas occupied by *Gopherus berlandieri* in Texas. Photo courtesy of TD Tuberville.

to desert pavement or to soils with high gravel content (i.e., 25%), which reduces their ability to burrow and limits growth of grasses and other forage species (Morafka 1982, Lieberman and Morafka 1988).

Gopherus berlandieri inhabits Tamaulipan brushlands in the plains of the Lower Rio Grande Valley (Kazmaier et al. 2001a; fig. 9.4), and prefers sandy, well-drained soils (Ernst and Lovich 2009). It occurs at elevations from sea level to 200 m elevation in southern Texas, and up to 884 m elevation in Coahuila, Mexico (Rose and Judd 1982). In coastal areas, the species is noted for its association with lomas—isolated clay or sand dunes surrounded by salt flats that may be periodically inundated with water (Auffenberg and Weaver 1969). Flats that are frequently flooded remain unvegetated, while the adjacent lomas create windbreaks that accumulate windblown silt. Lomas along vegetated flats are only temporarily stabilized; most of the time, however, more soil is eroding than is accumulating (Auffenberg and Weaver 1969). Lomas exhibit remarkable spatial and temporal heterogeneity, varying in size and proximity to one another. Because intervening salt flats are unsuitable for the species, lomas function as habitat islands, with populations associated with particular lomas and apparently little interchange of individuals among them (Bury and Smith 1986).

Gopherus polyphemus is restricted to the Coastal Plain of the southeastern U.S., where it is associated with deep, well-drained sandy soils (Auffenberg and Franz 1982, USFWS 1990a). Although the relative proportions of sand, clay, and loam components may vary, these soils generally have a sandy layer extending at least 1–1.5 m deep (USFWS 1990a, Jones and Dorr 2004). Jones and Dorr (2004) found active burrows to be positively associated with sandy soils and negatively associated with loamy soils in industrial forests of Alabama and Mississippi (fig. 9.5). The probability of the species occurring in the landscape (based on presence of their burrows) decreases with increasing clay content in surface soil horizons (Baskaran

Fig. 9.5. Open canopy Longleaf pine habitat occupied by *Gopherus polyphemus*. Photo courtesy of TD Tuberville.

et al. 2006). In the western portion of the species' range, deep sandy soils are limited, and individuals are associated with soils with higher clay content than elsewhere throughout the range (Means 1982).

VEGETATION

Gopherus agassizii is found in Mojave Desert scrub vegetation with the dominant vegetation alliances of creosote / bursage (*Larrea tridentate / Ambrosia dumosa*) scrub (Turner 1994, Germano et al. 1994; fig. 9.1). Vegetation changes predictably with elevation, and at lower elevations, individuals are found in areas dominated by saltbush (*Atriplex* sp.). In upper elevations they can be found in Mojave mixed desert scrub, which frequently includes *Yucca* species (*Y. schidigera, Y. brevifolia,* and *Y. baccata*) in a diverse mix of perennial plants. With increasing elevation, the dominant vegetation transitions to blackbrush (*Coleogyne ramosissima*) and eventually piñon (*Pinus monophylla*)–juniper (*Juniperus osteosperma*) woodlands (Turner 1994), and tortoises become scarce (Schamberger and Turner 1986, Bury et al. 1994). At higher latitudes, individuals can be found at times in sage- (*Artemisia* sp.) dominated areas in the transition between the Mojave and the Great

Basin (Nussear 2004). In the Colorado Desert of California, *G. agassizii* is found in areas with incursions of plants more typically associated with Sonoran Desert scrub (e.g., ocotillo [*Fouquieria splendens*] and ironwood [*Olneya tesota*]), in washes interspersed in a *Larrea/Ambrosia*–dominated landscape, and in rockier sites with cactus-scrub vegetation (Luckenbach 1982). Individuals actively seek specific species of perennial plants for shelter (Woodbury and Hardy 1948, Luckenbach 1982, Bury et al. 1994), presumably because of their thermal characteristics (Burge 1978, Hillard 1996, Nussear 2004). In addition to creosote and bursage in the Mojave Desert, *Ephedra* sp., *Lycium andersonii,* and *Yucca* sp. also are frequently chosen (Woodbury and Hardy 1948, Burge 1978, Nussear 2004).

The Arizona Upland occupied by *G. morafkai* has diverse species assemblages, including palo verde (*Cercidium* sp.), ironwood, and saguaro with many shrubs (e.g., brittlebush, California buckwheat) and cacti in the understory (Turner and Brown 1994, Van Devender 2002a). Washes and arroyos emanating from the uplands frequently contain mesquite trees and lowland shrubs (e.g., *Lycium* sp.; Van Devender 2002a). The Lower Colorado River Basin has a less diverse and less dense flora as a result of high temperatures and low precipitation (Turner and Brown 1994, Van Devender 2002a). Saguaro, palo verde, and ironwood are still present but restricted to washes. The flora is dominated by creosote and bursage (Turner and Brown 1994, Van Devender 2002a), much like in the Mojave Desert (Turner 1994). In Mexico, the species inhabits densely vegetated Sinaloan thornscrub, with as many as 2,000 plants per hectare, consisting predominantly of shrubs in the north and transitioning to trees in the south (Brown 1994). Individuals are sparsely distributed on the eastern border of the species' range, and diminish quickly as habitat transitions to Madrean evergreen woodlands (Fritts and Jennings 1994).

Gopherus flavomarginatus inhabits grasslands of the Chihuahuan desert (Morafka 1982). The dominant perennial vegetation is comprised largely of sparsely distributed shrubs and small trees such as mesquite (*Prosopis* sp.), creosote, tarbush (*Flourensia cenua*), guayule (*Parthenium argentatum*), and cacti, such as pencil cholla (*Opuntia* sp.) and prickly pear (*O. rastrera*) and *Agave* sp.—all of which are used for shade (Morafka 1982, Tom 1994) or for protection of burrows (Aguirre et al. 1984). Common grasses include grama grass (*Bouteloua* sp.; Aguirre 1994, Tom 1994) and tobosa grass (*Hilaria mutica*), the latter of which can comprise up to 23% cover in tortoise-occupied areas. Although individuals are strongly associated with tobosa grass, they are not common in lowlands where tobosa grass forms a monoculture: they are more prominent in areas with a mixed community of grasses, including *Sceleropogon brevifolis, Sporoholus* sp., and *Erioneuron pulchellus* (Lieberman and Morafka 1988). Ocotillo and the euphorb *Jatropha dioica* also occur on the desert pavement slopes above the grasslands (Morafka 1982).

Gopherus berlandieri is distributed in the Tamaulipan Biotic Province and the Rio Grande Plains Ecoregion (Auffenberg and Weaver 1969, Kazmaier et al. 2001b). In coastal areas, the species occupies isolated lomas embedded in a matrix of expansive, treeless grasslands or salt flats (Auffenberg and Weaver 1969, Bury and Smith 1986). The lomas have distinct, concentric vegetation zones with characteristic plant associations—brush, *Baccharis* sp. (an aster), and grass and cactus zones (Rose and Judd 1982). Elsewhere, individuals are most commonly associated with thornscrub savannas and woodlands characterized by dense patches of brush and cactus with scattered open grassy areas (Kazmaier et al. 2001a, fig. 9.4). The species favors open scrub habitats, but will use a diversity of scrub habitats varying in cover.

Individuals of *G. berlandieri* are strongly associated with prickly pear cactus (*O. lindheimeri, O. leptocaulis*), which provides both cover and food with high water content (Rose and Judd 1982). Dominant woody species in their habitat include honey mesquite (*P. glandulosa*), huisache (*Acacia farnesiana*), lotebush (*Condalia obtusifolia*), and yucca (*Y. traculeana*) (Auffenberg and Weaver 1969). Buffalo grass (*Buchloe dactyloides*) can be prevalent on the tops of lomas, with three-awn grass (*Aristida* sp.), short grass (*Bouteloua* sp.), and sand bur (*Cenchrus* sp.) more likely around the perimeter. Lomas have few forbs, and are typically surrounded by the coarse, bunchy sacahuista grass (*Spartina alterniflora*), although glasswort (*Salicornia* spp.) and seashore saltgrass (*Distichlis spicata*) may also occur (Rose and Judd 1982, Bury and Smith 1986). Dominant woody species in thornscrub are honey mesquite, hog plum (*Colubrina texensis*), and several acacia species (*Acacia rigidula, A. berlandieri, A. schaffneri*). The most common native grasses are species of *Setaria* and *Paspalum,* hooded windmill grass (*Chloris cucullata*), hairy grama (*Bouteloua hirsuta*), and fringed signal grass (*Brachiaria ciliatissima;* Hellgren et al. 2000).

Gopherus polyphemus is best known for its association with longleaf pine (*Pinus palustris*), but the species also occurs in coastal dune, scrub, and oak hammock communities. Longleaf pine forests occur along a soil-moisture continuum ranging from xeric upland sandhills typified by smaller, scattered pines and a diverse understory of oaks and other hardwoods to more mesic pine flatwoods with large, widely spaced pines and a grassy understory and no shrubby mid-story (fig. 9.5). Dominant tree species in the overstory frequently include longleaf, loblolly (*P. taeda*), and/or slash (*P. elliottii*) pines, and turkey (*Quercus laevis*), post (*Q. stellata*), bluejack (*Q. incana*), and/or laurel (*Q. laurifolia*) oak. Hickory (*Carya* spp.) and sweet gum (*Liquidambar styraciflua*) also may be present. Blueberry (*Vaccinium arboreum, V. stamineum*), fetterbush or staggerbush (*Lyonia lucida, L. fruticosa*), gallberry (*Ilex* spp.), and wax myrtle (*Myrica cerifera*) commonly occur in the shrub understory. Saw palmetto (*Serenoa repens*) can dominate at some sites, particularly in coastal areas. The longleaf pine forest is well known for its understory diversity (Walker and Peet 1984, Peet and Allard 1993). More than 1,000 plant species have been documented at individual sites (e.g., Drew et al. 1998, Sorrie et al. 2006), although species diversity can vary as a function of local edaphic conditions, site history, and contemporary management (Kirkman et al. 2001, Brudvig and Damschen 2011). The herbaceous understory typically is comprised of a combination of grasses and forbs, including wiregrass (*Aristida stricta, A. beyrichiana*), bluestem grasses (*Andropogon* spp.), asters, and legumes. Prickly pear cactus (*O. compressa*) is characteristic in some habitats.

Longleaf pine historically was widespread, constituting about 60% of the Coastal Plain. Few remnants of old-growth remain, however (<3% of historical distribution; Ware et al. 1993, Means 2006), with much having been converted to other land uses or replanted with faster-growing pine species. Where soils are suitable, *G. polyphemus* can inhabit ruderal habitats, including abandoned pastures and agricultural fields, pine plantations managed for timber production, powerline and transportation rights-of-way, military training areas, and other disturbed sites with open canopies and abundant forage (Hermann et al. 2002, Jones and Dorr 2004, Baskaran et al. 2006, Yager et al. 2007, Ashton et al. 2008).

PRECIPITATION

Global atmospheric circulation patterns result in warm, high-pressure air masses causing generally dry conditions at the latitudes where the western species of North American tortoises live. General precipitation patterns periodically are affected by the El Nino–Southern Oscillation Cycle, every seven to ten years. During positive cycling (El Niño), rainfall increases in western North America and decreases in the East, while the converse occurs during negative cycling (La Niña). In the West, mountain ranges create rain shadows that, coupled with cold marine currents, influence the North American hot deserts (Evenari 1985, Shmida 1985) occupied by the three western-most tortoise species—*Gopherus agassizii* (Mojave Desert), *G. morafkai* (Sonoran Desert), and *G. flavomarginatus* (Chihuahuan Desert) (Patterson 1982, Murphy et al. 2011, Morafka 1982). Each desert provides different opportunities for water (and consequently food; chapter 10) acquisition because of a precipitation cline overlaid on the latitudinal temperature gradient, running diagonally through all three deserts. The northwest Mojave Desert is normally dominated by winter rains from November through March. Summer storms of tropical origin move across the Mojave Desert each summer, but precipitation in the summer typically is low, increasing toward the southeast. Likewise, winter storm dominance diminishes toward the south and east. Precipitation within the distribution of *G. agassizii* ranges from 100 to 210 mm per year, with most falling as winter rain, especially in the western Mojave Desert (Germano et al. 1994, Tracy et al. 2004). Both summer and winter precipitation influence habitat suitability, with areas having winter rainfall of 50–180 mm and summer rainfall of 20–130 mm predicted to support more suitable tortoise habitat (Nussear et al. 2009).

The desert scrub habitat occupied by *G. morafkai* has low and unevenly distributed rainfall. The proportion of summer to winter rainfall is 35:70%, far greater than in the adjacent Mojave Desert (<10%; Germano et al. 1994). Average precipitation ranges from 0 to 170 mm in summer and 15 to 70 mm in winter for the Lower Colorado River Basin; 60–240 mm in summer and 40–140 mm in winter in the Arizona Upland, and 45–185 mm in summer and 20–60 mm in winter in the Plains of Sonora (Turner and Brown 1994). In each case, precipitation varies far more in summer than in winter; but summer rainfall is more reliable in the Sonoran Desert than in the Mojave Desert (Germano et al. 1994, Averill-Murray et al. 2002b). Tortoises generally are absent from areas with <100 mm of annual precipitation (Fritts and Jennings 1994). Drought can increase mortality rates and potentially decimate local populations of the desert-dwelling tortoises (Howland and Rorabaugh 2002, Averill-Murray et al. 2002b, Peterson 1994, Longshore et al. 2003). In contrast, *G. flavomarginatus* habitats receive the majority of their precipitation during summer months, as a result of tropical storms in late June through November. Annual precipitation ranges from 200 to 323 mm (Morafka and Berry 2002), and is concentrated during summer months (~70% in June through October; Morafka 1982, Adest and Aguirre 1995).

Annual rainfall is significantly higher for the two easternmost tortoise species than for the western species (fig. 9.6), because of moisture from the subtropical ridge of the Atlantic Ocean and Gulf of Mexico meeting with cooler continental air masses. These conditions create humid and subtropical environments for *G. berlandieri* and *G. polyphemus* (Rose and Judd 1982, Auffenberg and Franz 1982). Annual precipitation in *G. berlandieri* habitat averages 650–700 mm, but is seasonal and can vary dramatically from year to year. In coastal areas, nearly 70% of rainfall occurs in the "rainy season" (May–October), with peak rainfall during September–October, when a third of annual rainfall occurs (Auffenberg and Weaver 1969, Rose and Judd 1982). At the northern extent of the species' range, inland thornscrub sites have a bimodal rainfall pattern with a peak in May–June and a minor peak in September (Kazmaier et al. 2001b). Droughts are frequent, but even in drought years, relative humidity in coastal areas can average 88%, resulting in heavy dew (Auffenberg and Weaver 1969, Rose and Judd 1982).

Even though *G. polyphemus* occupies some of the most xeric sites found within its geographic range, humidity and water availability are much higher than in habitats used by its congeners (Douglass and Layne 1978), generally receiving more than ten times the precipitation of sites in the Mojave Desert, for example. Mean annual rainfall ranges from 1140–1650 mm (Minnich and Ziegler 1977). The majority of rainfall for the eastern species occurs during summer months, enhanced by tropical storms, although the northern part of the range experiences relatively even amounts of precipitation throughout the year (Germano 1994). In the humid subtropical region of south Florida, precipitation is markedly seasonal, with >60% falling during the four-month wet season (June–September; Douglass and Layne 1978).

TEMPERATURE

North American tortoise habitats range from hot deserts to subtropical regions. Tortoises are notably absent from the colder temperate regions in the United States, appearing to

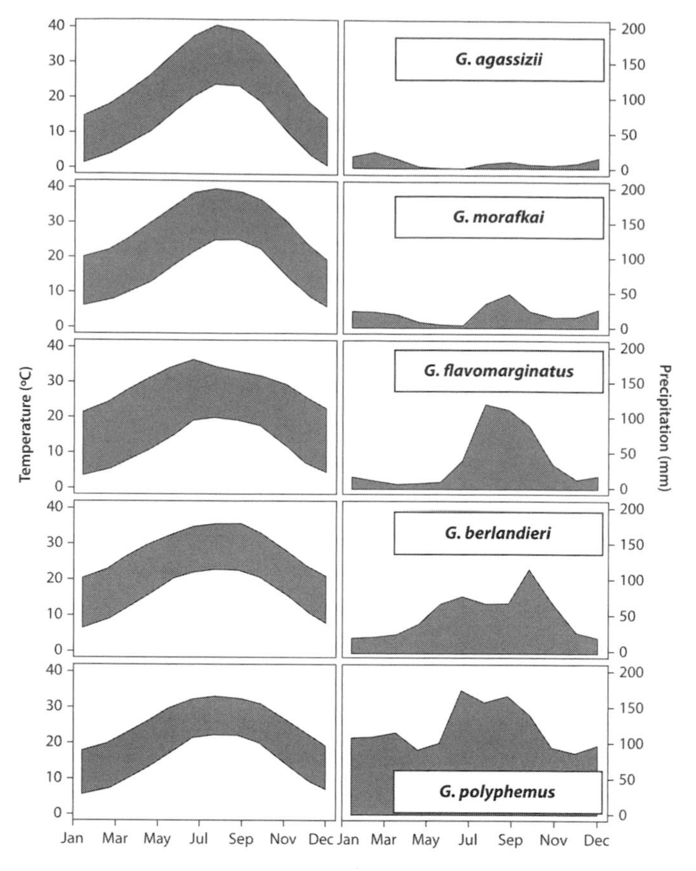

Fig. 9.6. Average high and low monthly temperatures (*left*) and precipitation (*right*) for localities within habitats for the five species of tortoises in North America. Data for *Gopherus agassizii* were compiled from 30 year normals of Inyokern, California; Baker, California; Twenty-Nine Palms, California; Las Vegas, Nevada; and St. George, Utah. Data for *G. morafkai* were for Needles, California; Phoenix, Arizona; Tuscon, Arizona; Ajo, Arizona; and Hermosillo, Mexico. Climate for *G. flavomarginatus* was represented by three locations in Mexico: Minas Nuevas, Sonora; Ciudad Delicias, Chihuahua; and Ciudad Jimenez, Chihuahua. Stations contributing to the climate data for *G. berlandieri* included Brownsville, Crystal City, and Del Rio in Texas and Sabinas, Coahuila; Monterrey, Nuevo León; and Ciudad Victoria, Tamaulipas, in Mexico. Climate data for *G. polyphemus* were taken from Yemassee, South Carolina; Hialeah, Florida; Tampa, Florida; Gainesville, Florida; Albany, Georgia; Atmore, Alabama; Hattiesburg, Mississippi; and Baton Rouge, Louisiana. All data for U.S. cities were acquired from www.weather.com, and data for Mexico were acquired from www.weatherbase.com.

be more limited by the presence of cold than hot temperatures. The combinations of precipitation and temperature that define their habitats vary greatly among the five species (fig. 9.6). In the western deserts, tortoises inhabit some of the hottest areas, with distributions largely limited by cold temperature extremes in the Mojave Desert (although the mechanism is not well understood). High temperatures may restrict tortoises to upland habitats in the Sonoran and Sinaloan Deserts, although in southern regions they may also be excluded from the lowlands due to the potential for flooding.

Temperature extremes in the Mojave and Sonoran Deserts are broad, with mean annual high and low temperatures for each ranging from –2 to 43°C (Germano et al. 1994). Preferred temperatures for *Gopherus agassizii* (20–38°C; McGinnis and Voigt 1971, Zimmerman et al. 1994) and *G. morafkai* (20–45°C; Averill-Murray et al. 2002a) are bracketed closely by these temperatures; however, the temperature experienced by the individuals (operative temperatures) are much greater. Operative temperatures are influenced by air temperature, solar radiation, and other inputs (Porter and Gates 1969, Tracy 1982) and for tortoises in deserts are estimated to be as high as 55–60°C. Tortoises avoid extreme temperatures by retreating to burrows (Zimmerman et al. 1994). Favorable operative temperatures for *G. agassizii* generally are available during midday in spring, in mornings and evenings during summer months, and again during the midday in fall (Zimmerman et al. 1994, Hillard 1996). Although summers typically are hot in the Sonoran Desert, seasonal monsoonal events stimulate activity as tortoises respond to the resulting increased forage availability (Averill-Murray et al. 2002a). *Gopherus agassizii* has the northernmost distribution of the North American tortoises, and its northern limit is associated with the boundary between the Mojave (hot) and the Great Basin (temperate) Deserts. Despite ample precipitation in the eastern Sonoran Desert, extended periods of cold and freezing temperatures exclude most of the Sonoran Desert perennial plant species and likely limit the distribution of *G. morafkai,* as well (Van Devender 2002a). The habitat of *G. flavomarginatus* is considered warm-temperate to subtropical (Lieberman and Morafka 1988), with annual temperatures typically averaging 20–35°C; however, in extreme years winter temperatures can fall as low as –7°C and summer temperatures can reach 39°C (Morafka 1982).

The two eastern tortoise species experience a more hospitable climate with respect to temperatures than the western species (fig. 9.6). *Gopherus berlandieri* occurs where summers are hot and winters short and mild (Kazmaier et al. 2001c, 2001d). Mean temperatures are relatively warm in all seasons, peaking in July when mean air temperatures are 28.7–30.4°C (Auffenberg and Weaver 1969, Kazmaier et al. 2001d). Like tortoises in the desert regions, *G. berlandieri* experiences temperatures that are sufficiently high to induce bimodal activity (morning and afternoon), which is more pronounced in the warmer months of June and July than in the cooler months of November and April (Auffenberg and Weaver 1969). Mean temperatures where the species occurs are 12.5–15.4°C in the coldest month (January), with an average daily minimum of 10.5°C. The distribution of *G. polyphemus* has a relatively narrow range of average daily temperatures (5–17°C in winter and 20–32°C in summer; fig. 9.6). Individuals are most active from March to April through October to November (Eubanks et al. 2003, Yager et al. 2007), with activity becoming bimodal during the hot summer months. Winter temperatures drive a brief period of overwintering (Douglass and Layne 1978). Although individuals may be active above ground in any month of the year, winter activity is more likely in the southern portion of their range. Even at the northernmost portion of the species' range, however, individuals may occasionally emerge to bask on warmer winter days (DeGregorio et al. 2012).

SHELTER

The burrow is a key feature of the habitat for all species of *Gopherus,* and is used to avoid temperature extremes both in summer and winter, and to provide protection from predators (Woodbury and Hardy 1948, Zimmerman et al. 1994, Nussear et al. 2007; chapter 9.4). Burrows are of such importance that friability of the soil can limit tortoise occurrence: in some areas, soils are too sandy to support burrows, and in other areas, soils are too rocky, or consist of exposed or shallow bedrock in which burrows cannot be constructed (Burge 1978, Averill-Murray 2002a). Tortoises, however, are quite adaptable, and individuals of some species take advantage of naturally occurring caves, rock crevices, or other shelter sites. For example, in the northern extent of the species' range, individuals of *G. agassizii* living in sandy soils use naturally occurring caves to provide needed shelter during winter hibernation (Woodbury and Hardy 1948, Nussear et al. 2007), and also use dens formed by erosion in the gravel layers within washes, especially in habitats with a shallow caliche layer beneath the soil (Woodbury and Hardy 1948, Germano et al. 1994). Individuals of *G. morafkai,* which live on rocky hillsides, also use caves under large rocks or boulders and caliche dens, and both desert tortoise species opportunistically use pack rat middens (Woodbury and Hardy 1948; Barrett 1990; Averill-Murray et al. 2002a, 2002b; Van Devender 2002b; Riedle et al. 2008). Individuals of *G. flavomarginatus* use burrows of other species, such as rodents, if available; but, unlike desert tortoises, are not known to use rock shelters (Lieberman and Morafka 1988, Tom 1994).

Some tortoise species are more closely tied to their burrows than others. For example, both *G. polyphemus* and *G. flavomarginatus* exhibit strong affinity to burrows. Individuals typically have one or two primary burrows which they occupy during periods of daily and seasonal inactivity (Aguirre et al. 1984). Both species dig extensive burrows extending 6 m to 10 m in length, with a depth of 2 m below the surface (Auffenberg 1969, Morafka 1982, Aguirre et al. 1984, Witz et al. 1991, Aguirre 1994, Smith et al. 2005). Burrows are reused among years (Guyer and Hermann 1997), but individuals of *G. polyphemus* can have high abandonment rates in disturbed habitats (22% per year; Aresco and Guyer 1999a). Interestingly, burrows of *G. polyphemus* in mesic habitats can be subject to winter flooding, but individuals usually continue to occupy them, and have even been observed submerged in water (Means 1982, Diemer 1992a). Individuals of *G. flavomarginatus* also can withstand periodic short-term flooding of burrows, but areas subject to extended periods of flooding are not inhabited (Lieberman and Morafka 1988).

THREATS TO HABITAT

Most threats to tortoise habitat generally can be attributed directly to anthropogenic influences, including habitat loss and fragmentation as a result of the establishment of urbanized areas and their associated infrastructure, agriculture, and the proliferation of paved and unpaved roads. Roads, in addition to causing direct mortality and injury to animals, impede movements of individuals among populations and facilitate the spread of invasive plants (Auffenberg and Weaver 1969, Bury and Smith 1986, USFWS 1994a, Howland and Rorabaugh 2002, USFWS 2011b). Effects of habitat loss and fragmentation are most pronounced for *Gopherus polyphemus,* which has been relegated to small, fragmented populations throughout much of its range (McCoy et al. 2007). Although many tortoises are successfully translocated after being displaced by development and other losses of habitat (e.g., Drake et al. 2012; Nussear et al. 2012; Tuberville et al. 2005, 2008), the net loss of suitable habitat continues to reduce the number of extant viable populations. In *G. flavomarginatus* habitat, the added threat of human predation on tortoises for consumption and export to the pet trade has depleted individuals from large areas, isolating populations and impeding gene flow (Morafka 1982, Bury et al. 1988, Lieberman and Morafka 1988, Aguirre 1994). For many *Gopherus* species, military installations represent some of the largest tracts of intact natural communities (Wilson et al. 1997), but increased training demands and/or increased training intensity could result in detrimental consequences for tortoise populations and their habitats.

In the western U.S., where some of the fastest-growing human populations in the country are located (U.S. Census Bureau 2010), livestock grazing has occurred in tortoise habitat for many years and directly and indirectly affects tortoises and their habitats. Effects of grazing include reduction and alteration of native vegetation and changes in soil nutrients and soil structure, which may take decades to recover (Morafka 1982, Lieberman and Morafka 1988, USFWS 2011b). *Gopherus berlandieri* appears to be somewhat more tolerant than the desert tortoises to grazing. Some managed grazing techniques may even be beneficial, particularly if they promote establishment of *Opuntia* patches (Auffenberg and Weaver 1969, Rose and Judd 1982). Even for this disturbance-tolerant tortoise species, however, overgrazing can significantly alter habitat structure by favoring shorter grass species and by contributing to brush encroachment. In response to encroachment, many managers have implemented "range improvement" techniques to control woody vegetation, such as "chaining," roller chopping, and root plowing (Kazmaier et al. 2001a, 2001c), all of which are extremely detrimental to *G. berlandieri* (Rose and Judd 1982). Furthermore, the benefits of these techniques to livestock grazing are short-lived, as these habitats become shrubbier and less diverse following brush removal (Kazmaier et al. 2001a). Fire has been suggested as a potential technique to control woody vegetation and create grassy openings on lomas, but its effectiveness is unknown (Bury and Smith 1986).

With increasing invasions of exotic grasses in western U.S. deserts, wildfires are becoming more common in desert tortoise habitats. For example, the grasses *Bromus madritensis* and *Pennisetum ciliare* can be in high densities in intershrub spaces (12–20% cover), connecting an otherwise sparsely vegetated shrub habitat with fine fuels (Brown and Minnich 1986, Brooks 1999, Esque et al. 2002, Van Devender 2002a). This situation has artificially increased fire frequency and intensity in the Mojave and Sonoran Deserts (Brown and Minnich 1986, Brooks and Esque 2002, Esque et al. 2002). For instance, fire destroyed 36,000 acres of critical habitat for *G. agassizii* in southern Nevada in 2005. Unlike ecosystems inhabited by the *G. polyphemus,* the Mojave and Sonoran Desert ecosystems have not evolved with fire, and thus wildfires have the potential to alter habitat drastically. After each fire, the prominence of exotic grasses typically increases, creating a dynamic in which native vegetation is replaced by exotic grasses, which in turn increases fire risk (Brown and Minnich 1986, D'Antonio and Vitousek 1992). Changes in habitat may be realized not only in a reduction of annual forage species, but also loss of cover sites, as fire causes mortality of many perennial species that provide shade resources for tortoises (Esque et al. 2002). In contrast to the desert species, fire suppression presents a threat for the *G. polyphemus.* For this species, prescribed fire is important for maintaining the open canopy conditions. Fire not only suppresses the hardwood mid-story, but also is critical for germination of many groundcover species. Thus, frequent application of fire shapes the habitat structure and contributes to the incredible plant diversity of longleaf pine communities (Walker and Peet 1984, Van Lear et al. 2005). In the face of encroaching development, however, it becomes increasingly difficult to implement prescribed fire, even in protected areas. Without fire, habitat once suitable for *G. polyphemus* can quickly become unsuitable, resulting in high rates of burrow abandonment (Aresco and Guyer 1999a). Sites with a long history of fire suppression will require multiple fire applications to be restored (Yager et al. 2007, Ashton et al. 2008). Alternative methods for controlling hardwoods, such as mechanical removal and herbicide application, are being used on some sites, but their effectiveness in promoting desired habitat conditions has not been thoroughly evaluated.

Alternative energy is a new threat to tortoises, especially in the western U.S., with approximately 70 projects being proposed at the time of this writing, and a combined 10,000 km^2 of habitat potentially affected. In combination with other disturbances, these developments will result in further loss and fragmentation of habitat. Desert tortoise populations exist as metapopulations, and gene flow and the possibility of demographic rescue depends on the ability of individuals to migrate among habitat patches (Hanski 1991, Averill-Murray et al. 2002b, Edwards et al. 2004a). With each additional loss and fragmentation of habitat, this capability is diminished and ultimately negatively influences the potential for continued survival of these species.

CONCLUSIONS

We have described habitat for the North American tortoise species, using vital constituents of the landscape that bound the factors that provide or constrain their opportunity to acquire resources and persist. These factors include the physiography and geography of the landscape, and the associated vegetation, precipitation regimes, and environmental temperatures that promote the well-being of tortoise populations. Collectively, these factors represent important components of habitat interacting with the needs of tortoises, ranging from suitable temperatures for activity and egg incubation, to substrates that facilitate their ability to dig burrows and find shelter, to the availability of sufficient rainfall to provide needed forage and water. The perennial vegetation associated with each of the species provides shelter from the typically warm environments they inhabit, and these associations also have specific soils, temperatures, and precipitation regimes that drive their distributions. The combinations of these habitat attributes not only have shaped the distributions of tortoises over time, but perhaps even the distinctions among species as they adapted to local habitats. Sadly, many of these habitats are threatened by human development, and thus North American tortoises, like most tortoises around the world, face many threats to their persistence, and require conservation efforts to ensure their continued presence.

Acknowledgement

Manuscript preparation by TDT was partially supported by the Department of Energy under Award Number DE-FC09-07SR22506 to the University of Georgia Research Foundation.

10 Water and Food Acquisition and Their Consequences for Life History and Metabolism of North American Tortoises

Todd C. Esque
K. Kristina Drake
Kenneth E. Nussear

Acquisition of sufficient food and water and the physiological consequences that occur when resource availability fluctuates are key to understanding the maintenance (Peterson 1996a, b), growth (Medica et al. 2012), reproduction (Henen 2002b), and health of tortoises (Jacobson et al. 1991) and wildlife populations generally (Robbins 1983). Diet information can inform studies on the nutritional requirements of herbivorous ectotherms (Nagy 1977, Christian et al. 1984, Bjorndal 1987, Pough 1973, Zimmerman and Tracy 1989, Troyer 1991, Oftedal et al. 2002, Tracy et al. 2006). Translating this information into nutritional status can be challenging, however, because of the tortoise's ability to persist for long periods without food or water and to amortize nutritional gains and losses over periods extending over years (Nagy and Medica 1986, Henen 1997, Peterson 1996b, Hellgren et al. 2000, Oftedal et al. 2002). In this chapter, we compare and contrast the acquisition of water, food, and their influence on nutrition in relation to tortoise life history and physiology.

Because of the paucity of information on smaller life stages of tortoises, the information here focuses on adults, unless specified. This distinction is important because of the challenges associated with the small body size of neonate and juvenile tortoises (hereafter called juveniles). Smaller body size has morphological, physiological, behavioral, and ecological consequences for juvenile tortoises relative to adults. For example, juvenile tortoises are smaller and weaker, cannot travel as far, and have reduced fields of view and relatively large requirements for body-building nutrients. Smaller tortoises also have body temperatures and hydric status that conform rapidly to ambient conditions; thus, internal stasis is challenging (Morafka 1994, Morafka et al. 2000). These factors tremendously influence requirements for drinking and accessibility to feeding opportunities that juvenile tortoises respond to by modifying their behaviors and increasing efficiencies.

Opportunities for water and food acquisition among North American tortoises are driven by environmental factors at global, regional, and site-specific scales (chapter 9), as the five species' ranges nearly span the southern extent of the continent, encompassing the driest to wettest regions. For example, *Gopherus agassizii* near Barstow, California, experiences an average of ~104 mm of precipitation annually, while *G. polyphemus* in southwest Georgia experiences 1270 mm (Turner 1994, McRae et al. 1981). *Gopherus agassizii* experiences predominately winter rainfall, *G. morafkai* experiences 50:50 summer-to-winter rainfall in their north-central range (near Tucson, Arizona), and *G. flavomarginatus* experiences predominantly summer rainfall. *Gopherus berlandieri* has moisture availability that is intermediate between the three western-most species and *G. polyphemus*. The majority of rainfall for the eastern species occurs during summer months, enhanced by tropical storms, although the northern part of the range experiences relatively even amounts of precipitation throughout the year (Germano 1994). Precipitation variability is greater where the three western species live than for the other two species. Habitats spanning this range of conditions would seemingly result in divergent challenges for water and food acquisition. Closer inspection of the microenvironments that tortoises occupy reveals the similarities and differences in how tortoises acquire water and nutrition and in the strategies they employ to overcome environmental variability.

WATER ACQUISITION

Hydration

Tortoises require free water for hydration, and opportunities for water acquisition vary considerably among North American tortoises. *Gopherus agassizii* has the fewest drinking opportunities among North American tortoises. Extreme aridity (100–200 mm/y) and infrequent precipitation events demand that they drink whenever possible, including winter. Winter drinking for *G. agassizii* occurs at temperatures as low as 6°C (T. Esque and K. Nussear, unpublished data). It is striking to discover this ectotherm exiting hibernation and actively drinking or wallowing below known hibernation temperatures of ~10°C (Nussear et al. 2007). Activity at such low temperatures is a regular occurrence for the species in the northern part of its range, but is mostly unnecessary for the other species. In contrast to *G. agassizii*, *G. flavomarginatus* drinks primarily during summer storms (Adest et al. 1989, Brown 1994). Geographically intermediate to the other desert-inhabiting tortoises, *G. morafkai* receives winter and summer hydration as a result of the bimodal rainfall in the Sonoran Desert (Turner 1994, Turner and Brown 1994). Even though *G. morafkai* drinks in two seasons, the annual precipitation received in Tucson is within 10% of that in *G. flavomarginatus* habitat (NCDC 2012a, 2012b), and roughly twice that available for *G. agassizii*. *Gopherus berlandieri* also experiences bimodal rainfall patterns, with peaks in late spring and late summer (Hellgren et al. 2000). Drinking opportunities for *G. berlandiaeri* are intermediate between species to the east and west, but most similar to *G. polyphemus*, for which tropical summer storms dominate total precipitation and provide more than 40 mm of rain per month in the region (e.g., Tampa, Florida). In some *G. polyphemus* habitats, the water table is so shallow that there are drinking opportunities at the bottoms of their burrows (Ashton and Ashton 2008), at least seasonally. Furthermore, the high precipitation and abundant surface and subsurface water results in a mesic environment where evaporative losses are reduced overall. Because of the smaller body size and greater thermal reactivity, juvenile tortoises may have even more flexibility to respond to what would be considered seasonal drinking opportunities than adults. Juvenile *G. agassizii* are increasingly active in the mid- to late-winter season (late November onward, Wilson et al. 1999a) and, in the western Mojave Desert, this is the rainy season. Thus, juvenile *G. agassizii* adapt to challenges in water balance by shifting their activity to cooler periods, when the risk of water loss is decreased and water gain is increased. It is possible that the other juvenile tortoises respond in similar ways. The timing and amount of available rainfall governs drinking opportunities, which influence physiological demands in relation to other activities and triggers tortoises to respond as they can, sometimes in novel ways.

Drinking opportunities are also affected by variation in substrate materials, which influence surface flow, permeability, and infiltration, creating ephemeral catchments of water where tortoises drink. In western North America, areas of exposed bedrock moderate the availability of surface water after rains, depending on whether the parent material is relatively permeable (e.g., sandstone) or not (e.g., granite, basalt, or limestone). The two eastern tortoises occupy areas where sandy hammocks, limestone formations, river bottoms, and inundated wetlands have significant influences on the availability of resources for tortoises (Auffenberg and Franz 1982; chapter 9), such that drinking opportunities may be less available than expected in this mesic climate regime (Diemer 1986). Sandhill habitats are the driest areas within that region (Garner and Landers 1981), because coarse sand and gravel soils are well drained, and provide fewer opportunities for drinking. In those areas, light rainfall percolates immediately as it falls and is largely unavailable as free water. In coastal areas, *G. berlandieri* occupies silt/clay-dominated lomas surrounded by bottomlands seasonally inundated by seawater. Further inland, the species occupies seasonally humid shrublands with sandy soils, where only heavy rains or patches of less permeable materials provide drinking sites (Rose and Judd 1982). Because of variable permeability among soil surfaces, intermittent puddling occurs in microsites that are patchy and widespread in the landscape.

Tortoises are not only opportunistic about drinking, but they employ active water acquisition, including environmental engineering and using their own bodies to form water catchments. *Gopherus agassizii* locates water catchments and revisits them as habitual drinking sites (Medica et al. 1980). This behavior is especially apparent on soils covered by desert pavement, a thin layer of interlocked gravel atop thin and semi-impermeable clay layers where the tortoises scrape gravel clear for drinking. *Gopherus agassizii* habitually uses dependable puddling locations and are known to gather in anticipation of precipitation on rocky outcrops, calcium carbonate hardpans, washes, and abandoned roadways (T. Esque, P. Medica, and K. Nussear personal observation). *Gopherus polyphemus* may drink in a manner that is no less novel by rising to their burrow entrances in response to rain, turning sideways, and extending their upslope forelimb to intercept the flow of water. Once individuals assume this position, the water pools on the apron of their burrows and they bend their heads over to drink from the pool (Ashton and Ashton 1991).

When rain pools or saturates soil surfaces, tortoises large and small drink by positioning their heads nearly perpendicular to the soil surface, pressing beak and nares toward the ground surface, and submerging their heads to the periocular region. Tortoises can also rehydrate from saturated soils by drawing moisture directly through their nares. This behavior is regularly observed among *G. agassizii* and *G. polyphemus*, and is undoubtedly important for other North American tortoises, especially those living in sandy areas (Ashton and Ashton 1991; chapter 9).

Response to Drought Conditions

Drought has many technical definitions, but we define drought relative to environmental conditions: tortoise movement, body condition, physiology, and survivorship (Peterson 1996b, Duda et al. 1999, Longshore et al. 2003). Our definition of a severe drought is a one-year period with <50 mm of precipitation. Climatic records indicate that these conditions occur fewer than 20 times over a 60-year period across the Mojave Desert and likely never in the ranges of the other tortoise species (NCDC 2012a, 2012b). Tortoises can respond to drought conditions by reducing energetic needs: for example, individuals of *G. agassizii* that were well hydrated and fed during a year of abundance (250% of normal winter precipitation) reduced their activity dramatically in a subsequent year with below-average precipitation (42 mm; Duda et al. 1999). During less intense dry periods, *G. agassizii* and

G. berlandieri both reduced movement and overall activity (Luckenbach 1982, Duda et al. 1999, Kazmaier et al. 2001c), presumably decreasing physiological demand. Dry conditions result in reduced activity and movement of most tortoises, making it even more difficult than usual to find individuals during surveys (R. Kazmaier, personal communication; Duda et al. 1999). Responses to drought conditions have not been published for *G. polyphemus,* but individuals may shift their activity areas toward more mesic sites during drought periods (Diemer 1986). It would be interesting to compare the results of water balance studies for the tortoises from more mesic environments during more stressful seasons or years with results on water balance in *G. agassizii* for comparative purposes.

Water Balance and Physiology

Tortoises drink to replace water lost to dehydration from body surfaces and respiration (Minnich 1977). The vegetation that they eat contains excess salts and other compounds (e.g., sodium, chloride, and potassium; Peterson 1996b) that must be eliminated with urea from the body, resulting in water loss through excretion. Although tortoises are not thought to have evolved in desert conditions (Van Devender 2002b), they appear to be somewhat behaviorally and physiologically adapted to drastic changes in environmental conditions, making adult tortoises extremely resilient to osmotic fluctuations and water imbalance. To maintain water balance, tortoises decrease evaporation by altering their behavior, reducing activity, and retreating to burrows or caves with cooler, humid microenvironments until environmental conditions improve (Morafka 1982, Peterson 1996a, Nagy et al. 1997). These behaviors result in cutaneous water loss for *G. agassizii* (1.5 mg/cm^2 per day) that is less than a third of eastern box turtles' (*Terrapene carolina*) (5.3 mg/cm^2 per day; Schmidt-Nielsen and Bentley 1966). Although the majority of water loss by *G. agassizii* is through the integument and respiratory processes, both sources of loss are reduced relative to those of turtles in more mesic climates (Schmidt-Nielsen and Bentley 1966). Although juvenile tortoises employ all of the behaviors and physiology described for adults, their small size, and thus increased surface-to-volume ratios, increase hydration challenges considerably (Berry and Turner 1986, Hillard 1996, Wilson et al. 1999b). To reduce losses, juvenile tortoises shift activity periods. Because of their small size juvenile tortoises can opportunely become active much earlier than adults, often emerging in late winter (Naegle 1976, Diemer 1992a, Wilson et al. 1999a). Cooler activity periods means lower evaporative losses, and for *G. agassizii,* earlier activity coincides with the rainy season, thus increasing drinking opportunities and potentially feeding opportunities. Conversely, extended dry periods without summer precipitation put juvenile *G. agassizii* at risk of dessication (Nagy et al. 1997), potentially explaining an important question about the species' demography: why are wild juvenile tortoises so rare? Perhaps the rigors of summer and generally dry conditions drive large losses of juvenile tortoises by dessication in most years. Although winter behaviors and water balances have been explored, research on late spring and summer survivorship through drought are difficult and costly to research and have not been fully explored.

During prolonged drought conditions, tortoises may lose up to 40% of their body weight, reducing their mean total body water volume by nearly 60% (Peterson 1996b). Drastic reductions in water volume in the bladder lead to increases in electrolyte concentrations in the remaining urine and increases in plasma osmolality (Nagy and Medica 1986). Reptiles use the urinary bladder, cloaca, and colon, rather than the kidney, to modify urine and plasma concentrations (Jorgensen 1998). The bladder plays a significant role in regulating blood osmolality, being highly permeable to water, urea, ammonia, and small ions, but not uric acid (Jorgensen 1998). This permeability allows tortoises in arid environments or those hibernating for weeks/months without fluid replacement to resorb water from the bladder and store highly concentrated wastes (e.g., uric acid), that can later be excreted quickly and completely when water becomes available for replacement.

Excreting excess salts (primarily potassium) requires balancing osmotic and ionic concentrations in urine and plasma using water or nitrogen. Excreting salts while conserving water requires removing nitrogen from the body that may be needed for other physiological processes (e.g., growth, reproduction). North American tortoises excrete waste mainly as uric acid or urates. One advantage of this process is that urates precipitate out of the solution in the bladder before fluids there are saturated. An active concentration process using water is therefore not needed, thereby avoiding water and energy losses. Tortoises must void the toxic bladder contents when opportunities arise and rehydrate, and tortoises can drink 11–28% of their total body mass (Minnich 1977, Nagy and Medica 1986, Peterson 1996b), simultaneously excreting materials ranging in consistency from syrupy tannin-colored water to granular grey wet-cement of a similar volume.

Because rainfall is tightly linked to food production for tortoises, periods exceeding one year with insufficient precipitation can cause declines in body condition (Peterson 1996b, Henen 1997, Nagy and Medica 1986), and periods longer than two years can result in high mortality for *G. agassizii* (Longshore et al. 2003). In retrospect, instances of drought-related mortality are widespread for desert-dwelling species: *G. agassizii* in Piute Valley, Nevada, in 1983 (P. Schneider, unpublished data); Goffs, California, in ~1996 (T. Shields, personal communications); and Ivanpah Valley, California (Peterson, 1994); and *G. morafkai* during the early 1980s in the Maricopa Mountains of central Arizona (82% decline; Wirt and Holm 1997). Given the dire consequences of prolonged dehydration in lands where seasonal dry periods are the rule, it is no surprise that tortoises do not pass up drinking opportunities when precipitation occurs, even if it interrupts hibernation.

FOOD ACQUISITION

Foraging

Foraging behavior varies among tortoise species based on how environmental conditions interact with the plant species available (fig. 10.1), and tortoise physiological condition. During years of low forage availability, *Gopherus berlandieri* and *G. agassizii* demonstrated greatly reduced activity (Kazmaier et al. 2001c; Ruby et al. 1994; Duda et al 1999; USFWS, unpublished data). North American tortoises tend to be relatively quiescent from November through March, in response to relatively cool temperatures (Adest et al. 1989, Nussear et al. 2007, Van Devender 2002a, Diemer 1992a, McRae et al. 1981; fig. 10.1). Depending on environmental conditions in the previous year, springtime can be a period of nutritional need for tortoises, to recover from weight loss during winter, and to address nutritional requirements for females developing egg follicles and shells (Garner and Landers 1981; chapter 5). Two of the three western species demonstrate seasonally bimodal activity, with a spring feeding period, followed by decreased activity in the heat of midsummer, and resumption of activity with the advent of summer rains (Henen 2002, Van Devender et al. 2002, Adest et al. 1989, Kazmaier 2000, Birkhead et al. 2005). Too little information exists about the springtime activity of *G. flavomarginatus* for us to generalize. For *G. agassizii,* peak foraging occurs March through June (Woodbury and Hardy 1948, Burge and Bradley 1976, Esque 1994, Jennings 2002), and the diet consists of plants that thrive on winter rainfall. Occasionally, autumn feeding opportunities occur (Oftedal 2002, Henen 2002a), resulting from localized summer rainfall germinating different annual flora than in the spring. Although *G. morafkai* feeds on spring annuals that are equally abundant as those in the Mojave Desert (Averill-Murray et al. 2002a), a broad survey of studies indicated that they feed primarily during summer and autumn (Van Devender et al. 2002; fig. 10.1). This pattern raises the question, why is *G. morafkai* not using the abundant spring vegetation that is available? Such a pattern is inconsistent with generalizations about tortoise opportunism in feeding and drinking. *Gopherus flavomarginatus* feeds from June through October (Morafka 1982), which is consistent with reliable precipitation and resulting food plants (Barbault and Halffter 1981; fig. 10.1). *Gopherus berlandieri* forages during late spring (April-May) and resumes feeding into the autumn (fig. 10.1). *Gopherus polyphemus* forages from April through September, coinciding with summer rainfall (fig. 10.1).

Tortoise foraging patterns are influenced by environmental conditions, habitat obstacles (e.g., topographic features and vegetation), cover site juxtaposition, and social interactions, but here we focus on variation in food availability. Most *G. agassizii* feeding occurs on sparsely vegetated, flat, or gently sloping outwash plains; however, some individuals forage on mountain slopes and boulder fields (chapter 9). Individuals meander during foraging bouts, frequently returning to their original cover sites, and sometimes making straight-line movements toward known cover sites (Esque 1994). *Gopherus agassizii* feeding has been described as nonrandom, because of repeated visits to the same feeding patches and because they follow washes (dry watercourses) in linear patterns, where "wash endemic" plant species are consumed (Jennings 1993). Tortoises also follow linear routes, along roadsides (P. Medica, personal observation) and cultivated areas where food is abundant, and frequent well-used trails through vegetation (McRae et al. 1981). Individuals of *G. polyphemus* repeatedly return to their original burrows, but may have short-range seasonal migrations from mesic lowlands to drier highlands, driven by food competition in localized areas (McRae et al. 1981). Similarly, individuals of *G. flavomarginatus* use regular feeding routes with "rigid repetition" (Morafka 1982), and these routes may be travelled to take advantage of the fresh growth that results from previous foraging, similar to descriptions of foraging in the Green Sea Turtle (*Chelonia mydas*) (Bjorndal 1980). Alternatively, forays by *G. polyphemus* may be interspersed by longer meanderings into new areas where particular food species may be found (McRae et al. 1981, Ashton and Ashton 2008).

Foraging bouts for *G. agassizii* averaged 30 to 75 minutes in a day in the northeast Mojave Desert, and individuals spent 6–22% of their time foraging (Esque 1994). *Gopherus berlandieri* spent 6% of their active time feeding (Kazmaier et al. 2001c). For *G. berlandieri* and *G. polyphemus,* home range sizes were larger where food plants were scarce or widely scattered (Rose and Judd 1975; chapter 11), but McRae et al. (1981) cautioned against using home range size alone as an index of foraging investment because individuals may spend more time and travel farther while engaged in social interactions than feeding. *Gopherus agassizii* spent a smaller percentage of overall time feeding, with shorter feeding bouts, during years when food availability was low (Esque 1994). Individuals of *G. polyphemus* have a feeding radius within 30 m of their burrows (McRae et al. 1981), and *G. flavomarginatus* feeding travels have been described as like those of *G. polyphemus,* with males foraging over larger areas than females (Lieberman and Morafka 1988).

Sustenance for Juveniles

Upon hatching or shortly thereafter (1–3 days), neonatal tortoises resorb the remaining egg yolk that is attached to intestines in a sac providing sustenance beyond the egg (Miller 1932)—a characteristic known as postparative lecithotrophy (Morafka et al. 2000). This process likely is crucial to survival in environments where feeding opportunities are unreliable for small tortoises. For example, emergent *G. agassizii* may find no feeding opportunities available on emergence from the egg, and such poor conditions frequently continue even into the spring activity season. As neonatal tortoises develop from lecithotrophy to herbivory, however, there is another crucial step in their ecology, requiring that they inoculate themselves with digestive symbiotic gut flora (Morafka et al. 2000). Inoculation may be accomplished by consuming soil or feces of adult tortoises containing the same symbionts (Troyer 1982).

Juvenile tortoises soon shift toward herbivory. It has been suggested that animal matter may play a significant role in nutrient acquisition, but few systematic studies have been conducted. Preference for insects over other diet items was demonstrated for *G. agassizii,* however, and consumption of animal matter has been noted for all North American tortoises (Okamoto 2002). Activity seasons for juveniles begin earlier in

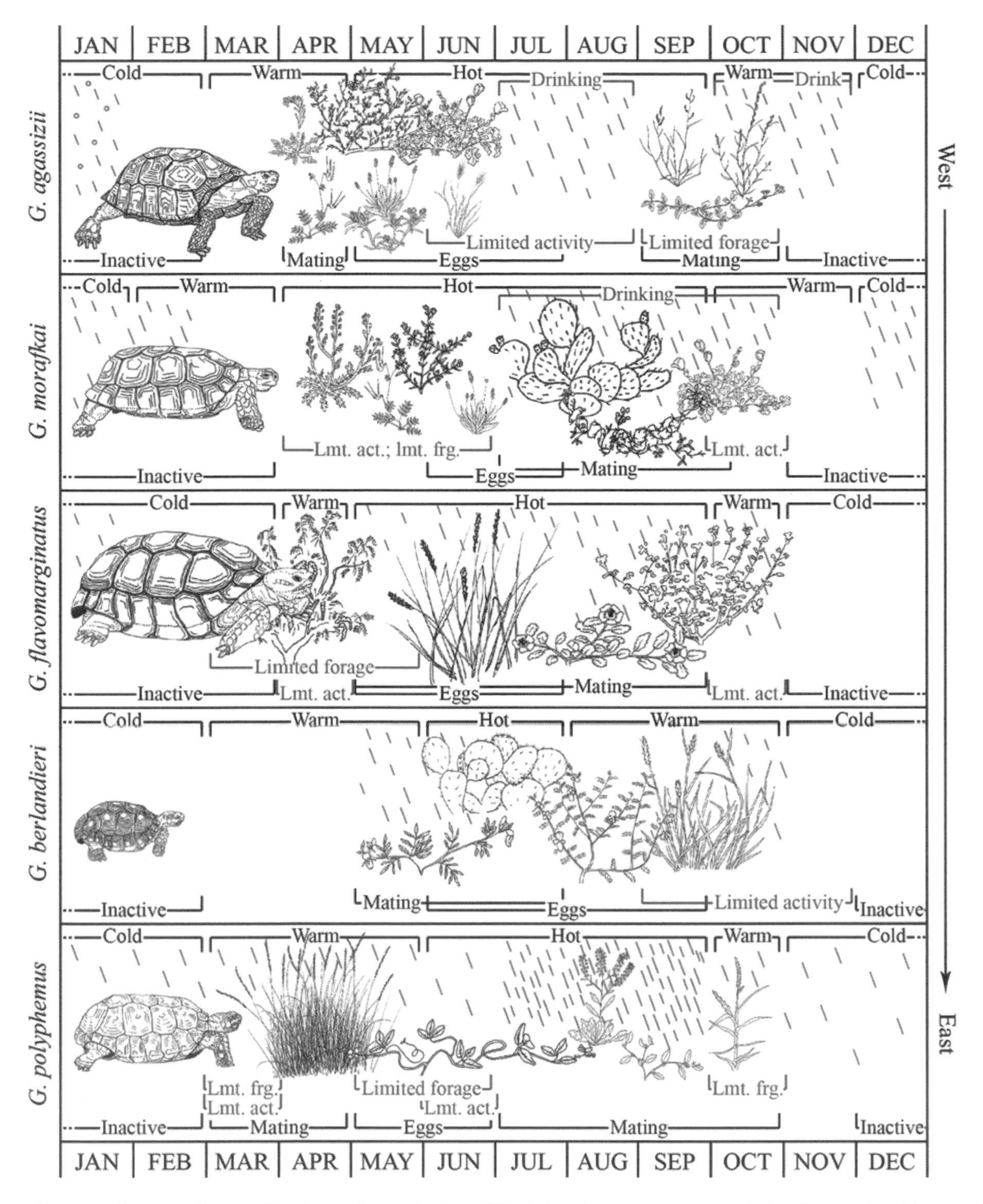

Fig. 10.1. Climate and timing of activity, foraging, and reproduction of North American tortoise species during the months of the year. Main precipitation periods are indicated with angled dashes (rain) and circles (snow). Foraging activity is indicated by the presence of plants specific to each species' diet. Sizes of tortoises are relative to each species' average MCL. Illustration by MA Walden, adapted from Morafka and Berry (2002).

the year than for adults. Interestingly, the activity shift means that some plants may be more accessible to the small tortoises with limited reach because during winter many herbaceous plants occur as basal rosettes that are easily in reach of the juveniles, prior to elongation (Morafka and Berry 2002). The fresh young plants likely contain fewer structural materials and thus could also be more readily digestible than coarser mature plants.

While foraging, juveniles of *G. polyphemus* move constantly over relatively short distances (25 m ± 41.5 SE m), the time outside their burrows is often brief (19.4 ± SE 10.3 min), and when disturbed they moved rapidly toward their burrows (Mushinsky et al. 2003, Halstead et al. 2007). Although the behaviors could provide protection from thermal extremes or have to do with feeding requirements or predator avoidance (Mushinsky et al. 2003), it was concluded that the "full stomach hypothesis" was the most likely driving factor, meaning that satiation took precedence over predator avoidance or temperature seeking behaviors (Halstead et al. 2007). Wild neonatal *G. flavomarginatus* spent 1% of their total time budget feeding, and "walk and feed" behaviors occupied 15 or 20% of their time in two different years (Tom 1988). During May and June, neonatal *G. flavomarginatus* foraged from 0900–1000 h, and during August through October they usually foraged from 1100 to 1200 h (Tom 1988). Seasonal consumption of dry food was estimated at 175g dry matter for 2-year-old *G. agassizii*, and body mass increased by 34–55 g (Nagy et al. 1997).

Because their small body size translates into body temperatures that track ambient temperatures, small tortoises may increase feeding opportunities outside seasons when adults are normally active (Wilson et al. 1999a). Observations of juvenile *G. agassizii* during winter indicate that feeding was rare during October to January, but increased during warmer days beginning at the end of January (Wilson et al. 1999a). In contrast, adults in the area are not expected to be actively feeding until late March or April (Esque et al., unpublished data). Although temperature certainly plays an important role in activity periods, we suggest that foraging may also be linked to availability of plants that also only begin to elongate with warming winter trends and adequate rainfall. While research has focused on cool season activity in juvenile tortoises, there is little information available on warm season activities. We suggest that *G. agassizii* juveniles may behaviorally modify their body temperatures by avoiding extreme temperatures and the inherent cost of water loss that could be life threatening. Therefore, juveniles may shift their activity periods toward cooler season temperatures, thus increasing their overall activity seasons. No research on juvenile winter activity was available for the other North American species, however.

Diet

Tortoise diet diversity is noteworthy. A compilation of reports throughout the range of *G. morafkai* yielded about 222 diet species (Van Devender et al. 2002) and, with sufficient observation, all tortoise species may prove to have similarly broad diets. *Gopherus agassizii* sampled about 50% (87 of 168 spp.) of the floras at each of two sites in the northeast Mojave Desert across five years. Three or four plant species comprised >70% of the diet at both sites in any year, however (Esque 1994, DeFalco 1995). In the western Mojave Desert, about 25% (44 of 173 spp.) of the flora at one site was sampled during a one-year study; ten species comprised >81% of the diet (Jennings 1993). Individuals switched diet species sequentially, following phenological patterns to acquire the freshest plants until only dry senescent plants were available (Jennings 2002). Individuals of *G. berlandieri* were observed foraging at one site for several years and 29 plant species were included in tortoise diets (R. Kazmaier, unpublished data). In Florida, the diet of 50 *G. polyphemus* at one site consisted of 68 plant genera, including all common plant species in the area, but only ten species comprised 75% of the diet (Macdonald and Mushinsky 1988). In southwest Georgia, *G. polyphemus* consumed 53 genera, and individuals were considered opportunistic frugivores with fruit seeds—particularly of *Rubus* sp.—prevalent in spring and fall diets (Birkhead et al. 2005). Dietary analyses for *G. flavomarginatus* are not available, but observations indicate that they eat mostly tobosa grass (*Pleuraphis mutica*), supplemented by prickly pear cactus (*Opuntia* sp.) fruits and various herbaceous annuals (Morafka 1982). We speculate that their diets are most diverse when the summer annual flora peak.

Considerable speculation exists that juvenile tortoises consume qualitatively different diets than adults (Morafka and Berry 2002), and analyses of diets present mixed results. The few observations of juvenile *G. flavomarginatus* feeding identified mesquite (*Prosopis* sp.) flowers and green grass shoots, and tobosagrass (*Pleuraphis mutica*) was found in the feces of a juvenile (Morafka 1982). Diet selection was observed among juvenile *G. polyphemus*, including some species not abundant in adult diets (Mushinsky et al. 2003). During an El Niño year of increased precipitation, juvenile *G. agassizii* selected plant leaves (>70% diet) from the herbaceous annuals brown-eyed primrose (*Camissonia claviformis*) and desert dandelion (*Malacothrix glabrata*), while the annual Mediterranean grass (*Schismus* spp.) was avoided (Oftedal et al. 2002). Selected plants were higher in water, protein, and PEP—an index of potential potassium excretion—but not lower in potassium than species that were bypassed (Oftedal et al. 2002). Juvenile *G. agassizii* fed single-species diets lost phosphorus, shell mass, and volume on grass diets, regardless of whether the grasses were native or exotic (Hazard et al. 2010).

Dietary overlap generally is not widespread among North American tortoise species, except for the two desert tortoises. These two species have contiguous ranges, separated only by the Colorado River, and the plants do not differ greatly there. A synthesis of diet plants found 39 of 43 species occurred in both the Mojave and Sonoran Deserts, and nine species were diet items for both tortoise species (Oftedal 2002). The wide latitudinal range of *G. morafkai* is sufficient to create mostly nonoverlapping diet lists between distal ends of the species' range, however. A similar pattern likely exists for *G. berlandieri*, but no data are available concerning its diet in Mexico. At least one native plant species (*Allionia incarnata*) is shared among the diets of the two desert tortoise species and *G. berlandieri*. Minimally, three species are shared between the diets of *G. berlandieri* and *G. polyphemus* (Macdonald and Mushinsky 1988; R. Kazmaier, unpublished data). *Gopherus flavomarginatus* also may have some dietary overlap with other species; however, data for a rigorous comparison are unavailable.

Food availability depends on the species richness, phenologies, and physical structures of plants, and the dietary preferences of the tortoises. Plant availability, palatability, and nutritional value differ among growth forms (annual forb,

Table 10.1. Percentages of diet eaten by North American tortoises within major plant growth forms

	Annual Forb	Annual or Perennial Grass	Perennial Forb	Woody Plant (vine, shrub, tree)	Succulent
Gopherus agassizii	52.9	12.6	19.5	11.5	3.4
Gopherus morafkai	*	30 to 53	7 to 28	6 to 22	**
Gopherus berlandieri	*	21.0	37.0	9.0	28.0
Gopherus polyphemus	20.0	7.0	46.0	20.0	**

Sources: G. agassizii: Esque 1994, DeFalco 1995; *G. morafkai*: citations in Van Devender 2002a; *G. berlandieri*: Scalise 2011; *G. polyphemus*: Macdonald and Mushinsky 1988.

* Not eaten

** Not determined

grass, perennial forb, woody shrub, tree, succulent) and plant parts (stems, leaves, flowers, fruits), partially because they differentially incorporate structural materials that can affect palatability and digestibility (McArthur et al. 1994). Fruits, on the other hand, may have particularly palatable parts that serve as attractants for animals, including tortoises, to encourage seed dispersal (Rick and Bowman 1961, Rose and Judd, 1982). Woody trees, vines, and shrubs play a role in tortoise diets when palatable parts such as twigs, leaves, flowers, and fruits are available (Van Devender et al. 2002). The presence of some woody plant parts in diets are difficult to explain. For example, pine needles and oak leaves ranked among the top five species present in the scats of *G. polyphemus* (Macdonald and Mushinsky 1988). It is hard to imagine that these plants provide nutritive sustenance—oaks have tannins that are difficult to digest, for instance—and the authors suggested alternative functions such as gut maintenance during hibernation. Perhaps these strange diet items fulfill an unknown function that will be revealed in the future.

Generally, perennial or annual forbs are the most important plant groups in diets of tortoises. Forbs are particularly important in *G. agassizii* diets, and moderately so for *G. morafkai* and *G. berlandieri* (Jennings 2002, Scalise 2011, Van Devender et al. 2002). Dietary differences between *G. agassizii* and *G. morafkai* may be related to the emergence of tortoises in the newly forming Mojave Desert from tortoises in the Sonoran Desert and their subsequent adaptation to conditions there (Van Devender 2002b). Forbs encompass the most diverse nutritional array and it is difficult to generalize about them; however, they have less structural fiber, which means that beneficial nutrients may be more accessible. Annual forbs and grasses are the most important items noted in *G. agassizii* diets (Woodbury and Hardy 1948, Hansen et al. 1976, Esque 1994, Oftedal 2002, Jennings 2002; table 10.1). Hohman and Ohmart (1980) found similar results, but with a seasonal shift to feeding on shrubs, from < 2% of diets in April to 20% in August. Surprisingly, although many *G. agassizii* diet species are available to *G. morafkai* (in some parts of their ranges), diets of the latter species are dominated by summer perennial forb species (Van Devender et al. 2002; table 10.1). Succulents (cacti) are particularly important components of the diet of *G. berlandieri* (Scalise 2011; Kazmaier, unpublished data; table 10.1). Perennial, rather than annual, forbs are dominant in the diet of *G. polyphemus*, but grasses are important in some areas (Macdonald and Mushinsky 1988, Douglas and Layne 1978). The range of observations demonstrates a great variability in the use of growth forms, but probably is influenced by presence or absence of various species and the other factors mentioned previously.

Questions about the benefit of grasses in the diets of North American tortoises stem from widespread invasion of annual and perennial exotic grasses, and consideration of their direct or indirect influence on tortoise health. Mediterranean annual grasses are fully integrated into western North American landscapes, and in the Mojave and Sonoran Deserts, and have the capacity to modify vegetation composition drastically. Although annual invasive grasses can have nutritional values similar to native grasses (Hazard et al. 2009, 2010), they can overwhelm ecosystems, reducing plant diversity and ultimately lowering overall nutritional availability. Invasive grasses also may affect tortoises directly: one study noted impaction of an invasive annual grass in tortoise mouths, for example (Medica and Eckert 2007). Jennings (2002) found few invasive grasses in the diet of *G. agassizii* during a one-year study. This finding apparently contrasts with findings from the northeast Mojave Desert, where invasive grasses were common in the tortoise's diet (Esque 1994): the principal forage species there included the several invasive grasses, such as brome grasses (*Bromus madritensis, B. tectorum*) and Mediterranean grass (*Schismus* spp.) (Hansen et al. 1976, Esque 1994). Jennings' (2002) observations were made during an extremely productive year, with a great diversity of nongrass plants available. In contrast to *G. agassizii, G. morafkai,* and *G. berlandieri* eat invasive (perennial) grasses, such as Lehmann lovegrass (*Eragrostis lehmanianna*) and buffelgrass (*Cenchrus ciliaris*), respectively (Morafka 1982; Kazmaier, unpublished data). Although the invasive perennial cogon grass (*Imperata cylindrical*) has very low nutritional value and is not eaten by *G. polyphemus,* it is a threat to habitat because it crowds native food plants and interferes with movement (Basiotis 2007).

Issues with invasive exotics aside, by most accounts grasses are important components in tortoise diets, and all North American tortoises eat substantial amounts of them. In addition, many African tortoises consume grasses as a large portion of their diets (Kabigumila 2001). Grasses are important diet components probably because of their high relative abundances at certain times: they are less important when forbs are readily available (Garner and Landers 1981, Macdonald and Mushinsky 1988, Esque 1994, Scalise 2011). The low nutritional and high indigestible fiber content of grasses may account for lack of preference or avoidance by tortoises (McArthur et al. 1994, Ramirez et al. 2004, Basiotis 2007). *Gopherus polyphemus* eats early spring grasses, and switches to broadleaf grasses and legumes as they become available; then, in the fall, as forbs senesce, they switch back to grasses in some areas (Garner and Landers 1981). Overall, wiregrass (*Aristida beyrichiana*) was less favored than broad-leaved grasses (Garner and Landers 1981). Where choices are available, annual grasses are used to the near exclusion of perennial grasses, potentially because of their higher N, Total Nonstructural Carbohydrates (TNC), mineral content, and low fiber (Esque 1994; data in McArthur et al. 1994). One study found native perennial grasses to be absent from the diet of *G. agassizii* in April, but comprising approximately 10% later in the season (Hohman and Ohmart 1980). Others found *G. agassizii* eating dry grasses once well hydrated after summer rains (Nagy and Medica 1986). Native perennial grasses may be more important than currently documented in the diet of *G. agassizii,* and certainly comprise measurable portions of the diets of *G. morafkai, G. berlandieri,* and, sometimes, *G. polyphemus* (Van Devender et al. 2002, Scalise 2011). In general, tortoises may use grasses when more nutritious items are unavailable. Perhaps if the tortoises have slower passage times during these times—as predicted by digestive optimality (Tracy et al. 2006), they are able to obtain more nutrients than not eating at all, and over time gain mostly energy.

Prickly pear cactuses (*Opuntia* spp.) are the most widely available succulent plants to North American tortoises, and can be important for some species, especially when other plants are not available (table 10.1). The role of prickly pear cactuses in tortoise diets is interesting and not entirely understood at this time. They are considered to be survival food for *G. agassizii,* a seasonal benefit as a mast crop to *G. morafkai,* a staple in the diet of *G. berlandieri*, peripherally available to *G. polyphemus,* and possibly important in the diet of *G. flavomarginatus* (Morafka 1982). Although consumption of large amounts of prickly pear cactus over long periods of time is hypothesized to compromise tortoise physiology by potentially affecting calcium balance (Hellgren et al. 2000), prickly pear cactuses also ranked highest for mineral and carbohydrates content among 15 plant species analyzed for nutrition (McArthur et al. 1994). Beavertail cactus (*O. basilaris*) and pencil cholla (*O. ramosissima*) are used by *G. agassizzi* during years when primary production is negligible (T. Shields, personal observation; T. Esque, personal observation; Turner et al. 1984). *Gopherus morafkai* also eats prickly pear cactus pads; and, in years of good fruit crops, the large fleshy fruits of Engelmann's prickly pear (*O. englemannii*) are eaten with gusto. While the fruit is available, tortoises may have bright purple stains on their beaks and fore limbs. X-radiography indicates that individuals may fill their entire guts with the fruits (T. Esque, personal observation). During times of low production, the diet of *G. berlandieri* also may be dominated by prickly pear cactus (Hellgren et al. 2000, Kazmaier 2000), including Texas prickly pear (*O. lindheimeri*) and desert Christmas cactus (*O. leptocaulus*), both of which are especially important components of the summer diet (Auffenberg and Weaver 1969, Scalise 2011). Both *G. flavomarginatus* and *G. polyphemus* also eat prickly pear cactus (Morafka 1982, Macdonald and Mushinsky 1988, respectively).

All young, growing animals have energy and nutrition requirements that surpass those of adults (adjusted for size), because of their need to synthesize growing bodies (Nagy et al. 1997, Morafka et al. 2000). Efficient foraging, by gathering high quality food items, can be essential to juvenile tortoises, because high quality diets result in faster growth, and increased size reduces predation risk. Faster growth also is commensurate with more rapid attainment of sexual maturity (e.g., in *G. polyphemus;* Halstead et al. 2007; chapter 7). Although some growing animals increase their metabolisms to accommodate growth needs, juvenile tortoises do not. Therefore, they must make up nutritional demands modifying behaviors to be more efficient, and there is speculation that they choose richer diets, although it has been noted in one experiment that nutritional quality of *G. agassizii* juvenile diet does not differ from that of adults. Other ways of increasing nutritional content include eating higher quality or more digestible plant parts. A demonstration of such preferences includes strong avoidance of wiregrass by juvenile *G. polyphemus.* Wiregrass has relatively low nutrition value compared to other forbs and some grasses (Mushinsky et al. 2003).

Nutrition and Metabolism

Tortoise nutrition is the sum of energy, protein and amino acids, minerals, vitamins, water, and other components of plants used as the building blocks for healthy tortoises. The majority of energy tortoises receive comes from carbohydrates and through fermentation of the fiber in the plants they eat (Bjorndal 1987). The time and distance tortoises forage is a function of forage availability, but as their guts become full, some digestion is required to make room for additional food prior to consumption. Food passage rates are temperature dependent (Bjorndal 1987, Troyer 1991, Dean-Bradley 1995, Zimmerman and Tracy 1989); thus, foraging behavior is intimately related to tortoise physiology and thermal biology. Tortoises also supplement their diets with nonvegetative materials such as animal matter for protein and supplemental mineral nutrition. Nutrition analyses of tortoise food plants indicate some important correlations. One study of plant nutrition found that although nutritive and mineral content of food plants varied between two soil types (i.e., sand versus carbonate parent materials), the availability of precipitation had a larger influence on plant nutrition than variation in soil parameters (McArthur et al. 1994). Generally, plant species that are high in protein and low in fiber are also high in energy content, calcium (Ca), magnesium (Mg), phosphorus (P), and potassium (K) (Tracy et al. 2006). Another study found that plants with higher water content were generally correlated with better nutritive and mineral content, and high protein was correlated with high fat content; protein was negatively correlated with Acid Detergent Fiber (ADF—primarily compounds for structural materials), and TNC and Acid Detergent Fiber were negatively correlated (McArthur et al. 1994).

In chelonians, the thyroid gland plays a central role in regulating metabolism and maintaining physiological balance. The structure of the thyroid gland and the hormones it produces are very similar across vertebrate taxa, and the hormones have a wide variety of functions. In reptiles, thyroid hormones such as tri-iodothyronin (T3) and tetra-iodothryonin (T4 or thyroxin) play an integral part in growth and development (Maher 1961, 1965), regulation of nutrient assimilation (Eales 1979, Kohel et al. 2001), reproduction (McNabb 1992), ecdysis (Lynn 1970), behavior (Steinberg et al. 1993), and metabolism (Maher 1961, 1965). Thyroid hormones bind to mitochondria and activate enzymes involved in glycolysis and ATP production, which increases the rates metabolism in target tissues. These hormones also influence a wide variety of peripheral tissues, by elevating oxygen and energy consumption and increasing heart rate, blood pressure, and sensitivity to sympathetic stimulation. Although thyroid functions are diverse, their primary role is to support energy-demanding activities that occur when environmental conditions are favorable (Eales 1979). A comparison of thyroid functioning in reptiles and mammals concluded that although the reptilian thyroid is active at high temperatures it is still considerably less active than the mammalian thyroid (Hulbert and Williams 1988). Desert tortoises were at the lower end of the range reported for thyroxin levels compared to other reptiles, indicating a very low metabolic rate.

Thyroid activity in North American tortoises is most important in metabolism and is positively influenced by temperature and seasonality (Kohel et al. 2001). For tortoises in temperate climates, we would expect thyroid hormone production to increase during seasons with increased metabolic activity, such as feeding, growth, and reproduction or surface activity, and decrease during periods of metabolic quiescence, such as hibernation (McNabb 1992, Gerwien and John-Alder 1992, Kohel et al. 2001). Studies examining the effects of plasma thyroxine (T4) levels in *G. agassizii* (Kohel et al. 2001) and the Painted Turtle (*Chrysemys picta*) (Licht et al. 1985) described distinct monthly and seasonal changes in thyroxine levels, as well as differences related to sex, size, and reproductive activity. Thyroxin levels were lowest during hibernation and increased during spring emergence, when tortoises increase feeding and activity, all of which have been implicated with activation of the hypothalamo-pituitary-thryoid (HPT) axis (Kohel et al. 2001). Thyroxin levels also are influenced by nutrient uptake. For example, individuals of *G. agassizii* that were fasted for a two-week period had decreased thyroxin levels (Kohel et al. 2001), but those levels increased within 36 hours after feeding, suggesting that nutrients activate the HPT axis (Eales 1988, MacKenzie et al. 1998). The thyroid response to food intake may be an important factor in stimulating increased metabolic activity for tortoises as they emerge from winter hibernation, supporting increased metabolic needs after emergence.

Nonvegetation Food Items

The diversity of diets appears to be driven in part, by sampling potential foods and accidentally ingesting materials that occur near target food items (Van Devender et al. 2002), and tortoises occasionally ingest manufactured objects such as human trash (Macdonald and Mushinsky 1988, Walde et al. 2007). Feeding observations have also documented tortoises consuming the feces of congenerics and other animals (Auffenberg and Weaver 1969, Garner and Landers 1981, Esque and Peters 2004, Walde et al. 2006). Observers are fascinated when tortoises ingest nonplant items, yet this behavior is widespread and possibly has a nutritional basis. Tortoises have widely been observed ingesting: soil at mineral licks, a variety of bones, snail shells, raptor pellets, charcoal, sand, and stones (Auffenberg and Weaver 1969; Sokal 1971; Hohman and Ohmart 1980; Marlow and Tollestrup 1982; Macdonald and Mushinsky 1988; Esque and Peters 1994; Murray 1997; Hellgren et al. 2000; Stitt and Davis 2003; Walde et al. 2007; Kazmaier, unpublished data 2012). Individuals of *Gopherus polyphemus* have been observed carrying bones to their burrows and eating carrion (Garner and Landers 1981). Insect matter was the fourth-most abundant material in *G. polyphemus* scats (Macdonald and Mushinsky 1988). Selection of animal matter by *G. polyphemus* may be motivated by low phosphorus, calcium, and protein concentrations in its food (Carr 1952), all of which are needed for egg formation (Garner and Landers 1981). Growing tortoises require minerals and female tortoises require additional mineral expenditures for eggshells. It has been hypothesized that female tortoises mobilize $CaCO_3$ from their own bones during eggshell formation, but also supplement dietary calcium when possible (Hellgren et al. 2000). Alternative hypotheses include nonfood items as geoliths for food mastication and vermifuge for the removal of internal parasites (Marlow and Tollestrup 1982, Macdonald and Mushinsky 1988, Esque and Peters 1994). Although intriguing the overall importance of animal matter in tortoise diets remains physiologically unexplained.

NUTRITIONAL ECOLOGY: DIET SELECTION

Diet selection is the preferential use of foods by individuals or populations and is usually quantified by the relative use of a food item with respect to its availability. Preferred plants are those eaten in greater abundance than if tortoises found them at random, while avoided species are eaten less frequently than their availability in the environment would indicate. Food items that are neither preferred nor avoided appear to be used randomly. Does diet selection occur for specific plant species or qualities of plants? If diet selection occurs, what are the common dietary characteristics of the plants that are selected (e.g., energy, minerals, vitamins, or water)? Diet selection analyses have documented preferences for the *Gopherus polyphemus* (Macdonald and Mushinsky 1988), *G. berlandieri* (Scalise 2011), and *G. agassizii* (Esque 1994, Jennings 2002, Tracy et al. 2006). Some items, although widely available and frequently investigated by these species, were never observed in diets; while other species that were rare in the environment were disproportionately represented in diets or were not quantified during vegetation surveys, even though they were consumed (Macdonald and Mushinsky 1988, Esque 1994, Jennings 1993). Although grasses are important diet components across species, it is probably because of their high relative abundances at certain times: they are less important when forbs are readily available and are consumed at the frequency with which they occur in the environment (eaten at random), and generally are avoided (Garner and Landers 1981, Macdonald and Mushinsky 1988, Esque 1994, Scalise 2011). The most frequently ingested food items (grasses) for *G. polyphemus* were eaten at random with respect to their abundance in the environment, but one highly preferred genus (*Richardia*) was among the most nutritional diet plants (i.e., high in water, fat, Ca, and other mineral content) (Macdonald and Mushinsky 1988). Diet and preference data pooled among five sites throughout the range of *G. berlandieri* within Texas indicated that it prefers cactus, avoids grasses, and eats forbs and woody plants at random (Scalise 2011). *Gopherus agassizii* also demonstrated preferences (Esque 1994, Tracy et al. 2006), but in contrast to the other two species. Diet selection was found to vary by the plants' phenological state, with fresh plants preferred and dried plants avoided until nothing else was left (Jennings 2002). No single species was consistently preferred or avoided among years in the northeast Mojave (Esque 1994). Both native and invasive species were among the preferred plants, and the patterns varied among years in relation to the number of species available.

Tortoise diet selection data can be used to support or refute theories about diet and foraging optimality (Emlen 1966, MacArthur and Pianka 1966, Schoener 1971, Stephens and Krebs 1986), nutritional wisdom (Westoby 1974), and functional response (Spalinger and Hobbs 1992). Diet selection data can be coupled with experimentation on nutrition and digestive efficiency to explain the relative amounts of nutrients in tortoise diets and what they contribute to tortoise health. Ultimately, such work can be placed in the theoretical context of physiological ecology to explain animal nutrition (Tracy et al. 2006). Attempts to describe observed diet and nutrition patterns mechanistically include many theoretical contexts: digestive efficiency costs (Nagy and Medica 1986); energy assimilation (Bjorndal 1987); self-selection (Waldbaeur and Friedman 1991); and the cost of switching food types (Tracy et al. 2006), all reviewed in relation to North American tortoises by Oftedal (2002) and Tracy et al. (2006). In this framework, Oftedal et al. (2002) and Tracy et al. (2006) integrated scientific theory, field observations, and laboratory experiments to generate alternative hypotheses for understanding tortoise nutritional ecology, specifically for *G. agassizii*. The Potassium Excretion Potential (PEP) Index hypothesis hinges on the need for tortoises to excrete excess K in the form of urates, resulting in a high ratio of water and nitrogen to potassium and requiring large amounts of nitrogen and water (Oftedal 2002). According to this hypothesis, tortoises lose so much nitrogen and water in the process of eliminating excess K that they must select plant species of high PEP Index to avoid nutritional deficits. Observations of growing tortoises in semiwild conditions and feeding on freely available plant species, indicated that they selected plants with a high PEP Index from among all the plants available, supporting the PEP Index hypothesis, although K was not avoided specifically. It is widely observed that several plants with high PEP values are eaten vigorously by tortoises when they are located, but that many of these plants such as locoweeds (*Astragalus* spp.) or evening primroses (*Oenothera* and *Camissonia* spp.) are usually uncommon; thus, they likely do not make up large portions of diets in most years. In contrast, Tracy el al. (2006) found that *G. agassizii* neither avoided plants with high K nor preferred plants with high PEP indices. They noted that the diets of individuals did comprise a subset of available plants, and that the preferred species were those that remained green for long periods of time: tortoises consumed these species until "switching" was necessary (Tracy et al. 2006). Based on these results, Tracy et al. (2006) developed the Integrated Resource Acquisition Hypothesis (IRAH) by integrating aspects of optimal foraging, optimal digestion, and cost-of-switching hypotheses. In the IRAH, nutritional value is increased by feeding on a few plant species that the tortoise finds readily available and that the symbiotic gut flora are already used to (i.e., switching would require adjustments of the gut flora, leading to inefficiencies in digestion), and feeds on them until they are unavailable, then switches to another green species, until finally feeding on cured materials. Interestingly, this is the pattern of foraging that was also described for an independent site in the northwest Mojave Desert (Jennings 2002). This exciting and relevant field of work demands future synthesis including other species of tortoises.

CONCLUSIONS

Although the availability of drinking water varies considerably among North American tortoises, acquisition of free-standing water is essential for tortoise survival. The basic physiological hardware to respond to this variability is similar among tortoises, but behavioral responses are surprisingly variable among species and between adults and juveniles. Arid conditions appear to be most challenging for juvenile tortoises, yet few systematic studies quantify juvenile performance during the most arid seasons. The diets of North American tortoises include a broad sample of all plant species available in their habitats, but a small subset of plant species usually comprise the majority of diets. Nutrition may be supplemented with minerals and protein from animal matter and juveniles may particularly benefit from nonvegetative diet items, but most of such observations are anecdotal and more work on this aspect of tortoise nutrition is warranted. Although grasses and cactus are widely prevalent among adult tortoise diets, grass-dominated diets may be nutritionally detrimental to juvenile tortoises and may have important consequences for tortoise populations in habitats where invasive grasses dominate plant communities. There is debate in the scientific community regarding the mechanistic underpinnings of tortoise foraging, diet, and nutrition. Additional research in this field would benefit the conservation and management of North American tortoises.

Acknowledgements

We thank R. Kazmaier for providing valuable diet information on Texas tortoises, M. Walden for adapting the graphical concept of tortoises throughout the year from the inspired work of D. J. Morafka and K. H. Berry first appearing in Van Devender et al. (2002). P. A. Medica, S. P. Jones, and the editors of this volume provided valuable feedback and discussion that improved the quality of the chapter. We also thank F. Chen, R. Lamkin, and M. Walden for providing technical assistance in the review and synthesis of literature associated with this chapter.

Any use of trade names or specific products is for descriptive purposes only and does not imply endorsement by the U.S. government.

11 Home Range and Movements of North American Tortoises

Joan E. Berish
Philip A. Medica

Home range and movements of North American tortoises are influenced in both dramatic and subtle ways by climatic factors, topographical features, burrowing substrate, forage availability, social interactions, anthropogenic disturbances, and the physical structure of vegetation. Home range has been defined by Burt (1943) as "that area traversed by the individual in its normal activities of food gathering, mating, and caring for young." A number of methods used in determining "home range" exist, and the merits of each have been discussed elsewhere (Calhoun and Casby 1958, Jennrich and Turner 1969, Anderson 1982). Descriptions of habitat characteristics for each species are covered in chapter 9, and details of basking, overwintering, social interactions, and egg-deposition sites are discussed in chapters 4, 12, and 13. Although we overlap with some of those topics, our objectives are to explain why, when, and how far tortoises move. Here, we discuss all North American tortoises (genus *Gopherus*), following nomenclature of Crother (2008) for *G. polyphemus*, *G. flavomarginatus*, and *G. berlandieri*. We use the recent division of the desert tortoise (Murphy et al. 2011) into *G. agassizii* and *G. morafkai*.

HOME RANGE SIZE AND DISTANCES MOVED

Although generalities do exist regarding movements of North American tortoises, considerable individual and site-related variation has been observed in home range size and distances moved for *Gopherus polyphemus* (McRae et al. 1981, Diemer 1992a, Wilson et al. 1994, Butler et al. 1995, RB Smith et al. 1997, Mitchell 2005), *G. berlandieri* (Auffenberg and Weaver 1969, Rose and Judd 1975, Judd and Rose 1983, Kazmaier et al. 2002), *G. flavomarginatus* (Aguirre et al. 1984), and both species of desert tortoise (Burge 1977, Hohman and Ohmart 1980, Turner et al. 1980, Medica et al. 1981, Barrett 1990, Martin 1995, Duda et al. 1999, Averill-Murray et al. 2002a, Franks et al. 2011, Nussear, unpublished data). Judd and Rose (1983) summed it up by noting: "Perhaps the single most striking aspect of the home range ecology of *G. berlandieri* is the great variability among individuals in home range size." This behavioral plasticity and intersite variation are both intriguing and confounding, especially when trying to estimate minimum area requirements for preserves (Eubanks et al. 2002, Mitchell 2005).

In general, both species of desert tortoise make longer movements and have larger home ranges than the other three species of North American tortoises; however, Kazmaier et al. (2002) documented relatively large home ranges for the diminutive *G. berlandieri*. Overlapping home ranges tend to be the norm for North American tortoises (McRae et al. 1981, Aguirre et al. 1984, Diemer 1992a, O'Conner et al. 1994b, Averill-Murray et al. 2002a, Kazmaier et al. 2002). On average, adult males of all five species tend to travel more frequently, move longer distances, and have larger home ranges than adult females (McRae et al. 1981, Aguirre et al. 1984, Diemer 1992a, O'Conner et al. 1994b, RB Smith et al. 1997, Averill-Murray et al. 2002, Kazmaier et al. 2002, Eubanks et al. 2003, Franks et al. 2011). This difference between sexes is not surprising, considering the intensive mate-seeking that males undertake (McRae et al. 1981, Douglass 1986, Diemer 1992a, Averill-Murray et al. 2002a, Kazmaier et al. 2002, Boglioli et al. 2003, Eubanks et al. 2003). An exception to the above trend has been documented at two sites in southern Nevada (Bird Spring Valley and Lake Mead), where adult female *G. agassizii* had home ranges nearly double that of males (Nussear, unpublished data). Furthermore, some immature tortoises have larger home ranges than some adults (Aguirre et al. 1984, Diemer 1992a, Butler et al. 1995, Kazmaier et al. 2002, Pike 2006).

Taking into account variations in habitat type, climatic factors, study duration, sample size, and data analysis, mean home ranges (minimum convex polygons) for adult and immature North American tortoises are shown in table 11.1. In some cases, tortoises moved only between two burrows

Table 11.1. Home range sizes (minimum convex polygons) of North American (*Gopherus*) tortoises

Species	Number by sex / size class	Mean Home Range (ha)	Minimum/Maximum Home Range (ha)	Duration of Study (yr)	State/Habitat type	Reference
G. polyphemus						
	8 M	0.45	0.06–1.44	<1	GA sandhill	McRae et al. 1981
	5 F	0.08	0.04–0.14	<1	GA sandhill	McRae et al. 1981
	68 M	1.10	0.00–4.80	1	GA longleaf pine	Eubanks et al. 2003
	51 F	0.40	0.00–3.40	1	GA longleaf pine	Eubanks et al. 2003
	43 F	1.86	0.01–13.49	2	GA sandhill	Mitchell 2005
	7 M	1.05	0.28–2.17	<1	FL W. Indian scrub	McLaughlin 1990
	6 F	0.06	0.01–0.12	<1	FL W. Indian scrub	McLaughlin 1990
	6 M	0.88	0.23–2.88	1–2	FL planted pine	Diemer 1992b
	5 F	0.31	0.00–1.18	1–2	FL planted pine	Diemer 1992b
	8 F	0.48	0.00–1.44	1	FL sandhill	Smith 1995
	6 F	0.11	0.00–0.48	1	FL old field	Smith 1995
	10 M	1.90	0.30–5.30	1–2	FL coastal scrub	Smith et al. 1997
	4 F	0.60	0.30–1.10	1–2	FL coastal scrub	Smith et al. 1997
	9 M	0.32	0.13–0.63	<1	FL beach dunes	Lau 2011
	11 F	0.42	0.01–2.94	<1	FL beach dunes	Lau 2011
	11 M	1.95	0.63–4.89	1–2	MS burned pine	Yager et al. 2007
	9 F	1.07	0.11–2.46	1–2	MS burned pine	Yager et al. 2007
	9 M	1.30	0.71–2.43	1–2	MS unburned pine	Yager et al. 2007
	11 F	1.90	0.21–7.65	1–2	MS unburned pine	Yager et al. 2007
	4 S	0.50	0.01–0.25	1–2	FL planted pine	Diemer 1992b
	7 J	0.01	0.00–0.25	1–2	FL planted pine	Diemer 1992b
	9 J	0.07	0.01–0.36	1	FL sandhill	Wilson et al. 1994
	9 H	0.25	0.06–0.42	1–2	FL sandhill	Butler et al. 1995
	7 H	1.95	0.01–4.81	1	FL coastal scrub	Pike 2006
G. flavomarginatus						
	5 M	5.35	—	1–2	MX desert grassland	Aguirre et al. 1984
	6 F	4.72	—	1–2	MX desert grassland	Aguirre et al. 1984
	9 J	1.10	—	1–2	MX desert grassland	Aguirre et al. 1984
	5 J	22.3 m^2	—	1	MX desert grassland	Tom 1988
	4 J	28.4 m^2	—	1	MX desert grassland	Tom 1988
	— M	3.1	—	—	MX desert grassland	Adest 1994
	— F	2.5	—	—	MX desert grassland	Adest 1994
	— J	0.4	—	—	MX desert grassland	Adest 1994
G. berlandieri						
	31 M	2.57	—	2–5	TX salt flats, lomas	Judd and Rose 1983
	26 F	1.42	—	2–5	TX salt flats, lomas	Judd and Rose 1983
	9 F	6.8	1.5–21.6	2	TX mesquite/catclaw, ungrazed	Kazmaier et al. 2002
	7 M	31.8	9.2–130.7	2	TX mesquite/catclaw, ungrazed	Kazmaier et al. 2002
	13 F	5.0	1.0–19.8	2	TX mesquite/catclaw, grazed	Kazmaier et al. 2002
	7 M	9.5	4.8–23.2	2	TX mesquite/catclaw, grazed	Kazmaier et al. 2002
G. agassizii						
	— M	53	39–77	—	CA creosote/bursage	Berry 1974
	— F	21	8–46	—	CA creosote/bursage	Berry 1974
	3 M	26	20–38	1.5	NV creosote/bursage	Burge 1977
	3 F	19	11–27	1.5	NV creosote/bursage	Burge 1977
	5 M	23	2–34	2	AZ creosote/bursage	Hohman and Ohmart 1980
	3 F	11	1–29	2	AZ creosote/bursage	Hohman and Ohmart 1980
	2 S	3.5	3.4–3.5	2	AZ creosote/bursage	Hohman and Ohmart 1980
	4 M	1.3	0.22–3.4	1	AZ creosote/ bursage/blackbrush	Esque 1994
	6 F	3.5	1.2–11.0	1	AZ creosote/ bursage/blackbrush	Esque 1994
	10 F	5.6	0.3–11.85	1	UT creosote/ bursage/blackbrush	Esque 1994
	10 F	10.3	0.11–14.6	2	UT creosote/ bursage/blackbrush	Esque 1994
	1 M	4.41	—	1	UT creosote/ bursage/blackbrush	DeFalco 1995
	4 F	3.65	3.0–4.5	1	UT creosote/ bursage/blackbrush	DeFalco 1995
	17 M	26.6	2.9–88.6	1	CA creosote/bursage	Turner et al. 1980
	38 F	19.0	2.9–88.6	1	CA creosote/bursage	Turner et al. 1980
	25 M	24.5	1.6–72.7	1	CA creosote/bursage	Medica et al. 1981
	52 F	23.7	1.6–2.7	1	CA creosote/bursage	Medica et al. 1981
	13 M	7.65	3.8–16.9	1	CA creosote/bursage	Duda & Krzysik 1998
	16 F	7.49	1.0–15.9	1	CA creosote/bursage	Duda and Krzysik 1998

(*continued*)

Table 11.1. (cont.)

Species	Number by sex / size class	Mean Home Range (ha)	Minimum/Maximum Home Range (ha)	Duration of Study (yr)	State/Habitat type	Reference
	14 M	3.11	0.0–14.4	1	CA creosote/bursage	Duda and Krzysik 1998
	15 F	0.91	0.01–3.7	1	CA creosote/bursage	Duda and Krzysik 1998
	34 F	22.24	2.22–109.7	4	NV creosote/bursage	Nussear unpub. data
	22 M	11.95	1.70–33.83	4	NV creosote/bursage	Nussear unpub. data
	5 S	32.92	5.79–102.45	2–3	NV creosote/bursage	Nussear unpub. data
	7 F	24.45	1.97–124.35	2	NV creosote/bursage	Nussear unpub. data
	9 M	14.27	1.09–38.61	2	NV creosote/bursage	Nussear unpub. data
	7–16 F	5.3–7.2	—	1–2	CA creosote	Franks et al. 2011
	5 M	16.2	—	<1	CA creosote	Franks et al. 2011
	18–22 F	7.6–9.1	—	1–2	CA creosote	Franks et al. 2011
	6 F	1.6	—	<1	CA creosote	Franks et al. 2011
	5 M	9.2	—	<1	CA creosote	Franks et al. 2011
	4 F	2.1	—	<1	CA creosote	Franks et al. 2011
	4 M	5.8	—	<1	CA creosote	Franks et al. 2011
	8 M	20.9	7.7–40.0	<1	NV creosote/bursage	O'Connor et al. 1994b
	7 F	9.0	5.9–13.6	<1	NV creosote/bursage	O'Connor et al. 1994b
G. morafkai						
	9 M	9.2	1.0–22.3	2	AZ Arizona upland	Averill-Murray et al. 2002a
	4 F	4.7	2.7–7.5	2	AZ Arizona upland	Averill-Murray et al. 2002a
	4 M	21.7	15.2–31.5	2	AZ interior chaparral	Averill-Murray et al. 2002a
	6 F	23.3	3.4–51.5	2	AZ interior chaparral	Averill-Murray et al. 2002a
	10					
	F	12.8	2.3–50.7	2	AZ Arizona upland	Averill-Murray et al. 2002a
	5 M	25.8	5.6–53.4	1.5	AZ palo verde / cacti mixed shrub	Barrett 1990
	9 F	15.3	2.8–48.5	1.5	AZ palo verde / cacti mixed shrub	Barrett 1990
	4 M	11.0	3.9–22.2	2	AZ desert grassland	Averill-Murray et al. 2002a
	4 F	2.6	1.1–6.0	2	AZ desert grassland	Averill-Murray et al. 2002a
	3 M	16.3	3.3–25.2	2	AZ semidesert grassland	Martin 1995
	3 F	13.1	8.1–21.6	2	AZ semidesert grassland	Martin 1995

Note: For the number by sex / size class column, M = males, F = females, J = juveniles, S = subadults, and H = hatchlings.

during the study period, and their home ranges could not be computed. Home ranges may be overestimated by minimum convex polygons, which can include significant areas that tortoises may not use; however, this analysis has been widely employed, and its pros and cons have been discussed in the literature (O'Conner et al. 1994a, Kazmaier et al. 2002, Pike 2006, Harless et al. 2010). Home ranges derived using this method are perhaps best considered as indicators and estimators of movement scales and patterns (O'Conner et al. 1994a) and may be most useful in interpreting home ranges for habitat conservation purposes (Pike 2006, Harless et al. 2010).

Some researchers studying *G. polyphemus* have divided movements into those associated with feeding and those related to social interactions (McRae et al. 1981, Smith 1995). Mean feeding radius from the burrow in a southwest Georgia sandhill was 13 m, with 95% of feeding activity occurring within 30 m of the burrow (McRae et al. 1981). In north Florida sandhills, Smith (1995) reported that female *G. polyphemus* foraged within 17 m of their burrow, and Ashton and Ashton (2008) described a similar radius as the primary forage area around the burrow and 100 m as a secondary forage radius that is less frequently used. Auffenberg and Iverson (1979) cited the connection between foraging distance and food availability/quality, and noted that food is so plentiful in much of the range of *G. polyphemus* that most individuals usually move < 50 m from their burrows to feed. *G. polyphemus* also undertakes longer movements (e.g., 1–3+ km), however, apparently associated with seasonal depletion of preferred forage or highly selective foraging (McRae et al. 1981, Ashton and Ashton 2008). In some habitats, *G. polyphemus* commonly feeds along grassy roadsides and uses roadways and fire lanes as travel corridors during social encounters, thereby elongating home ranges (McRae et al. 1981, Douglass 1990, Diemer 1992a, RB Smith et al. 1997).

In contrast to *G. polyphemus, G. flavomarginatus* does not appear to increase movements in response to decreases in quantity and diversity of forage; instead, the tendency for *G. flavomarginatus* is to decrease the distance of movements during seasons of less forage abundance and increase them during the season of major productivity (Aguirre et al. 1984). For *G. agassizii,* home ranges in the same habitat may vary from one year to the next with changing environmental factors, such as rainfall. The overall mean home range of *G. agassizii* in Ivanpah Valley, California, varied from 19.0 ha for females and 26.6 ha for males in 1980, a year of high ephemeral plant production (7.3–10.0 g/m^2) (Turner et al. 1980), to 16.5 ha for females and 23.0 ha for males in 1981 (a drought year with ephemeral production of 0.04–0.11 g/m^2) (Medica et al. 1981). Similarly,

G. agassizii home ranges at both the Sand Hill and Pinto Basin sites at Twentynine Palms, California, were much larger in 1995, a productive year for winter annuals, than the drought year of 1996 (Duda and Krzysik 1998, Duda et al. 1999).

Movements of hatching *G. polyphemus* have been studied on several sites in Florida and Mississippi. Mean distance from the nest to the first burrow is generally ≤15 m (Butler et al. 1995, Epperson and Heise 2003, Pike 2006). Dispersal from the nest is random, and hatchlings move farther from the nest over time (Pike 2006). Mean daily movements are relatively short, averaging 8 m (Butler et al. 1995, Epperson and Heise 2003, Pike 2006), but some individuals make longer movements (e.g., 80 m in a 24-hour period; Epperson and Heise 2003). Mean daily movements for older juveniles are similar (McRae et al. 1981, Diemer 1992a, Wilson et al. 1994), with some longer forays. *Gopherus flavomarginatus* juveniles exhibit relatively high motility and also expand their activity area as they become larger in body size (Aguirre et al. 1984).

Comparing movements within and among tortoise species is complicated by how tracking data are collected. In general, daily maximum movements of *G. polyphemus* within home ranges (e.g., 200–400 m; Diemer 1992a, Smith 1995) are shorter than those of *G. flavomarginatus* (>900 m; Aguirre et al. 1984) and other western species. Emigrating *G. polyphemus,* however, have been known to move 0.7–6.4 km over periods ranging from about four days to more than three months (Diemer 1992a, Eubanks et al. 2003). *Gopherus agassizii* commonly travel 470–823 m/day (Berry 1974), and males are known to cover >1000 m/day within their home range. Some individuals have traveled longer point-to-point distances (1.4–7.3 km) outside their usual activity area over periods ranging from 16 days to five years (Berry 1986a); similarly, Kazmaier et al. (2002) documented an 11-km dispersal by a juvenile *G. berlandieri* over one year. One of the longest movements was by a radio-telemetered female *G. morafkai* that traveled 32 km from the Rincon Mountains south to the Santa Rita Mountains in Arizona. This tortoise's journey over anthropogenic barriers (railroad tracks and interstate highways) was facilitated by researchers and it was even temporarily adopted and held in captivity by private citizens on several occasions (Edwards et al. 2004a). *Gopherus morafkai* historically dispersed between mountain ranges and despite the challenges posed by anthropogenic habitat modifications, such interpopulation movements may be critical to the persistence of small tortoise populations (Edwards et al. 2004a).

Erratic behavior and movements prior to long-distance dispersal have been reported for both *G. polyphemus* (Diemer 1992a) and *G. flavomarginatus* (Aguirre et al. 1984). Diemer (1992a) noted that a subadult *G. polyphemus,* which eventually emigrated 0.7 km from its natal colony over about four days, moved back and forth along a forestry windrow for several days before its dispersal, even crawling up onto the debris and precariously balancing on logs. Dispersing immature and adult *G. flavomarginatus* exhibited erratic movements within relatively small areas (0.2–0.7 ha) before making extended, almost straight, advances (1–6 km), generally in a single direction. These individuals were not transients at the time of their displacement and had been members of their colonies for at least two years before making their long-distance dispersal (Aguirre et al. 1984).

Gopherus polyphemus movements and home ranges are also influenced by natural vegetation succession, forestry practices, and land management practices, such as controlled burning (Landers and Speake 1980; Auffenberg and Franz 1982; Diemer 1986, 1992a). *Gopherus polyphemus* in thickly planted pines or in areas that do not receive controlled burns may move to roadsides or natural openings in the forest (Landers and Buckner 1981, Diemer 1992a). Anthropogenic disturbance also can prompt movements to different burrows or excavation of new burrows, and may provide tortoises with alternative burrowing terrain such as berms and windrows (Diemer 1992a). Although disturbance can prompt movement by *G. polyphemus,* it does not necessarily result in significant changes in home range (Mendonca et al. 2007, Yager et al. 2007). In contrast to the response of *G. polyphemus* to disturbance, Kazmaier et al. (2002) reported significantly larger home ranges for *G. berlandieri* in ungrazed pastures than in grazed pastures, but noted that research design and analysis method could have affected this finding.

DAILY ACTIVITY AND SEASONALITY OF MOVEMENTS

Gopherus polyphemus is almost exclusively diurnal, although rare nocturnal activity by both adults and juveniles occur (Alexy et al. 2003; Ashton and Ashton 2008; T. Radzio, personal communication). Ashton and Ashton (2008) observed individuals foraging as late as midnight in northern Florida during July to September when daytime temperatures exceeded 36°C. Both *G. polyphemus* (T. Radzio, personal communication) and *G. flavomarginatus* (Adest et al. 1988) are known to emerge during nocturnal rain events. Nocturnal emergence by *G. flavomarginatus,* with or without rainfall, may be related to elevated burrow or soil temperatures and may occur more widely than was once thought (Adest et al. 1988). Nocturnal activity in *G. agassizii* has been observed at Lake Mead National Recreation Area in southern Nevada (elevation ~500 m), where temperatures in summer reach ~49°C in the daytime and commonly are still greater than 38°C at sunrise (Nussear 2004). At a higher elevation (~1020 m) at Rock Valley, Nevada, a large male *G. agassizii* was observed in its burrow at dusk before a nocturnal thunderstorm; it exited the burrow during the night and was observed at sunrise under a shrub surrounded by a fresh moat of dry mud (P. Medica, personal observation). Modern technology (e.g., remote video cameras) can provide 24-hour coverage of burrows (Alexy et al. 2003; T. Radzio, personal communication), improving future understanding of tortoises' responses to diel and climatic factors.

Daily activity by *G. polyphemus* generally is unimodal, with a midday peak, and follows the daily temperature cycle (Douglass and Layne 1978, McRae et al. 1981, Alexy et al.

2003), especially during the spring months. In southern Florida, the activity peak was during the hottest hours, 1300–1600, throughout the year (Douglass and Layne 1978). McRae et al. (1981) noted a bimodal activity pattern during July and August in southwest Georgia, however, with feeding forays more common during 1000 to 1200 and 1600 to 1800; activity returned to a unimodal pattern following the intense summer heat. A bimodal activity pattern in July–August occurs in *G. flavomarginatus* (Morafka et al. 1981). *Gopherus berlandieri* on the coast had two diel activity periods in spring, with a primary peak during 1000 to 1100 and a secondary peak during 1700 to 1800 (Bury and Smith 1986). Seasonal variation in the daily activity of *G. agassizii* is observed, with the pattern unimodal in spring and fall, but bimodal during the warmer summer months (Woodbury and Hardy 1948, Luckenbach 1982, Zimmerman et al. 1994).

Seasonal activity of *G. polyphemus* varies latitudinally, depending on the severity of winter temperatures. In much of its range, *G. polyphemus* is most active May–August and least active December–February (Douglass and Layne 1978, McRae et al. 1981, Diemer 1992a). In south-central Florida, some activity occurs outside the burrow every month of the year (Douglass and Layne 1978). On the southeast Florida coast, *G. polyphemus* engages in courtship year-round; there is no winter dormancy, although activity may wane during late January into early February (Moore et al. 2009). Morafka (1982) reported that winter dormancy of *G. flavomarginatus* generally extends from November through April. Activity of both species of desert tortoise also varies greatly with latitude and elevation. Activity extends into early December on Tiburon Island, Sinaloa, Mexico (Bury et al. 1978). The onset of hibernation for *G. agassizii* at four sites in Nevada and Utah during winter 1998–1999 was from 15 October to 10 November and was dependent upon the cooler fall temperatures at the study site (Nussear et al. 2007). In an extensive four-year study at Yucca Mountain, Nevada, involving 365 individuals, most entered hibernation during October or the first half of November, with 98% entering hibernation by 15 November with some variation based upon size and sex (Rautenstrauch et al. 1998). Eighteen percent of the juvenile *G. agassizii* at Fort Irwin, California were active above ground on winter days when temperatures were above 10°C (Wilson et al. 1999a).

Nesting tends to dictate late spring movements of adult female *G. polyphemus,* and relatively long-distance movements can occur as females search for open-canopied, sunlit egg-deposition sites during May and June (Landers 1980, McRae et al. 1981, Diemer 1992a, Smith 1995). Female *G. berlandieri* also made long-distance forays up to 0.8 km during May–June (Kazmaier et al. 2002). Similarly, Averill-Murray et al. (2002a) observed a female *G. morafkai* moving 0.75 km from its usual home range in June presumably to lay eggs. An extended movement of 1.3 km took place for a nesting female *G. agassizii* in Ivanpah Valley, California, between June 1980 and March 1981 (Medica et al. 1981).

Increased late summer and early fall movements by male *G. polyphemus* are well documented in both Georgia and Florida (McRae et al. 1981, Douglass 1990, Diemer 1992a, Eubanks et al. 2003) and coincide with the period of active spermatogenesis (Taylor 1982a, Ott et al. 2000). The outermost points of the home range of a male *G. polyphemus* may represent the burrows of preferred females; thus, female locations are important in determining the limits of males' home ranges (Douglass 1990; J. Diemer, personal observation). Male *G. berlandieri* and *G. morafkai* also make relatively long forays during late summer (Averill-Murray et al. 2002a, Kazmaier et al. 2002), although a long movement (3.1 km) by a male *G. agassizii* was observed in Ivanpah Valley, California, during 14–30 May 1980 (Turner et al. 1980).

Juvenile *G. polyphemus* (one–four years old) in west-central Florida were active during every month of the year (Wilson et al. 1994), with activity peaking in spring. Winter movements were limited, but juveniles basked more often on the burrow mound in winter than in other seasons. On the east-central Florida coast, Pike (2006) documented year-round activity in hatchling *G. polyphemus,* but noted that they moved the most during and immediately following the time of hatching (August–November). Butler et al. (1995) found consistently high activity of hatchling *G. polyphemus* from mid-April through early November in northeast Florida; activity then waned until late March.

BURROW/SHELTER-SITE USE

Unlike *Gopherus berlandieri* and the two species of desert tortoise, *G. polyphemus* and *G. flavomarginatus* almost universally excavate relatively deep burrows for shelter (Hansen 1963, Auffenberg 1969, Morafka et al. 1981, Morafka 1982). Ashton and Ashton (2008) recorded their longest *G. polyphemus* burrow as 20.5 m. Exceptions to the typically deep burrows are found in areas with lime rock or high water tables, or in shell-sands of the coast and off-shore islands; in these situations, burrows may be quite shallow (Auffenberg 1969, Ashton and Ashton 2008). Both adult and juvenile *G. polyphemus* occasionally use shallow depressions known as pallets (Auffenberg 1969, Diemer 1992a, Butler et al. 1995, Ashton and Ashton 2008), but not nearly as frequently as do the primarily non-burrowing *G. berlandieri* (Rose and Judd 1982). Both species of desert tortoise use a variety of burrows, dens, pallets, and other shelter-sites (Luckenbach 1982). In the northern extreme of the geographic range on the Nevada Test Site, *G. agassizii* burrows can extend 7.5 m under caliche overhangs in the sides of washes (Germano et al. 1994). Some tortoises in Nevada also shelter under shrubs overnight during July–September (Bulova 1994). *Gopherus morafkai* excavates soil burrows but also uses caliche caves in washes, rock crevices, vegetation, and even nests of wood rats (*Neotoma albigula*) as shelter-sites (Martin 1995, Averill-Murray et al. 2002a).

Individual and site-related variations are also reflected in number of burrows used by *G. polyphemus.* Depending on season, this species uses multiple burrows (McRae et al. 1981, Diemer 1992a, RB Smith et al. 1997, Eubanks et al. 2003). In longleaf pine (*Pinus palustris*) habitat in Georgia, females used an average of five burrows and males used ten burrows in one year (Eubanks et al. 2003), similar to earlier findings (females: four; males: seven; McRae et al. 1981) in comparable habitat for a slightly shorter period (April–December). In a Florida pine plantation, mean number of burrows used was three for females and six for males (April–December); males averaged eight burrows over two years (Diemer 1992a). Mean number of burrows used over one to two years was larger in a Florida coastal scrub habitat: nine for females and 17 for males (RB Smith et al. 1997). Eubanks et al. (2003) noted that most studies were not long enough to fully elucidate the number of burrows that a particular *G. polyphemus* uses during its lifetime; previously unoccupied burrows are continually being added to an individual's roster. Like *G. polyphemus,* male *G. flavomarginatus* also use a greater number of burrows (n = 4), on average, than females (n = 3) (Aguirre et al. 1984). Although male *G. agassizii* may use an average of 17.6 shelter-sites in a year and females may use 19, the majority of those used by males are burrows rather than nonburrows (Burge 1977). On another Nevada study area, *G. agassizii* used an average of 9.1 shelter-sites (range, 3–18) during July to October (Bulova 1994). *Gopherus morafkai* in southern Arizona may use as many as 27 shelter-sites (average, 21) over a two-year period (Martin 1995).

Hatchlings and other juvenile *G. polyphemus* may use pallets, abandoned adult burrows, or shelter under vegetation (Diemer 1992a, Butler et al. 1995, Ashton and Ashton 2001, Pike 2006); they also dig new burrows as they disperse from the nest area (Butler et al. 1995, Pike 2006). Depending on the site and tracking duration, this youngest class of *G. polyphemus* may use an average of one to five burrows (Wilson et al. 1994, Butler et al. 1995, Epperson and Heise 2003, Pike 2006). Hatchlings and older juvenile *G. flavomarginatus* use an average of two burrows (Aguirre et al. 1984), and like *G. polyphemus,* they shelter in pallets and under vegetation (Tom 1994).

Although burrows generally contain only one *G. polyphemus* at a time, co-occupation during the day or even overnight has been documented for both adults and immatures (McRae et al. 1981, Douglass 1990, Diemer 1992a, Pike 2006). Burrow or other shelter-site co-occupation has also been observed in other species of *Gopherus* (Rose and Judd 1982, Aguirre et al. 1984, Averill-Murray et al. 2002a). Cohabitation of burrows by both males and females is infrequently observed among *G. agassizii;* particularly during late summer and early fall, males have been frequently observed within burrows with females at Rock Valley, Nevada (P. Medica, personal observation). Burrow defense or usurpation may be associated with these social encounters; individuals of *G. polyphemus* will turn sideways within burrows and block the forward motion of tortoises trying to enter (Diemer 1992a). Similarly, individuals of *G. flavomarginatus* and *G. morafkai* defend their burrows against intruders (Aguirre et al. 1984, Averill-Murray et al. 2002a), and female *G. morafkai* have even been observed defending their burrows containing nests against Gila monsters (*Heloderma suspectum*) (Barrett and Humphrey 1986). Because *G. berlandieri* does not excavate deep burrows, individuals generally use pallets on a "first-come, first-served" basis and do not defend these depressions; moreover, these pallets do not serve as the central focus of *G. berlandieri* activity, as do the burrows of the other *Gopherus* species (Rose and Judd 1982).

CONCLUSIONS

The home range and movements of North American tortoises are strongly influenced by both environmental and anthropogenic factors, and therefore vary considerably among species, geographic locations, and individuals. In general, male *Gopherus* tortoises have larger home ranges than females, with a few exceptions documented for *G. agassizii.* Movements are closely tied to foraging needs and social interactions, but there are unknown factors that influence relatively long-distance dispersals. Although primarily diurnal, *Gopherus* species do occasionally engage in nocturnal activity. Two of the five species, *G. polyphemus* and *G. flavomarginatus,* generally excavate relatively deep burrows; the other three species use a variety of created or existing shelter-sites that include burrows, pallets, dens, and caves under rocks or caliche. Despite intensive research efforts over the last four decades, these tortoise species remain, in some ways, enigmatic.

12 Social Behaviors of North American Tortoises

Craig Guyer
Sharon M. Hermann
Valerie M. Johnson

Tortoises are thought to have inhabited North America for 50 million years (Le et al. 2006) and the genus *Gopherus* has been around for at least 30 million of those years (Bramble 1974; chapters 1 and 2). Although assemblages of tortoises in the past apparently harbored two or more species involving one large genus (*Hesperotestudo, Manouria,* or *Stylemys*) and one small genus (*Gopherus;* Le et al. 2006), all modern species belong to *Gopherus* and occupy areas in which they are the sole representative of this turtle radiation (Morafka 1982). So, tortoises seem to have evolved within assemblages in which they interacted with larger competitors and these interactions seem likely to have resulted in behavioral characteristics of living taxa that are adaptations to competition that no longer exists. Additionally, the genus is characterized, in part, by the presence of subdentary glands that are used in chemical communication (Le et el. 2006), a diagnostic feature that opens the door to the possibility that the earliest members of the genus were as highly social as the living descendants that will be discussed in this chapter. Because living and fossil members of the radiation show a strong affinity for grassland savannas and deserts, the development of extensive social structure in North American tortoises appears to be an adaptation of these creatures to dry open habitats. In this chapter, we review social behaviors of living tortoises and synthesize them with a view toward understanding their historical origin. Similar reviews have been published (e.g., Berry 1986a) and these invariably have noted the great degree to which all species share behaviors. We will focus, therefore, on the shared features of tortoise behavior and not on the relatively minor differences among species. We will point out the unique behavioral features that characterize adaptations of *G. agassizii* to the extreme aridity of the Mojave Desert, however, as well as shared behavioral features associated with the unusually deep burrows that characterize *G. flavomarginatus* and *G. polyphemus.*

North American tortoises live in aggregations that can be measured from the physical locations of individuals or of their burrows, generally over a season of activity (e.g., Styrsky et al. 2010). A variety of terms is available to describe this feature of tortoise biology, but the one most frequently encountered is the term "colony." Because burrows are relatively easy to discover and map, delimitation of colonies is frequently based on these structures. Designating tortoises as colonial creates an obvious comparison of tortoises to other colonial vertebrates, such as ground squirrels, that are presented in most ethology texts as classic examples of colonial social organization. Members of a tortoise aggregation typically have the potential to socialize, as evidenced by direct observations or by mapping the spatial arrangement of burrows and using estimates of long distance movements to infer which sets of burrows contain animals that are likely to socialize with one another. Because tortoise aggregations are so similar to ground squirrel colonies, we accept colony as a reasonable term for tortoise social structure. We note, however, that populations of tortoises are also delimited by spatial aggregations of burrows (McCoy and Mushinsky 2007, Styrsky et al. 2010), placing the terms colony and population in jeopardy of becoming synonyms. It would be preferable to have a colony represent some identifiable component of a population. Fortunately, advancements in the analysis of social networks provide a mechanism that allows examination of colony attributes in a fashion that separates them from attributes of the population to which the colonial organisms belong.

Colonial social organization in vertebrates has benefits that outweigh costs (Alcock 2009). These benefits generally derive from reduced predation associated with an increased ability of a group to detect approaching predators or the increased diversity of mating opportunities. North American tortoises are capable of recognizing predators and responding aggressively toward them (Barrett and Humphrey 1986), and individuals can vocalize in ways that some have interpreted as alarm calls (Nichols 1953). No tortoise species has been demonstrated to exhibit behaviors consistent with the types of behaviors used by colonial mammals, like ground squirrels, to detect a predator and use vocal cues to warn other colony members of its approach (e.g., Boellstorff and Owings 1995), however. Instead, published accounts of antipredatory adaptations of tortoises are associated with behaviors that individuals use largely to escape detection by a predator (Ruby and Niblick 1994). Because predator avoidance may not be the driving mechanism in social organization in tortoises, we focus on mating systems to better explain the dynamics of social organization. Because behavioral components of sexual selection, male–male competition and female choice, are known from tortoises, these features will guide our presentation of social behaviors of North American tortoises.

MALE–MALE AGGRESSION

Male–male competition has been described for all species of *Gopherus* and appears to involve key features that are consistent across taxa. Two categories of encounters are known. In type 1 encounters, physical combat occurs between males, ending with a signal of submission by the subordinate male to the dominant male (move away from dominant or withdraw into shell; Weaver 1970). In type 2 encounters, dominant and submissive roles are established previously and no combat occurs because aggression is avoided by the actions of the subordinate (Weaver 1970). Interestingly, the roles played by a pair of interacting males may change between encounters (Rose and Judd 1982). This change is interesting because it would appear to require an ability of individuals to recognize social context and make appropriate decisions based on those contexts.

For males, type 1 encounters generally are rare and involve visual signals associated with head-bobbing (Camp 1916, Douglass 1986, Eglis 1962, Weaver 1970). The height and duration of these bobs may signal the level of aggression of an individual (Ruby and Niblick 1994). Approaching males may then engage in physical combat in which the combatants attempt to push each other or attempt to hook their gular projections under the plastron of the rival and use their hind legs to jump forward and upward along a vector that overturns the rival (Hailman et al. 1991). Combat may continue even after a rival is overturned since in some cases the upright male continues to ram the rival, thereby righting that individual (Douglass 1986). When overturned, a tortoise will move his front limb in a circular fashion, attempting to contact the ground with enough force to right its body. Some have speculated that this waving of the forelimb provides a visual stimulus that maintains aggression by the upright male (Ruby and Niblick 1994) and that this may make righting of the overturned rival a consistent feature of type 1 encounters. Males exhibiting type 1 encounters may attempt to mount their rival, a behavior that is thought to signal dominance either as an ultimate behavioral event during type 1 encounters or as a misinterpretation of a signal of submission by the rival (Niblick et al. 1994).

Type 2 encounters apparently are frequent and occur in a variety of forms. In some cases a subordinate male responds to the visual stimulus of an approaching dominant male by turning and walking away at a faster speed than the dominant male (Rose and Judd 1982). In other cases, the subordinate male retracts his head and limbs, a signal that appears to reduce the level of aggression of the dominant male (Ruby and Niblick 1994). In still other cases, subordinate males detect and avoid chemical cues in the feces of dominant males (Bulova 1997). This reaction is complicated by the reaction of males to secretions from the subdentary glands, however. These glands are found in both males and females, and both sexes spread secretions across their forelimbs by wiping the chin across these appendages (Mason and Parker 2010). Males are more likely to enter burrows or approach tortoise shells marked with secretions from subdentary glands, but equally so for secretions of males and females (Bulova 1997). So, tortoises may mark trails as the secretions rub off during locomotion between burrows. Males may use these trails to follow females directly to their burrows or to follow rival males as they travel to burrows of females. In cases in which males travel to burrows occupied by rival males, the high concentration of repulsive chemicals associated with feces of the rival at his burrow likely prevents the trailing male from entering the burrow of a dominant male. Thus, the chemicals deposited from the subdentary glands appear to be used for species recognition and the chemicals deposited in the feces to identify sex and perhaps individual, because dominant males are avoided. For other reptiles, chemical cues allow rival neighboring males to recognize each other via pheromones and to reduce levels of aggression toward neighbors with which a resident has already established a relationship (dear enemy effect; Mason and Parker 2010). This system appears to characterize North American tortoises.

FEMALE–FEMALE AGGRESSION

Emerging evidence from all species of *Gopherus* suggests that interactions between females also affect the mating system. In fact, despite the impressive activities of males when they engage in type 1 encounters, the far-more-subtle activities of females may be the key to unraveling the mating system of North American tortoises. Female–female aggression is known from tortoises and is of the same two types described for males: type 1 involving physical combat and occasionally mounting of a female by another female (Douglass 1986, Switak 1973, Weaver 1970) and type 2 involving avoidance behaviors. Although rare, female–female aggressive encounters are known. Lee (1963) noted increased aggression of caged *G. agassizii* females during the nesting season and *G. polyphemus* females have been observed to mount other females during interactions at burrow entrances (personal observation). The primary evidence of avoidance behavior is an aversion of some females to the approach of other females in cage situations (Douglass 1986). Coupled with these observations, are reduced rates of burrow sharing by females with other females, compared with rates of male–male and male–female burrow sharing, which appear to be roughly equal in frequency for North American tortoises (Bulova 1994, Johnson et al. 2009). Finally, females are unlikely to enter burrows marked with chemicals found in feces of other females. The effect of these behaviors should be the creation of widely spaced females whose receptivity to mating appears to be unknown to males. The peak in male movement activity seen in all species of North American tortoises (chapter 11) may result from this dispersion of an unpredictable limiting resource (receptive females). The tendency of females, especially large, more fecund ones, to limit movements in the fall (Eubanks et al. 2003, Rose and Judd 1975) may increase the distances males must move to monitor females.

FEMALE CHOICE

In all *Gopherus* species, males travel more frequently and over longer distances than females, especially in the fall (chapter 11). Although mating may take place at any time of the year, mating attempts peak in the fall for *G. polyphemus* (Johnson and Guyer 2007) and are thought to peak at this time for other species, because mate-seeking is the consistent explanation for the long fall movement distances of males (Eubanks et al. 2003, Trápago et al. 2000, Woodbury and Hardy 1948). In addition to this well-recognized aspect of the tortoise mating system, females play an active role in mate choice, because females move to share burrows with males (about as frequently as males move to share burrows with females; Johnson et al. 2009) and their associations with males in cage settings suggest pair bonds may develop during a season of activity and that these bonds may differ from year to year (Nichols 1953, 1957). When offered a choice between a large and a small male in an experimental arena in which contact between sexes was prevented by fencing, *G. agassizii* females positioned themselves near the larger male during their first encounter, but near the smaller male during their second encounter (Niblick et al. 1994). Females may have learned from the first encounter that the preferred large male was unavailable. These results suggest that females process information about available mates and learn to investigate other options if their first choice is not available.

Descriptions of courtship suggest the approach of a male toward a female serves as a visual cue that causes the female to approach the male, move away from him toward her burrow, or turn sideways in her burrow to block his entry (Auffenberg 1966b, Black 1976). If the female is receptive, then courtship activities involve a stop-start approach by the male toward the female and may involve a steady approach by the female toward the male (Auffenberg 1966b, Black 1976). While stopped, the male extends his head and neck and bobs them up and down, the first of a series of behaviors that are similar to actions associated with type 1 male–male aggressive encounters, but that are given with less intensity in a courtship setting. Thus, visual signals from the male to the female provide one avenue of communication that allows females to select from among possible mates. The head bobbing activity also is thought to provide a chemical cue, wafted from the male to the female via the male's motions. Finally, tactile cues are provided by males to receptive females; these cues are in the form of ramming of the female by the male and by the male's nibbling of the head, forelegs, and anterior margin of the carapace of the female. In response to these advances, the female either remains motionless or backs away from the male along a semicircular path that presents the hind end of the female to the male. The male then mounts the female, a relatively precarious act, and, for successful copulations, the female elevates her hind end to allow the male to insert his erect penis into the cloacal opening of the female. Even at this late stage, females apparently are capable of effecting mate choice by failing to elevate the hind end. During copulation, males are known to vocalize, but this is not known to provide any communicative signal.

The descriptions above were generated largely from caged animals in which a single pair interacted without disturbances from other individuals, either because no other individuals were present or because dominance of one male over all others prevented such interactions. In field settings, females experience visitations from an average of at least three males during a season of activity (Boglioli et al. 2003) and visitations from these males may be a daily experience of females during the fall (Johnson and Guyer 2007). When three tortoises, a large male, a small male, and a female were allowed to interact in an experimental arena, the large male was more likely to mount the female. In fact, in only one instance did a small male in the experimental arena attempt to copulate and this was from the side, a position that could not possibly lead to insemination (Niblick et al. 1994). That these behavioral preferences of females lead to increased siring of offspring by large males have been documented in free-ranging (Moon et al. 2006) and translocated (Tuberville et al. 2011) populations of *G. polyphemus*. Large-male advantage, however, is not observed in mounting rates of females at their burrows in a habitat that mimics the ancestral landscape to which this species is adapted (Boglioli et al. 2003). Additionally, in the study of Tuberville et al. (2011), males that had longer residency times within a translocated population were much more likely to be selected by females than males that were recent releases. This preference for males with longer residency was equally strong for females with long residency times and recently released females. Because recently released males generally were larger than males with long residency times, factors other than size affect female choice in tortoises (Tuberville et al. 2011).

From the observations listed, we infer that females use a variety of cues in making mate choices and that the long-term social bonds that were created in the ancestral landscapes of North American tortoises resulted in patterns of female choice that were not dominated by male size, but instead by social experiences of females with particular males. Unsuccessful mountings of free-ranging males are well known and frequently involve attempts by a male to mount the female from the side. Such mountings can lead to ejaculation by the male onto the side of the carapace of the female (evidenced by photographs associated with data from Johnson et al. 2009). These observations leave a strong impression that male tortoises do attempt to force copulations upon females. Berry (1986a) has suggested that these forced copulations occasionally might be successful, but Niblick et al. (1994) noted that females seem to have enough behavioral options to thwart such attempts. In fact, a female can exhibit choice even if a male has mounted her because she can aggressively attack the male (Beltz 1954), position her plastron against the ground (Niblick et al. 1994), or enter her burrow (knocking off any male that may have mounted her) and block it at its entrance (Douglass 1986). A fatigue factor by females in the face of constant attempted matings by males has been suggested for sea turtles (Lee and Hays 2004) and may warrant examination in mating systems of North American tortoises. Nevertheless, the spatial arrangement of females, pattern of male movements in which individual males monitor several females more or less simultaneously, and frequent occurrence of multiple males monitoring a particular female are features that are most consistent with scramble competition polygyny as the mating system of North American tortoises.

DISPERSAL

Dispersal of individuals from the natal social group to other social groups is a key feature of mating systems. Documentation of this aspect of behavior of North American tortoises has remained elusive. In fact, all age and sex groups have been presented in the literature as the dispersing group. So, either this represents reality—all groups contain individuals that do disperse—or additional techniques in population genetics need to be applied to clarify how genes flow within and among tortoise groups. Berry's (1986a) definition of dispersal as being relatively linear movements away from a home range with no apparent tendency to return to it was used by Eubanks et al. (2003) to document two cases of dispersal in adult male *Gopherus polyphemus* who moved the equivalent of 10–13 mean adult male home ranges during a season of activity. There was no obvious tendency for subadults to be forced to the periphery of the colony in that study, contrary to suggestions by McRae et al. (1981); but, Diemer (1992a) noted a single subadult disperser, a finding supporting the suggestions of McRae et al. (1981). Pike (2006) argued that the slow tendency of juveniles to center annual home ranges farther and farther from the natal home range represented the dispersal stage of *G. polyphemus.* In the same species, Tuberville et al. (2005) noted that translocated females are more likely to disperse from a release site than other tortoises and that adults are more likely to disperse than juveniles, observations that may mirror dispersal tendencies of unmanipulated individuals. In *G. flavomarginatus,* Aguirre et al. (1984) observed both adult and subadult individuals to disperse, and in *G. agassizii,* Berry (1986a) observed dispersal in both males and females. So, a consistent pattern of dispersal has yet to emerge from studies of North American tortoises.

INTERSPECIFIC INTERACTIONS

Historically, interspecific interactions between tortoise species were likely to have been important. Because large competitors appear to have died out across the range of the living species (Preston 1979), and living species do not have geographic ranges that overlap, behavioral adaptations to such competitive interactions may have no current context for free-ranging individuals. Aspects of past competitive environments may emerge in caged animals, however. Anecdotal observations of animals in cages suggest a continuum of interspecific aggressiveness among living species such that *Gopherus flavomarginatus* is the most aggressive (consistent winner of caged aggressive interactions with all other species; Rose and Judd 1982), *G. agassizii* is next most aggressive, and *G. berlandieri* and *G. polyphemus* are the least aggressive (rarely responding to other congeners (Rose and Judd 1982). Because *G. flavomarginatus* is the largest of the living species, its apparent dominance over other species may merely be a size effect and have nothing to do with past competition. The observation that its ramming can kill individuals of other caged tortoise species but not other individuals of its own species (Rose and Judd 1982), however, seems excessive for a simple size effect. Also, because *G. flavomarginatus* appears to react more aggressively to its sister taxon, *G. polyphemus,* than to more distant relatives suggests that interspecific interactions have a historical context beyond mere size, perhaps related to evolutionary processes leading to isolating mechanisms associated with previous zones of sympatry.

SOCIAL NETWORKS OF NORTH AMERICAN TORTOISES

Highly social vertebrates develop distinctive patterns of communication that affect dispersion patterns, define mating systems, and shape strategies of predator avoidance. Analyses of social networks allow for visualization of the structure of such social groupings and for dissection of their components. No such analysis has been performed for any turtle species despite acknowledgement that such structure must exist, especially for tortoises (Berry 1986a). In this section we provide a preliminary analysis of the social network of *Gopherus polyphemus* for the Wade Tract, a private conservation reserve located in south Georgia and an important research site because it contains an apparently stable population of tortoises in old-growth longleaf pine forest, the likely ancestral habitat for the species (Guyer and Hermann 1997). At this site, a contiguous cluster of individuals was followed via radio telemetry for an entire field season, and visitations of individuals to females was monitored with camera devices (see Johnson and Guyer 2007, Johnson et al. 2009, Guyer et al. 2012 for details of the study site and methods). Data gathered at this site allowed

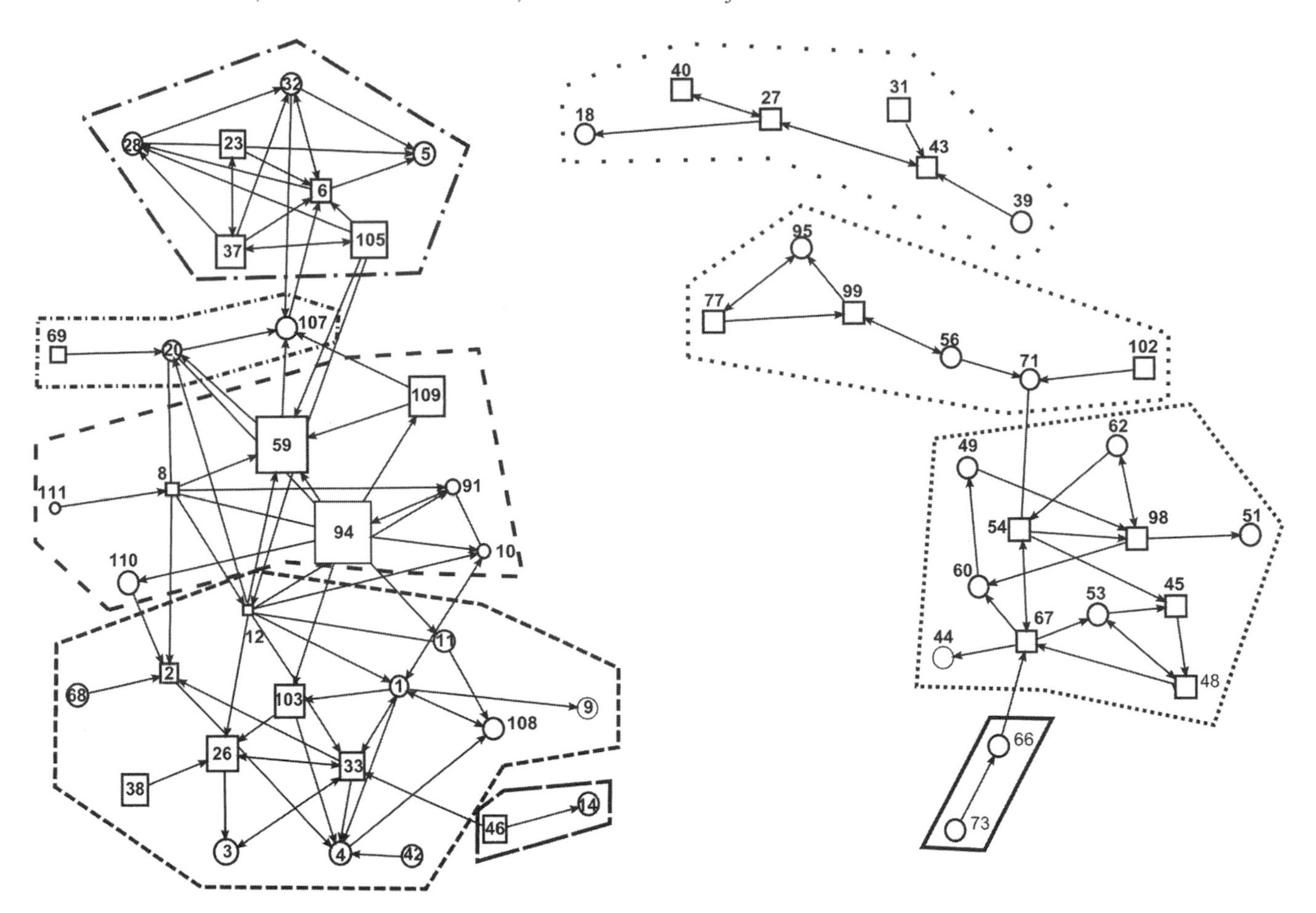

Fig. 12.1. Social network structure based on *A* (*above*), network centrality and *B* (*opposite*), network power, for *Gopherus polyphemus* during a season of activity (2002) at the Wade Tract in south Georgia. Squares are males and circles are females. Solid and dashed lines enclose cliques discovered via the method of Girvan and Newman (2002); numbers are ID numbers. Arrows indicate direction of a social interaction.

creation of an individual × individual matrix documenting the number of interactions initiated by each individual to any other individual. Because behaviors are so highly conserved within North American tortoises, we use this example to draw inferences about the likely social structure of the entire genus.

Social structure in *G. polyphemus* is decidedly nonrandom (fig. 12.1). In fact, of the individuals followed at the Wade Tract, interactions occur in three separate social groups within the population, and two of these groups contain three and five interconnected subgroups, respectively. Each social group contains both males and females in roughly equal numbers that total 6–12 individuals. Males dominate the social structure, if dominance is measured by centrality (fig. 12.1). Because females in high quality habitats, like the Wade Tract, move about half as frequently and over about half the area as males (Eubanks et al. 2003), however, the centrality of females is relatively large and remarkably uniform. This pattern results both from the activities of individual males, who visit one to three females on a more-or-less daily basis, and the activities of females, who commonly initiate interactions with males (Johnson et al. 2009).

This preliminary analysis documents substructuring of social interactions, and this finding generates a need for a term for the smallest subgroup. The term "pod" has been used to denote a social group within a population of tortoises (Ashton and Ashton 2008). Unfortunately, this term is widely accepted for matrilineal groups of toothed whales that are associated with each other during a majority of observations and that maintain cohesion through pod-specific vocalizations (Yurk et al. 2002). Because tortoises are unlikely to possess these features of social organization, we cannot recommend use of this term. Instead, we propose the term "clique" for the smallest subgroup within the social network of tortoise societies. This term is consistent with terminology associated with analysis of social networks (Borgatti et al. 2002). Additionally, because the Wade Tract subgroups overlap in geographic space (fig. 12.2), use of the term clique is consistent with use of this term in human societies, in which cliques share physical space but not social space (Freeman 1996).

Assuming that social structure in North American tortoises is conservative among species, the clique structure of *G. polyphemus* indicates several key features of social organiza-

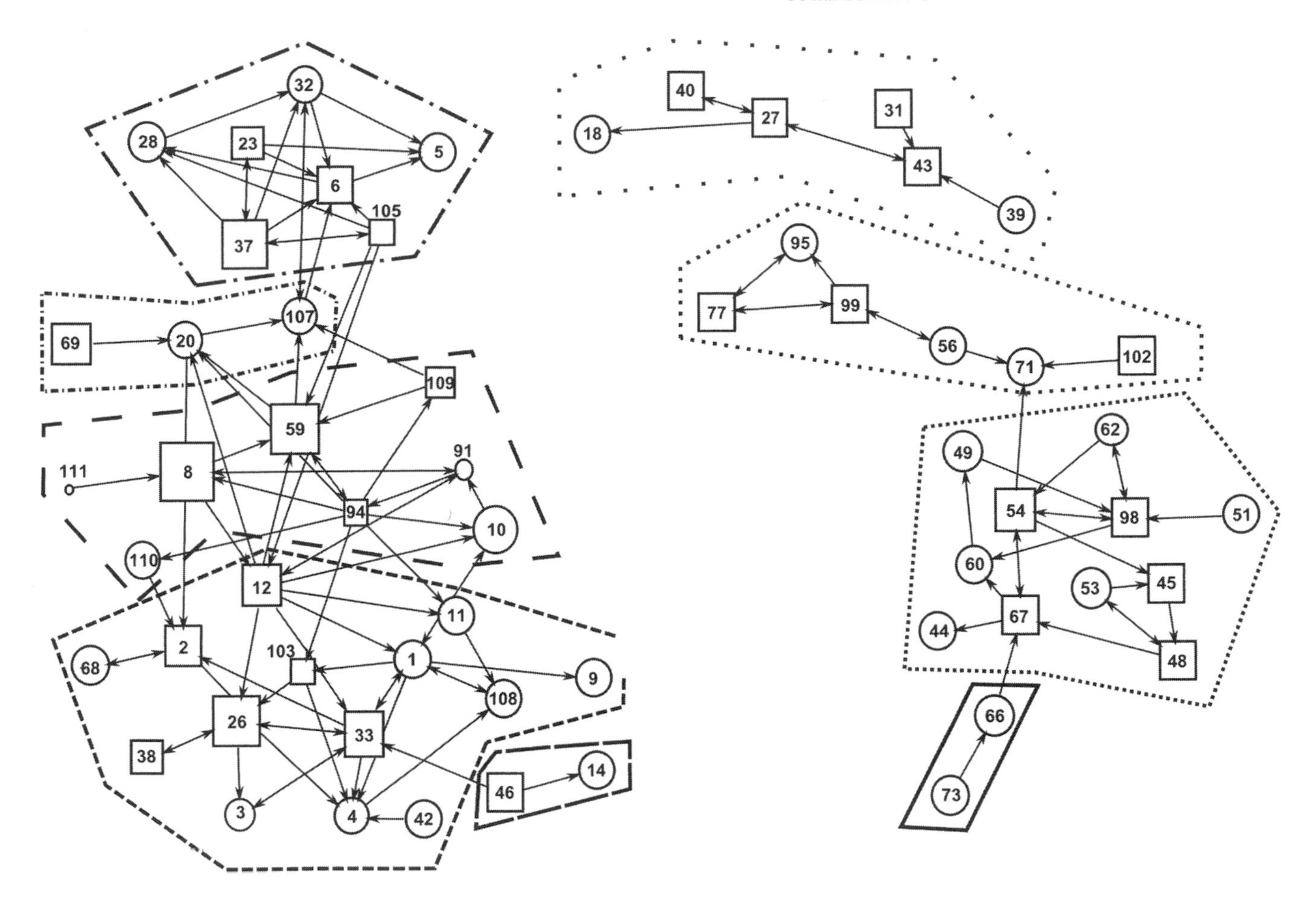

tion in tortoises. First, some cliques are remarkably isolated, in behavioral terms, from others. The three major Wade Tract groups were unconnected by edges between them. Although it is likely that some real links between groups were missed by our methods, the suggestion that tortoise social structure involves associations that include strong avoidances of some individuals despite close proximity of those individuals appears to be real. Such structuring requires individual tortoises to have strong spatial memory of not only their own burrows but also those of neighbors with whom they socialize. In addition to spatial memory, tortoises must recognize individuals so that some are avoided. Current evidence suggests that olfaction is the most likely mechanism of individual recognition (Bulova 1997), but visual cues might also play a role. Because both sexes initiate interactions with other individuals of either sex, both males and females appear to participate in maintaining clique structure. This description is consistent with observations by Douglass (1986), who noted that individuals of *G. polyphemus* may walk past nearby burrows in order to visit more distant ones. Because individual burrows can be used by up to 13 individuals during a season of activity (Eubanks et al. 2003) and these individuals might belong to different cliques, it is likely that tortoises walk by burrows that they know to exist, because they are occupied by attractive

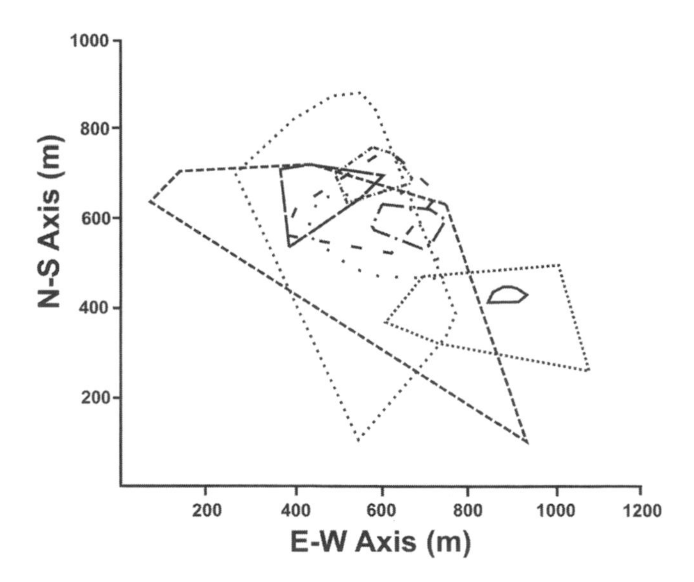

Fig. 12.2. Distribution of cliques on the Wade Tract in South Georgia. Ordinate and abscissa are latitude and longitude of burrow locations used by members of each clique. Solid and dashed lines create polygons that correspond to areas occupied by cliques of figure 12.1.

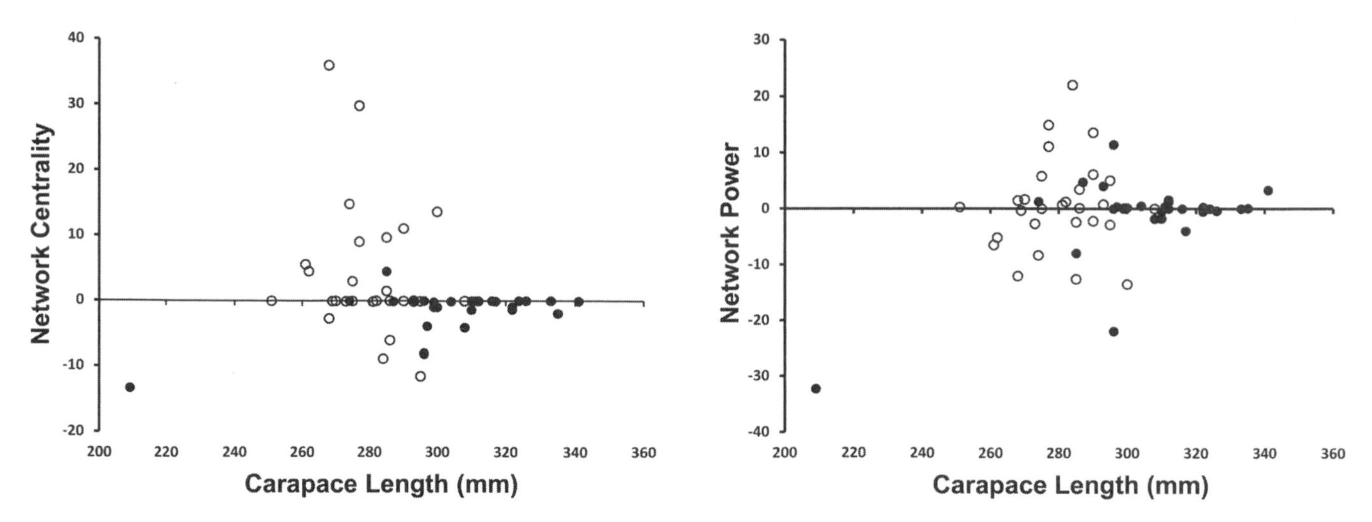

Fig. 12.3. Relationship between carapace length and (*left*) network centrality and (*right*) network power. Open dots are males; solid dots are females.

individuals during some times of year and by repulsive individuals at other times of year.

The network structure of Wade Tract tortoises is inconsistent with some published accounts of social organization of *G. polyphemus.* In particular, Landers et al. (1980) suggest that subadult males are forced to the periphery of tortoise colonies by aggressive interactions of large, dominant males. Similarly, Douglass (1986) described the mating system of *G. polyphemus* to be an incipient harem defense polygyny because large males had priority access to large females. These observations suggest that females should be centrally located within tortoise networks and that centrality should increase with body size for males. The observed network for the Wade Tract was characterized by greater centrality of males than females and of no relationship between degree of centrality and body size for either males or females (fig. 12.3A). In fact, centrality appears to increase for smaller males relative to larger ones. This lack of association between male body size and the position of males in a social network is repeated if position is based on capacity to exert power within the network (fig. 13.3B). Landers et al. (1980) and Douglass (1986) worked on lands altered by human activities, which may alter the strength of social networks because of a need for individuals to move to open areas when the landscape is fire-suppressed or clear-cut. We offer a similar explanation for the apparent disparity between information in our network analysis and observations by Tuberville et al. (2011), who noted a large male advantage in matings of repatriated *G. polyphemus.* We suggest that our lack of a size effect on centrality or power of males within social networks results from the long time period over which individuals on the Wade Tract have occupied burrows, because of its history of frequent fire and retention of old trees (Guyer and Hermann 1997). We argue that this arrangement has stabilized the social network for long time periods and has allowed females to make mate choices based on other features than size.

COMPARATIVE SOCIAL BEHAVIOR

The impression that one gets from the varied literature on behaviors of North American tortoises, some representing extensive, long-term studies, much representing tidbits of observations of free-ranging and captive individuals, is that these animals are quite intelligent and extremely social. Both terms, intelligence and sociality, are difficult to use for turtles because so many of the examples of extreme levels of intelligence and sociality come from mammals and birds. Nevertheless, tortoises show noteworthy intelligence in their spatial memory. All species appear to remember where five to ten burrows are located during an annual cycle of activity. These burrows include ones that are used over relatively short periods of time for specific purposes (nest burrows of females), as well as those that contain neighbors who are visited regularly. Additional burrows, whose locations apparently are remembered over longer periods of time, are also part of the lifetime home range of a tortoise. The identity of neighbors can be determined by chemicals in feces (Alberts et al. 1994) and probably by unique physical and behavioral features of individuals. Because of apparent long-term memory and responses to chemical cues, tortoises interact more amicably with known neighbors than with intruders from outside the social group (Alberts et al. 1994, Rose and Judd 1982). Desert tortoises learn and remember sources of water. Penned animals learn to anticipate timed sprinkler systems and use this knowledge to their behavioral advantage (Ruby et al. 1994). Female *G. agassizii* apparently can recognize Gila monsters (*Heloderma suspectum*) as a nest predator (Barrett and Humphrey 1986). Translocated animals apparently can reestablish burrow and neighbor recognition systems within a year (Field et al. 2007, Tuberville et al. 2005). Females can alter mate choices based on information gathered during an initial encounter (Niblick et al. 1994). All of these features document a remarkable ability of North American tortoises to learn as-

Table 12.1. A comparison of essential features of sociality in North American ground squirrels and tortoises

Social Feature	Ground Squirrels	Tortoises
Group size	80 (mean of populations in Fairbanks and Dobson 2007)	200–400 (range from McCoy and Mushinsky 2007; Styrsky et al. 2010)
Spatial structure of age/sex groups	Males and especially females establish home ranges near natal maternal den (Boellstorff and Owings 1995)	Tendency for females to avoid each other and for males to monitor females
Group cohesion	Certain individuals play strong role in cohesion of social network (Manno 2008); perhaps based on kin recognition (Sheppard and Yoshida 1971)	Based on centrality, certain males may play strong roles in cohesion of a social network
Degree and pattern of connectedness	Coteries; daily socialization	Cliques; daily visitations by males to adult females
Permeability to other groups	Groups not mobile	Groups not mobile
Compartmentalization	Dominant and submissive individuals in kin group	Dominant and submissive individuals likely to occur in each clique
Differentiation of roles	Dominant, subordinate, sentry	Dominant, subordinate
Integration of behaviors within groups	Dominants use behaviors that differ from subordinates'	Dominants use behaviors that differ from subordinates'
Information flow	Visual, vocal, chemical, and tactile cues; information identifies social status, group membership, approaching predators, and mating readiness	Visual, vocal, chemical, and tactile cues; information identifies social status, group membership, and readiness to mate
Fraction of time in social behavior	Large	Small

pects of their environment, especially the social environment, and to use this knowledge in ways that appear to be advantageous. For these reasons, we conclude that North American tortoises deserve recognition for displaying intelligence that is as advanced as many birds and mammals.

North American tortoises also have a remarkable level of social organization. For the ten essential features listed by Wilson (1975) for social animals, tortoises show comparable characteristics to those known for ground squirrels (table 12.1). In fact, the two societies appear to differ only in the presumed mechanism by which spatial structuring of age and sex groups is accomplished, the breadth of information communicated within each society, and the amount of time spent in social interactions. In ground squirrels, the tendency of offspring to establish home ranges near the natal burrow creates kin groups that socialize frequently and in different ways than they do with nonkin. In tortoises, social groups appear to be based on aversion of reproductive females to each other and the tendency of males to move frequently to monitor these widely separated females. In this way, the societies differ in the mechanism that creates social structure but not necessarily in the complexity of that social structure. In squirrels, vocal, visual, chemical, and tactile cues are used to communicate information within the social group and these cues indicate social status, group membership, willingness to mate, and approaching predators. Tortoises use the same set of cues and differ in types of messages only by lacking any known mechanism by which tortoises as a group ward off predators (but see Nichols 1953). Based on these observations, it appears to us that one major factor prevents wider appreciation of the degree of sociality present in North American tortoises. That factor is the difference in amount of time available for social behavior for endotherms versus ectotherms (Pough 1983). Ground squirrels are relatively unconstrained in the amount of time that they have available for social interactions, whereas tortoises are constrained by their need to spend significant time in thermoregulatory activities. Nevertheless, we present North American tortoises as an example of the remarkable degree of social organization that can be achieved given the behavioral constraints of ectothermy.

CONCLUSIONS

North American tortoises have many notable attributes, including complex social structure. This feature of these creatures has been revealed because of vastly improved mechanisms for following individuals. As remote video technology expands, we anticipate even more refined data that may clarify how social structure develops and is maintained within groups of tortoises. This expanded knowledge will be important to understanding the evolution of social structure within vertebrates. By expanding complex social structure to include the group Testudines, new light may be shed upon possible patterns of social structure in the earliest amniotes. Similarly, a refined understanding of tortoise social structure will be of obvious benefit to conservation of these animals. Models used to project disease transmission within tortoise populations will benefit from an improved understanding of the web of connections within tortoise social groups. Finally, an expanded understanding of social organization will likely improve methods used to retain tortoise populations on current conservation sites and repatriate populations to lands that have lost these special organisms.

13 Nesting and Reproductive Output among North American Tortoises

Roy C. Averill-Murray
Linda J. Allison
Lora L. Smith

Information on reproduction is highly variable among the five species of North American tortoises. Few quantitative data from wild populations exist for *Gopherus flavomarginatus* and none from Mexican populations of *G. morafkai,* while at least some data are available from more than 20 populations of *G. polyphemus.* The types of data from published studies vary among species as a result of differences in research methods. For example, reproduction in the desert tortoises *G. agassizii* and *G. morafkai* has been studied primarily by following radio-marked individuals throughout one or more reproductive seasons and by using x-radiography to determine clutch sizes and frequencies and egg sizes. Information on *G. polyphemus* comes more often from capture-recapture surveys and radiographs from the resultant samples or from systematic surveys of burrows for nests. In this chapter, we summarize information on nesting in wild populations and conduct a comparative analysis of reproductive output among the five species.

NESTING

Nesting Season

Both *Gopherus morafkai* and *G. polyphemus* produce a maximum of one clutch annually, but individuals of the remaining three species may produce two clutches per year, with occasionally three in *G. agassizii* and possibly *G. flavomarginatus* (see references in table 13.1). Each species typically nests during spring and summer. *Gopherus agassizii* begins nesting in mid-April, and females may lay subsequent clutches into July (Turner et al. 1986, Wallis et al. 1999, McLuckie and Fridell 2002, Baxter et al. 2008). The nesting season of *G. polyphemus* extends from April through June. Shelled eggs have been detected by radiograph in late April (Taylor 1982a, McLaughlin 1990). Peak nesting across most of the range occurs from mid-May to mid-June, however (Landers et al. 1980, Wright 1982, Diemer and Moore 1994, Butler and Hull 1996, Epperson and Heise 2003).

May and June also accounted for 80% of all nesting observations of *G. flavomarginatus,* but clutches may also be laid in April (Adest and Aguirre 1995). Nesting by *G. berlandieri* has been reported between April and August, but most observations have been in June and July (Auffenberg and Weaver 1969, Judd and McQueen 1980, Rose and Judd 1982, Judd and Rose 1989). Shelled eggs may be present as early as March and have been found in females as late as November (Auffenberg and Weaver 1969).

Gopherus morafkai appears to have the latest nesting season, beginning only by mid- to late June and occasionally extending through the end of August (Wirt and Holm 1997, Averill-Murray 2002), but primarily during June and July (Averill-Murray et al. 2002b). Shelled eggs are typically not detectable prior to early June, but they have been observed as early as mid- to late April (Averill-Murray, unpublished data), and individuals occasionally may retain eggs over winter (Averill-Murray 2002). Assuming incubation durations similar to those for the closely related *G. agassizii* (68–89 days; Spotila et al. 1994, Rostal et al. 2002), hatchlings from eggs laid in late summer probably overwinter in the nest, or they would otherwise emerge in November or December.

Nest-site Selection

Use of burrows for nesting varies among and within species. *Gopherus berlandieri* typically does not dig extensive burrows, and eggs also are often found on the ground surface or only partially covered, presumably due to limits of egg-chamber capacity or ground-surface hardness (Judd and Rose 1989). Most nests are constructed as shallow, covered scrapes in soil near the edge of bushes, where multiple nests may be clustered (Auffenberg and Weaver 1969). About half of *G. flavomarginatus* nests are associated with burrows, with nesting also occurring in the open or under shrubs or trees (Adest and Aguirre 1995).

Most *G. agassizii* nests are associated with burrows, although females may occasionally nest under shrubs not associated with burrows (Bartholomew 1993). Hampton (1981) found 15 nests at the entrance of burrows, 12 under creosote bushes (*Larrea tridentata*), and three in the open; one of the latter was on the bank of a wash. Only one of 26 nests found by Roberson et al. (1985) was not associated with a burrow, whereas most were inside burrows within 0.6 m of the burrow opening. Baxter et al. (2008) found that female tortoises laid their nests an average of 0.7 m into the tunnel of their burrows, irrespective of the total length of the burrow.

Nests of *G. morafkai* are almost exclusively associated with burrows, either under shrubs or boulders near the burrow entrance (Murray et al. 1996, Averill-Murray 2002). Reproductive females tend to remain at their nest-site burrows for several weeks subsequent to egg laying, while nonreproductive females are rarely found in the same burrow during consecutive weekly observations (Murray et al. 1996), a rare behavior in chelonians apparently directed toward defense against egg predators such as the Gila monster (*Heloderma suspectum;* Barrett and Humphrey 1986, Zylstra et al. 2005). Averill-Murray et al. (2002b) report similar defensive behavior of postreproductive females against human observers in which females actively block the entrance of their burrows to the intruder while limiting their own exposure.

Gopherus polyphemus females generally select nesting sites in open sunny areas, with little ground-cover vegetation, on the mound of sand at the entrance of a burrow (Landers et al. 1980, Wright 1982, Butler and Hull 1996). Like those for other tortoise species, however, direct observations of nesting are rare; most studies have focused on locating nests by searching on or around burrow mounds during nesting season. In southwestern Georgia, Landers et al. (1980) found that although most females nested within a meter of their burrow, 15% nested at alternative sites, including dirt roads, wildlife food plots, and a borrow pit as far as 134 m from the burrow. In Florida, nesting has also been observed along roads, in a clear cut, in a grassy field, and in firebreaks (Douglass and Winegarner 1977, Diemer and Moore 1994, Butler and Hull 1996). At a site in Alabama, Marshall (1987) found that nine of 11 nests were found away from the burrow mound (mean distance = 10.9 m), whereas all four nests located at a second site were within 0.72 m of the burrow mound. Evidence also exists that females may nest at recently unused burrows; Wright (1982) found two of 23 nests (9%) at old, abandoned burrows in South Carolina, and Smith (1995) found 3% of nests at old burrows at a north Florida site.

REPRODUCTIVE OUTPUT

Reproductive traits, including measures of egg size, clutch size, and annual clutch frequency, tend to be strongly correlated with female body size among turtle species (Iverson 1992a). Negative correlations between egg size and clutch size across genera and families, when corrected for body size, suggest an evolutionary trade-off in which females devote energy to large eggs in small clutches or to small eggs in large clutches (Elgar and Heaphy 1989, Iverson et al. 1993). Likewise, an inverse relationship between size-adjusted clutch frequency and clutch mass suggests a trade-off whereby turtles annually produce either several small clutches or few large clutches (Iverson 1992a). Age at maturity is negatively correlated with clutch frequency and annual reproductive output (annual clutch mass); early-maturing turtles produce more clutches and greater total clutch mass per year than later maturing turtles (Iverson 1992a). Latitude also correlates with various reproductive traits across turtle species (positive: clutch size; negative: clutch frequency) (Iverson 1992a, Iverson et al. 1993).

Increasing latitude is interpreted to reflect increasing seasonality and predictability of conditions as a function of season. Correlations of latitude in these relationships therefore point to a "bet-hedging" model of reproductive output where species at higher latitudes and shorter reproductive seasons lay fewer clutches per year, but offset this by producing more eggs per clutch compared to species at lower latitudes (Iverson 1992a). Large clutch sizes are concomitantly offset by smaller egg sizes (Iverson et al. 1993). Turtle species in highly variable and less predictable environments, irrespective of latitude, should demonstrate correspondingly variable reproductive output or more frequent reproductive bouts in order to increase the chance that at least some recruitment coincides with unpredictable resources (Iverson 1992a).

Members of the genus *Gopherus* occupy a relatively narrow latitudinal range (approximately 8°; Bury 1982, Iverson et al. 1993), but they vary dramatically in body size (225% difference between average carapace length in adult females of *Gopherus berlandieri* and *G. flavomarginatus*; Germano 1994). While these tortoise species live at similar latitudes, their environments differ in predictability. Therefore, comparison of the North American tortoises provides an opportunity to investigate correlates of, and potential trade-offs between, reproductive traits among closely related species. Based on the interspecific analyses cited above, we expect most reproductive traits to be positively correlated with female size, but expect clutch frequency to be negatively correlated with body size. Larger tortoise species should produce larger, less frequent clutches each year than smaller species. We expect smaller clutches to be comprised of larger eggs. Finally, we expect species that produce more clutches per year to live in less predictable environments and begin reproducing at earlier ages than species that produce few clutches per year.

Data Compilation

We compiled average values of reproductive traits from tortoise populations from the literature and unpublished data (table 13.1). These studies were not characterized by a consistent set of measured variables, and the same general characteristic could be measured in more than one way. Nonetheless, in the interest of having more populations and years represented, we took the following steps to create comparable measures between reported studies. If anything, this approach should increase sampling error and obscure patterns, so reported differences emerged from potentially high levels of background noise. We did not conduct comparisons with age

Table 13.1 Reproductive traits for North American tortoise populations

State	MeanS	MinS	CS	EW	EV	CV	CF	AV	AF	%R	Source
Gopherus berlandieri											AP = 674.6, P = 0.293, C = 0.131, M = 0.162
TX	170	147	2.7	34.1	25.3	65.7	0.64	41.7	1.7	0.63	Judd and Rose 1989
TX	144	131	2.1				1.34		2.8	0.55	Hellgren et al. 2000
TX	160	155	4.3								Rose and Judd 1982
Mean	**158**	**144**	**2.6**	**34.1**	**25.3**	**53.4**	**0.99**	**41.7**	**2.2**	**0.59**	
Gopherus agassizii											AP = 136.2, P = 0.391, C = 0.272, M = 0.119
CA	235	203	4.1	37.5	33.2	137.7	1.72	235.8	7.0	0.83	Wallis et al. 1999
CA	220	176	2.7	36.1	30.5	82.4					Griffith 1991
CA	212	189	4.4	34.2	24.8	106.0	1.72	178.5	7.6	0.96	Turner et al. 1986; Wallis et al. 1999
CA	215	188	3.9				1.72		6.7	0.90	Karl 1998
NV	247	209	4.7				1.50		7.1		Mueller et al. 1998
UT	225	192	5.2	37.2	32.1	169.9	0.86	143.1	4.5	0.67	McLuckie and Fridell 2002
Mean	**226**	**193**	**3.6**	**36.2**	**30.2**	**123.2**	**1.50**	**185.8**	**6.6**	**0.84**	
Gopherus morafkai											AP = 249.4, P = 0.408, C = 0.204, M = 0.204
AZ	256	220	5.6				1.00		5.6	1.00	Wirt and Holm 1997
AZ	236	226	3.3	32.4	22.2	73.3					Averill-Murray 2002
AZ	242	237	5.0				0.33		1.7	0.33	Wirt and Holm 1997
AZ	240	225	5.8	34.5	24.8	145.8	0.51	82.9	3.2	0.51	Stitt 2004; Averill-Murray unpublished
AZ	253	232	5.1	36.1	30.8	156.7	0.62	100.3	3.2	0.62	Averill-Murray 2002 unpublished
Mean	**245**	**228**	**5.0**	**34.5**	**26.0**	**125.3**	**0.62**	**91.6**	**2.7**	**0.62**	
Gopherus polyphemus											AP = 1373.3, P = 0.424, C = 0.275, M = 0.149
FL	313	300	11.1	41.6	37.7	418.4	0.64	266.3	7.1	0.64	Moore et al. 2009
FL	274	215	6.3								Ashton et al. 2007
FL	242	227	5.8	41.0	36.1	209.8	0.73	152.5	4.2	0.73	Small and Macdonald 2001
FL	270	246	6.0	42.3	39.6	233.1	0.85	195.2	5.0	0.85	Diemer and Moore 1994; Berish unpublished
FL	305	284	7.8								Ashton et al. 2007
FL	255	225	5.7								Smith 1995
FL	261	235	5.6	41.7	38.0	232.1	0.88	203.1	5.0	0.88	Diemer and Moore 1994; Berish unpubl. data
FL	265	187	8.0	41.0	36.1	288.7	0.46	133.7	3.7	0.46	Small and Macdonald 2001
FL	261	245	6.5	41.7	38.1	247.5	1.00	180.7	6.5	1.00	Diemer and Moore 1994; Berish unpublished
GA	290	261	4.5								Rostal and Jones 2002
GA	306	261	6.5								Rostal and Jones 2002
GA	290		7.0	44.8	47.1	329.6					Landers et al. 1980
LA	275	257	5.5				0.70		3.9	0.70	Smith et al. 1997
MS	278	250	5.6				0.55		3.1	0.55	Smith et al. 1997
SC	221	153	3.8	43.3	42.5	161.5	0.70	113.1	2.7	0.70	Wright 1982
Mean	**274**	**239**	**6.4**	**42.2**	**39.4**	**265.1**	**0.72**	**177.8**	**4.6**	**0.72**	
Gopherus flavomarginatus											AP = 328.4, P = 0.380, C = 0.153, M = 0.227
Durango	**348**	**315**	**5.4**	**44.1**	**48.5**	**261.9**	**0.61**	**159.8**	**3.3**	**0.44**	Adest et al. 1989; Adest and Aguirre 1995; Aguirre et al. 1997; Aguirre unpublished

Note: Values are averages across all respective years of study. Species means are weighted by number of years of study for each population. MeanS = mean size of reproducing females (mm), MinS = minimum size of reproducing females (mm), CS = clutch size, EW = egg width (mm), EV = egg volume (cm^3), CV = clutch volume (cm^3), CF = annual clutch frequency, AV = annual clutch volume (CV × CF; cm^3), AF = annual fecundity (CS × CF), %R = proportion adult females reproducing, AP = annual precipitation (mm), P = predictability, C = constancy, M = contingency (Germano 1993).

at maturity due to even greater variation in estimating this trait among studies than for other traits.

We included only populations for which the mean size (MeanS) of reproducing females was available. Estimates of minimum reproductive size (MinS) are a function of the range of female sizes included in a particular study and may be biased high; this bias will be more pronounced in cases where researchers studied larger adult females in the population. We estimated MeanS for *G. flavomarginatus* as the midpoint of the range reported by Adest and Aguirre (1995). We estimated MeanS within the *G. berlandieri* population studied by Hellgren et al. (2000) as that from all adults greater than 5 years of age in their table 4.

Measures of individual reproductive output include mean clutch size, mean egg width, and mean clutch volume. Mean clutch size (CS; number of eggs in a clutch) is the most consistently reported trait across studies. Egg width (EW) is more commonly reported in the literature than egg mass, so we calculated mean egg volume (if not reported directly) for each population and multiplied it by CS to derive clutch volume (CV) as an index of reproductive investment per clutch. For elongate eggs, EW becomes an increasingly poor measure of egg size relative to reproductive investment (Hailey and Loumbourdis 1988), so for most species we calculated egg volume (EV) with the equation $EV = \pi(EW^2)(L)/6$, where L is egg length (Coleman 1991). If egg length was not reported in the original study, we estimated it from the average ratio of egg length to width from available data for that species. For *G. polyphemus,* we calculated EV as a sphere (Ernst and Lovich 2009) for which $EV = 1/6\pi(EW)^3$. Diemer and Moore (1994) reported regression equations for mean egg diameter for three populations relative to population-specific body sizes, which we used to estimate unreported egg widths in the equation above using mean carapace lengths from the relevant populations.

Measures of reproductive output across adults in each population include annual clutch frequency, annual clutch volume, annual fecundity, and proportion of females reproducing. Annual clutch frequency (CF) is the number of clutches produced in a year divided by all adult females. Estimates of the proportion of adult females reproducing each year (%R = CF for species producing a single clutch) may be biased low in studies that relied upon capture-recapture instead of serial monitoring of radio-marked individuals, and we do not include such estimates if based on single capture periods. Proportion of *G. berlandieri* females reproducing from Hellgren et al. (2000) is taken from the maximum seasonal proportion gravid shown in their figure 13.3. We estimated annual clutch volume (AV) for the population as CV × CF and annual fecundity (AF) as CS × CF.

Data Analysis

We $\log_e$-transformed CS, EV, CF, and AV upon examination of residuals to improve assumptions of homogeneous variances and normality; the remaining variables did not require transformation. We investigated differences among species with general linear models, using MeanS (midline carapace length, mm) as a covariate. Species and MeanS were highly correlated ($r^2 = 0.79$, $p < 0.001$), and interactions between these main effects were nonsignificant in every case, except for the dependent variable EV. We excluded EV from further analysis based on this interaction and the information it shares with EW (size constraints) and CV (volumetric investment).

The strong correlation between the predictor variables MeanS and Species prevents straightforward interpretation of statistical significance in the main-effects models. If MeanS had a strong correlation with a reproductive trait, Species usually had only a slightly weaker correlation; these statistical differences should not be interpreted to mean that the biological importance of trait differences between species is only a function of mean reproductive size. We therefore report both single-factor correlation coefficients between each reproductive trait and MeanS and Species. To consider whether these correlations are describing the same association, we also account for the effect of MeanS before testing for the residual correlation of Species with each reproductive trait, and we account for the effect of Species before examining the residual correlation with MeanS. Where p-values indicated differences in a reproductive trait between Species, we performed pairwise comparisons to specifically characterize how the species differed. The linear models used for these tests included MeanS if this effect was indicated at $\alpha < 0.05$. On the same basis, interaction terms between the two predictors were not included. Because of limited sample sizes and to highlight any possible differences, we made no adjustment for multiple comparisons.

Following exploration of relationships between reproductive traits and size or species, we investigated potential trade-offs between CS:EW, CF:CS, and CF:CV. We removed the effects of size and species by using residuals from the full (two predictor) main-effects models for each reproductive trait. The resulting bivariate correlations between traits are partial correlations.

We graphically explored average values of traits for each species against average annual precipitation and average predictability measures for precipitation in the range of each species (Colwell 1974). Mean predictability (P) of rainfall reflects mean constancy (C; evenness of rainfall among months) and mean contingency (M; seasonality of rainfall within a year), all three of which were calculated for each species by Germano (1993) (table 13.1). Maximum uncertainty regarding precipitation would involve rainfall amounts that are independent of season and that fluctuate to the greatest degree possible during the course of the year. Rainfall is strongly correlated with primary production, although asymptotically (Pianka 1988), and represents a meaningful proxy for resources important for reproduction. Given variability across the ranges of each species, statistical comparisons among species would ideally make use of population-specific data, but such an investigation was beyond the scope of this preliminary analysis. Correlations with only five data points must be quite strong to be statistically significant at $\alpha = 0.05$, so to consider patterns to indicate paths for further investigation we highlight any relationships on the basis of an $r^2 > 0.66$ rather than on the basis of statistical significance. We note how these patterns may inform particular bet-hedging options exhibited by each species and may warrant further investigation as explanatory factors of tortoise reproduction and life history.

Relationships with Body Size

In contrast to the results of Iverson (1992a) for all chelonians, body size was not a reliable predictor of reproductive traits among North American tortoises (table 13.2). MeanS explained substantially more variation than Species only for MinS. MeanS explained similar amounts of variation to Species for CS and CV, where larger individuals (and species) tend to produce larger individual clutches, but relatively smaller volumes, than smaller ones. Even for those traits in which MeanS explained some degree of variation in a single-effect model, partial correlation analysis revealed that, compared to Species, MeanS explained additional variation only for MinS.

Even though physical constraints of a female tortoise's shell limit the maximum number of eggs she can carry (Hailey and Loumbourdis 1988, Wallis et al. 1999), female size usually

Table 13.2. Relationships of body size and species as predictors of reproductive traits within North American tortoises

Trait		MeanS		Species		
		Main	Partial	Main	Partial	Description of significant results, including post-hoc tests
MinS	Minimum size	0.86	0.12	0.65	ns	Larger species reproduce at larger MinS
CS	(ln) Clutch size	0.57	ns	0.54	ns	Larger species produce larger CS
EW	Egg width	0.46	ns	0.90	0.71	Larger EW in *Gopherus polyphemus* than *G. agassizii*, *G. berlandieri*, *G. morafkai*; *G. morafkai* EW is smaller than that of *G. flavomarginatus*
CV	Mean clutch volume	0.65	ns	0.64	0.75	*Gopherus flavomarginatus* produces smaller CV than other species; *G. morafkai* produces smaller CV than remaining species; *G. berlandieri* produces larger CV than *G. agassizii*
CF	(ln) Clutch frequency	ns	ns	0.59	0.53	*Gopherus agassizii* greater CF than *G. flavomarginatus*, *G. morafkai*, *G. polyphemus*
AF	Annual fecundity	ns	ns	0.54	0.57	Greater AF in *Gopherus agassizii* than other species
AV	(ln) Annual clutch volume	0.33	ns	0.79	0.87	*Gopherus agassizii* produces larger AV than other species; *G. polyphemus* produces larger AV than *G. flavomarginatus*, *G. morafkai*
%R	% reproducing	ns	—	ns	—	Highly variable
		ns	ns	0.49	0.49	Removal of anomalous population of *Gopherus morafkai*: higher %R in *G. agassizii* than *G. flavomarginatus*, *G. morafkai*; higher %R in *G. polyphemus* than *G. morafkai*

Note: Coefficients of determination (r^2) are listed for single-factor linear models (Main) and partial correlation analyses ($p < 0.05$; ns = not significant). MeanS = Mean reproductive carapace length. Descriptions summarize pairwise comparisons (uncorrected $p < 0.05$) among species from main-effect (two-factor) models or partial correlations to control for body size.

explains very little variation in clutch size within tortoise populations, even though the statistical test may carry a small *p*-value (e.g., Landers et al. 1980, Turner et al. 1986, Mueller et al. 1998, Averill-Murray 2002, Rostal and Jones 2002, Ashton et al. 2007). The size of the first clutch, but not the second, was correlated with female body size in one population of *G. agassizii* (Wallis et al. 1999), but the reverse pattern was observed in another (Karl 1998). Clutch size varied nonlinearly with body size in one population of *G. polyphemus;* intermediate-sized adult females produced the largest clutches, suggesting that larger (older) females in this population become senescent, although additional research is necessary to confirm the hypothesis (Ashton et al. 2007).

Other studies of individual populations have found significant positive correlations between female body size and EW (Wallis et al. 1999, Averill-Murray 2002), CV (Wallis et al. 1999), CF (Turner et al. 1986, Karl 1998, Wallis et al. 1999), and AF (Karl 1998, Mueller et al. 1998, Wallis et al. 1999). In particular, Wallis et al. (1999) found that smaller *G. agasizii* females laid fewer clutches and laid them later in the spring, presumably because their smaller body reserves required them to derive more nutrients directly from spring forage to support reproduction. Similarly, Averill-Murray (2002) found that smaller *G. morafkai* females were more likely to forego reproduction following drier winters.

Average MinS among species was not correlated with mean annual precipitation, a result similar to that reported by Germano (1994) for upper size limits of North American tortoises. The exceptionally large MinS for *G. flavomarginatus* prevented a positive correlation with rainfall predictability (fig. 13.1). *Gopherus berlandieri,* in particular, lives in an environment with the most variability between months in a year, but this variability is only weakly seasonal. This describes a much less predictable environment than those of the larger species, and *G. berlandieri* attains the smallest sizes overall and

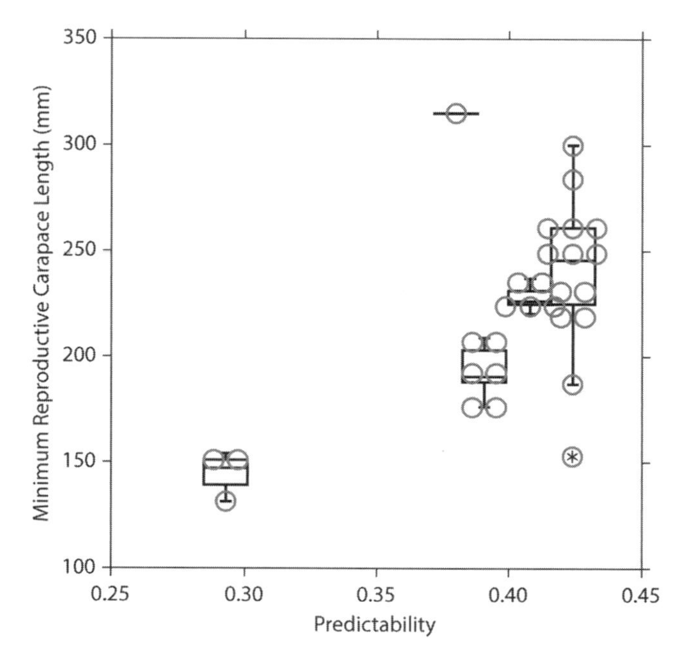

Fig. 13.1. Minimum reproductive size (midline carapace length) in relation to rainfall predictability. From left to right, species are *Gopherus berlandieri*, *G. flavomarginatus*, *G. agassizii*, *G. morafkai*, and *G. polyphemus*.

at maturity (fig. 13.1), with age at maturity occurring as young as five years (Hellgren et al. 2000).

Differences among Species

By itself, Species explained >50% of variation in all traits except %R (table 13.2). The strong correlation between Species and MeanS was most evident in MinS and CS, in which Species explained no additional variation to MeanS. *Gopherus agassizii* produced more eggs in total each year (AF) than the other

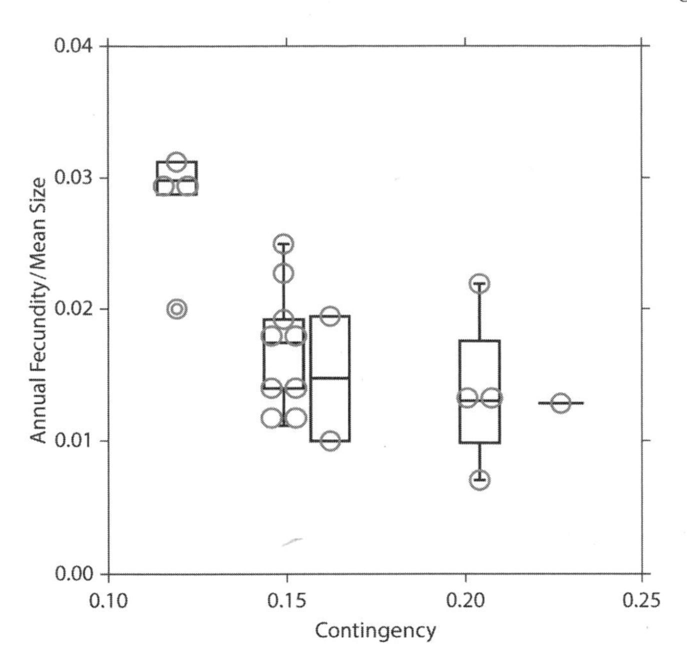

Fig. 13.2. Relative annual fecundity in relation to rainfall contingency (seasonality). From left to right, species are *Gopherus agassizii*, *G. polyphemus*, *G. berlandieri*, *G. morafkai*, and *G. flavomarginatus*.

species as a result of greater CF (tables 13.1 and 13.2), and this greater AF corresponded to low seasonality (contingency) compared to the other species (fig. 13.2). *Gopherus agassizii* also experiences the lowest overall annual precipitation compared to the other species (table 13.1).

Average CS within *G. polyphemus* populations appears to be relatively constant between years (Diemer and Moore 1994, RB Smith et al. 1997, Epperson and Heise 2003), but it increases with productivity (evapotranspiration) across populations (Ashton et al. 2007; see also Godley 1989 and Macdonald 1996 for examples of translocated tortoises on high-productivity, reclaimed phosphate mines). Size of second clutches was larger with higher rainfall between two populations of *G. agassizii*, and AF was correlated with annual plant biomass up to an asymptotic level above which other constraints (e.g., body size, nutrient reserves) apparently limited egg production (Wallis et al. 1999). Mean annual CS in a population of *G. morafkai* showed a strong positive correlation with spring rainfall, but the relationship was not statistically significant across four years of study (Averill-Murray 2002). Body size-adjusted CS may be negatively correlated with seasonality, as estimated by the coefficient of variation of mean monthly actual evapotranspiration among *G. polyphemus* populations (Ashton et al. 2007).

Gopherus polyphemus produced larger (wider) eggs for its body size than the other species, except *G. flavomarginatus* (table 13.2). In contrast to a previous observation by Averill-Murray (2002), our full analysis did not indicate that *G. agassizii* eggs were significantly larger on average than *G. morafkai* eggs. If we consider these two species in isolation, the pairwise comparison did indicate greater EW in *G. agassizii*, however ($p = 0.005$). Mean relative EW was not correlated with measures of rainfall predictability for each species.

Reproductive effort as measured by CV and AV appeared to have a slight tendency to increase with body size, but this effect was not present when the effect of Species was partitioned out (table 13.2). *Gopherus flavomarginatus* produced relatively smaller clutch volumes than all species, and *G. morafkai* produced smaller volumes than the remaining three species. The smallest species, *G. berlandieri*, produced similar relative clutch volumes to *G. polyphemus*, and exceeded those of the remaining species. *Gopherus agssizii* produced greater AV than the other species, and *G. polyphemus* produced greater AV than the remaining species, except *G. berlandieri*. For *G. agassizii*, this large volume resulted mostly from the consistently higher CF; for *G. polyphemus*, the effect resulted from larger eggs (EW) and more eggs per clutch (CS). Relative AV increased with increasing constancy ($r^2 = 0.76$), suggesting that species in more "even" rainfall environments, like those of *G. agassizii* and *G. polyphemus*, invest relatively more in reproduction for their body sizes than those in less even environments.

Neither MeanS nor Species significantly influenced %R (table 13.2). Data from a population of *G. morafkai* in which every individual reproduced during a one-year study (Wirt and Holm 1997) was not typical of other reports for this species, however (studentized residual = 2.35); so, we also considered the results without this data point. There is no evidence that this study provided invalid data, but with a small sample size ($n = 5$ for *G. morafkai*), such an observation can obscure more general patterns. In this case, data from all other studies indicated a significant Species effect on %R. Greater proportions of *G. agassizii* and *G. polyphemus* reproduced each year than the other species. Smaller proportions of females reproduced each year in *G. morafkai* and *G. flavomarginatus*, which are subjected to more strongly seasonal rainfall patterns (i.e., greater contingency; $r^2 = 0.84$; fig. 13.3). Greater proportions of *G. agassizii*, overall, reproduced each year in the least seasonal environment occupied by North American tortoises (table 13.1). The relationship between each species' average CF and contingency was less strong ($r^2 = 0.68$).

While *G. morafkai* and *G. polyphemus* females may produce either one or no clutches, *G. agassizii* females may reduce the number of clutches produced in a year while still remaining reproductively active. Winter rainfall and subsequent spring annual-plant production can positively influence CF in *G. agassizii* (Turner et al. 1986) and *G. morafkai* (Averill-Murray 2002). In *G. morafkai*, which lays its clutches later than *G. agassizii*, relatively small incremental increases in spring rainfall may also result in a large increase in the number of females reproducing (Averill-Murray 2002). Conversely, while *G. agassizii* females may produce fewer clutches under less favorable conditions, *G. morafkai* females are more likely to forego reproduction completely. In an exceptional low-frequency case for *G. agassizii*, CF and %R was smaller than expected given the preceding winter rainfall, potentially as a result of other, unknown factors contributing to population declines in the area (table 13.1; McLuckie and Fridell 2002).

Trade-offs between Reproductive Traits

Clutch frequency was not significantly correlated with CS among species ($r = -0.249$, $p = 0.828$), but was negatively correlated with CV ($r = -0.648$, Bonferroni $p = 0.036$). The latter relationship suggests a trade-off in which tortoise species that reproduce more frequently produce smaller-volume clutches, and species that reproduce less frequently produce larger-volume clutches, consistent with the general chelonian trend reported by Iverson (1992a). Clutch size was not correlated with EW (fig. 13.4; $r = -0.164$, $p = 1.000$), indicating that a trade-off does not exist across North American tortoises between investing in individual offspring (EW) versus investing in number of offspring in a given clutch. Correlations with CS were significant across populations of *G. polyphemus*, however (fig. 13.4; $r = -0.821$, $p = 0.012$). Within a *G. morafkai* population, CS also was not correlated with EW (Averill-Murray 2002), but body size-adjusted egg length and volume were strongly negatively correlated with CS in *G. agassizii* (EW was weakly correlated; Wallis et al. 1999).

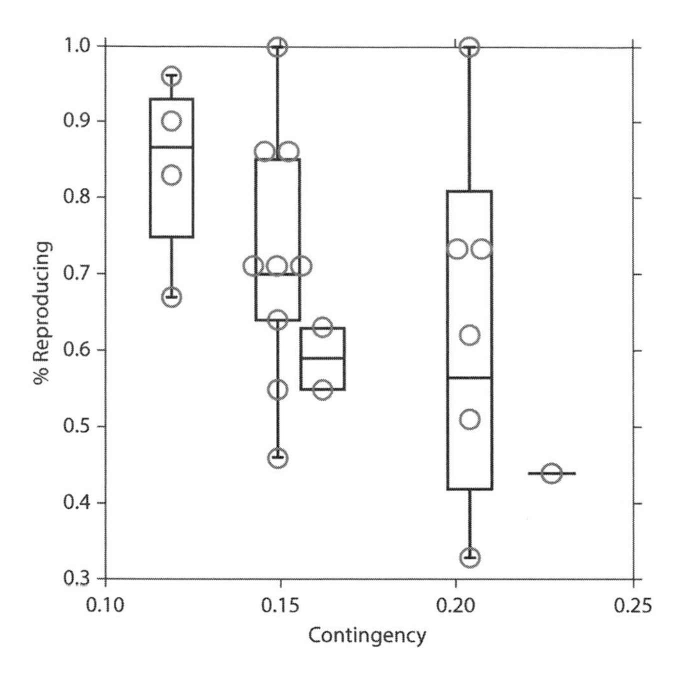

Fig. 13.3. Percentage females reproducing in relation to rainfall contingency (seasonality). From left to right, species are *Gopherus agassizii*, *G. polyphemus*, *G. berlandieri*, *G. morafkai*, and *G. flavomarginatus*.

CONCLUSIONS

North American tortoises occupy a narrow range of temperate latitudes. Nesting in all species typically occurs during spring and summer; however, the environments occupied by each species are quite variable, and nest-site selection also varies among species with respect to proximity to burrows or more open areas. One species (*Gopherus morafkai*) nests in strong association with burrows, potentially facilitating its rare chelonian behavior of active nest defense.

Variation in reproductive traits among North American tortoises differs in comparison with patterns previously documented among turtles. Body size among *Gopherus* species explains relatively little variation in reproductive traits, and there is no evidence of a trade-off between clutch size and egg size across species. A trade-off is apparent between having the ability to produce multiple annual clutches and investing in individual clutches (clutch volume), however. Previous studies have reported intraspecific patterns that deviate from those we report among species. For example, an apparent trade-off does exist between clutch size and egg size across the relatively wide latitudinal range of *G. polyphemus*, and clutch size was negatively correlated with egg size in populations of *G. agassizii*.

Even though our analysis of environmental factors was

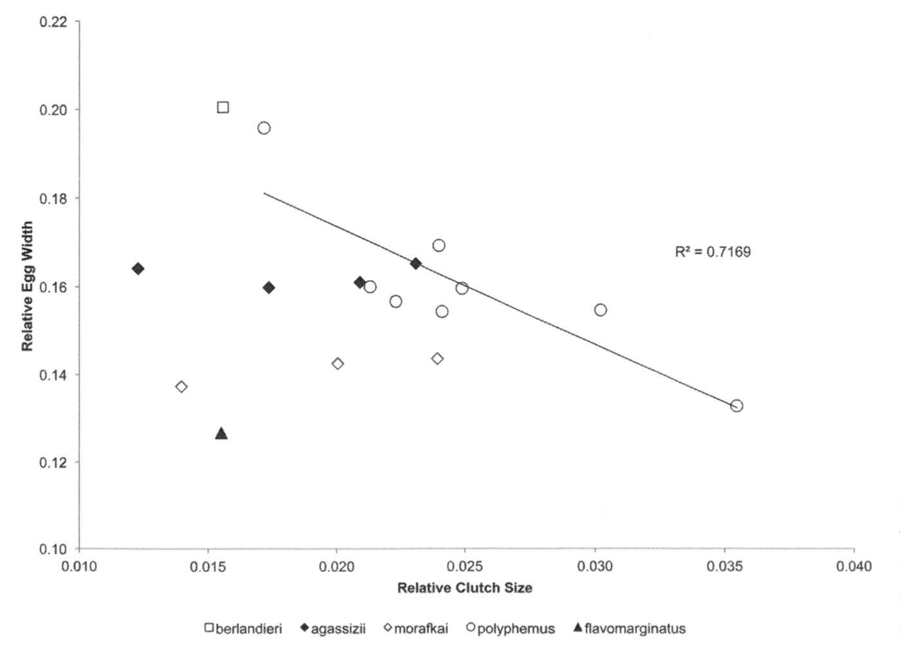

Fig. 13.4. Relative egg width (mean egg width / mean reproductive size) in relation to relative clutch size (mean clutch size / mean reproductive size). Regression is shown for *Gopherus polyphemus*.

coarse-grained, correlates of environmental variability support some predictions of a bet-hedging model of reproductive output across *Gopherus* species. As expected, even within a narrow latitudinal range encompassing all the species, those in more predictable environments tend to reproduce at larger sizes but less often than those in less predictable environments. *Gopherus berlandieri* occurs in a much less predictable environment than the other species and stands out as the dramatically smallest, and apparently youngest maturing, species. It appears to begin diverting energy toward reproduction as soon as it reaches the minimum size necessary to produce viable eggs/hatchlings. Notably, *G. berlandieri* possesses kinetic attachments between elements of the carapace and plastron, allowing passage of relatively large eggs (Rose and Judd 1991). Our analysis indicates that mean egg width in *G. berlandieri* is comparable to that in other *Gopherus* species, but it produces relatively large clutch volumes for its size. Also contrary to expectations, *G. berlandieri* produces a similar or smaller number of clutches of similar or smaller volume than the other later maturing species.

The two species that produce a maximum of one clutch per year, *G. morafkai* and *G. polyphemus,* live in the most predictable environments. *Gopherus polyphemus* experiences much greater overall annual precipitation than the other species, and the high predictability associated with its range derives from high constancy between months. The most fecund species, *G. agassizii,* occurs in the least seasonal environment, and lower seasonality of rainfall in particular is probably associated with greater proportions of females within a population reproducing annually and producing more eggs overall. The very low absolute amount of annual rainfall experienced by *G. agassizii* probably contributes to the production of multiple clutches in a year, even though less frequent reproduction would be expected based on the moderate predictability of rainfall the species does receive.

Habitat of *G. morafkai* receives strongly seasonal summer precipitation, and this species reaches maturity at a larger average size and reproduces less often than the closely-related *G. agassizii,* with individuals often foregoing reproduction altogether, especially in dry years. Similarly, the annual proportion of reproductive *G. flavomarginatus* is relatively low in its highly seasonal environment. Contrary to bet-hedging trade-offs, however, *G. agassizii* produces both a larger size and number of eggs compared to *G. morafkai.* Investment by *G. morafkai* of its entire annual reproductive output in a single clutch of relatively small eggs suggests that a more productive posthatching environment during the typical monsoon season may contribute to higher average juvenile survival than for *G. agassizii* (Averill-Murray 2002).

Tortoise species also apply the bet-hedging strategy differently under similar environmental patterns. For example, the ranges of both *G. agassizii* and *G. polyphemus* are characterized by relatively even annual rainfall patterns, and populations of those species produce a greater volume of eggs per year on average, relative to their body sizes, than the other species. *Gopherus agassizii* accomplishes this by producing several relatively small clutches per year, while *G. polyphemus* females produce one large clutch each year, if any. This difference is the result of the large absolute difference in rainfall and forage resources within the habitat of each species. Adult body condition and water balance fluctuates with rainfall much more in *G. agassizii* than in *G. polyphemus* (McCoy et al. 2011), reflecting a strong physiological connection to infrequent rain events that is even greater for hatchlings and juveniles (see Wilson et al. 2001). Multiple clutches by *G. agassizii* increase the chance that at least some offspring emerge during infrequent periods of water or forage availability. In contrast, *G. polyphemus* hatchlings from only a single clutch are more certain to emerge in favorable conditions.

The caveats noted regarding the compilation of reproductive-output data for our comparative analysis may have produced spurious patterns. More likely, limited data for *G. berlandieri* and *G. flavomarginatus* have obscured biological patterns with statistically nonsignificant results. Additional reproduction and environmental data from these two species and longer-term data within populations of all species will further elucidate factors that influence reproductive output within and among the North American tortoises. This point is also relevant to southern populations of *G. morafkai* from Sinaloa, Mexico (which may be a different species; Murphy et al. 2011). Sinaloan tortoises experience the greatest rainfall predictability ($P = 0.479$) and seasonality ($M = 0.253$) compared to all other North American tortoises (Germano 1993). From these observations and the results of our current analysis, we predict that Sinaloan tortoises begin reproducing at larger sizes, produce smaller eggs, and reproduce less often relative to their congeners.

Acknowledgements

We gratefully acknowledge J. Berish and G. Aguirre for providing unpublished data and two reviewers for their comments on a previous draft of this chapter.

The findings and conclusions in this article are those of the authors and do not necessarily represent the views of the U.S. Fish and Wildlife Service.

14 Abundance of North American Tortoises

LINDA J. ALLISON
EARL D. MCCOY

A species' commonness in an area can be described in several ways. *Abundance* is one quantity of interest, as is the *density* of individuals, which is abundance per unit area, and is useful as a currency for comparing areas, habitats, or even species. When individuals are dispersed over large areas, reporting the amount of available habitat that is occupied may be more practical. All of these measures have been employed to describe population size and status of tortoises.

Based on understanding of regional and/or local patterns in abundance, all but one species of North American tortoise are under some form of conservation management (chapter 18). Paradoxically, our understanding of the true abundance for any of these species is approximate and unsatisfactory, in part because sampling effort often is uneven across the range of a given species. Information on abundance of tortoises in Mexico is especially sparse.

Gauging true abundance through enumeration also is complicated because all of the species are variably active, both diurnally and seasonally, and cryptic in their landscapes. Individuals of *Gopherus berlandieri,* for example, are concealed by thornscrub vegetation over much of the species' range. Use of burrows for large portions of the day and year conceals individuals of the other North American tortoises from observation; therefore, it is necessary to account for tortoises that are present but not observed in order to accurately estimate abundance.

No one method of estimating abundance has proven effective for all species. Most surveys use grids and transects to describe whether tortoises are present and to generate a starting number of individuals, however. Various techniques have been applied across all of the species to address issues of detectability of individuals by accounting for tortoises that are present but unobserved. Among the methods used to convert simple counts of individuals into reliable estimates of population abundance are radio telemetry to assess availability for detection, distance sampling to estimate the proportion of available (here, equivalent to "visible") individuals that are not seen because they are cryptic, and mark-recapture to account for both undetectable and detectable-but-cryptic individuals. Each abundance estimate we encounter must be weighed against the limitations of the approach that was used. These "limitations" are the conditions or assumptions that must be met for the particular estimation method to perform at its best. The extent to which tortoise life history and study design meet those assumptions informs our interpretation of the abundance, density, and distribution estimates that are reported.

USING SIGNS AS AN INDEX OF TORTOISE ABUNDANCE

Sufficient individuals for direct enumeration cannot always be found reliably, and signs (surrogates, such as burrows, shells, and scat) are used as an *index* to abundance or are multiplied by a *correction factor* to calculate the number of individuals. *Gopherus flavomarginatus* lives in a setting in which almost constant shelter is required. Contemporary established populations have only been identified in the Bolsón de Mapimi, a 40,000 km^2 internal basin in central Mexico. Individuals may emerge from burrows only during a three-month wet season, when the Bolsón is less traversable by humans (Bury et al. 1988). For this species, sign evidence of recently active burrows provided the earliest estimates of total abundance. Based on locating seven active burrows on a one ha plot, Morafka (1982) also used this as the best density of individuals in a handful of identified upland areas. This density is an order of magnitude higher than known for local plots for *G. agassizii* and *G. morafkai.* Morafka (1982) also reported fewer than a single burrow per hectare in the basin floors themselves. Aguirre et al. (1984) and Aguirre (personal communication) concur with the description, estimating that most individuals occur at densities of about 10/km^2 or lower.

Observation of burrows has also been the primary method

of estimating abundance of *G. polyphemus* populations (Auffenberg and Franz 1982). Individuals inhabit burrows for extensive periods, and the depth of most burrows typically makes individuals undetectable from the surface. Therefore, although burrows can be readily observed when they are present, tortoises cannot. One cannot assume all burrows are occupied, however, or that each individual uses a single burrow, or that every burrow holds no more than one individual. Use of burrow counts as a relative indicator of tortoise abundance, or to convert burrow counts into an estimate of tortoise abundance, assumes that there is a predictable correlation between the number of burrows and the number of tortoises. This is the definition of an index. From intensive surveys at one site, Auffenberg and Franz (1982) proposed that 61.4% of active burrows at any site could be assumed to be inhabited. A later study using the same logic from seven sites proposed that 50% of burrows are inhabited (Ashton and Ashton 2008). Clearly, estimates will vary based upon the experience of observers, weather conditions, and other factors, but is it reasonable to assume these sources of variability are small so that the use of indices and correction factors is reasonable? In fact, actual burrow use is known to vary widely among sites (Burke 1989a, Breininger et al. 1991, McCoy and Mushinsky 1992a, Smith et al. 2005). McCoy and Mushinsky (1992a), for example, demonstrated that within a group of 26 sites, a more accurate regression relationship existed between log-transformed counts of individuals and of active burrows than the multiplicative relationship proposed by Auffenberg and Franz (1982), which consistently overestimated the actual number of individuals for the group of sites. The more reliable conversion factor of McCoy and Mushinsky (1992a) will also provide fairly imprecise untransformed abundance estimates in large populations, so that specific site estimates may vary widely from the true abundance. The authors suggested that different burrow use patterns among sites may reflect factors such as available forage and local abundance. These exceptions, and others that arise in time and space, are at odds with the assumptions of an index.

Because individuals of *G. agassizii* and *G. morafkai* can reliably be found on the surface outside of burrows in spring, live sightings were used to gauge abundance beginning with the earliest, localized surveys (Woodbury and Hardy 1948). Sign surveys for the *G. agassizii* became popular as regional questions about abundance became pressing, however, and timing and scale of these surveys could no longer be optimized to coincide with primary tortoise activity seasons. Luckenbach (1976, 1982) surveyed a nonrandomly selected set of 137 transects in California at times when individuals were not active by making use of signs to record presence at a site. Another early effort to describe distribution and abundance in California (Berry and Nicholson 1984) used sign counts on large numbers of strip transects, 1.5 mi long and walked in a triangle, to develop a map of relative abundance classes for tortoises.

The use of signs to estimate absolute abundances of *G. agassizii* and *G. morafkai* is problematic. Luckenbach (1982) applied correction factors of one or two burrows per individual, based on data from the western Mojave Desert, whereas Woodbury and Hardy (1948), working in Utah, reported that a single winter den was used by as many as 17 individuals. In the Sonoran, Mojave, and Colorado Deserts, tortoises make consistent use of preformed shelters in rock (Zylstra and Steidl 2009), where the number of shelters and of possible occupants per shelter is constrained by rock structure. Application of correction factors in the common situation where tortoises in one area use both soil and rock shelters would require adjustment for this fact.

Alternatively, adjusted correction factors were developed by relating sign counts from 1 mi^2 "calibration plots" to estimates of tortoise abundance from the same plots. These relationships would then be used to predict tortoise abundance on sign survey plots elsewhere in the Mojave and Sonoran Deserts. To accommodate any regional differences in the sign-to-tortoise ratios, calibration plots were generally in the same region as the sign survey plot. It was also recognized that there would be between-observer differences in detecting these signs, so effort was made to calibrate for each observer. The difficulty of accurately estimating abundance of individuals on calibration plots meant they were surveyed only in a few years, however, and usually not the same year as sign survey plots.

Duda et al. (2002) and Krzysik (2002) explored an alternate use of tortoise signs. Whereas accurate estimates of abundance are only possible on larger plots, relatively more common tortoise signs on the same large plot might be used to partition a single abundance estimate into separate abundances for each subplot. Krzysik (2002) found that the ration of tortoises to signs differed almost two-fold in the same years between large plots that were only 64 km apart, calling into question use of correction factors that are not developed for the specific time and place.

Estimation of the proportion of burrows inhabited by tortoises may be improved by monitoring burrow use with mechanical devices or cameras at burrow mouths, or by locating individuals in the burrow with a scope (an infrared video camera attached to a fiber-optic tube). Use of a scope offers the ability to verify presence of individuals in each encountered burrow; however, the method is prone to misclassification (Smith et al. 2005). Stober and Smith (2010) reported that over five separate survey periods, presence of an individual of *G. polyphemus* could not be determined using a scope for 6.7 to 16.4% of burrows, and can be even higher in certain environments (Kinlaw 2006; McCoy, personal observation). The method also may be prone to an unknown number of false positives and cannot be used for very small burrows.

Their relative size and opportunistic use of burrows excavated by larger individuals mean that juvenile tortoises of all North American tortoise species are more easily overlooked than adults, and their abundances may be drastically underestimated. The behavior of prereproductive animals can result in a different seasonal activity period than adults, and reproductive status is only loosely associated with size or age.

Especially for wide-spread species, use of size as a surrogate for age, reproductive status, or both can only further complicate comparison of size-specific estimates of abundance, and may explain why operative size classes for each species have changed over time and for different regions. Occasionally, life tables have been used to estimate abundance indirectly in size classes that were not sampled (e.g., Turner et al. 1987a, McCoy and Mushinsky 1995). Some promising work has been done using dogs for nonvisual searching (Cablk and Heaton 2006), but this method has not been used in surveys to estimate abundance of small individuals.

CORRECTING COUNTS OF INDIVIDUALS WITH ESTIMATES OF DETECTION PROBABILITY

Some of the above techniques to estimate abundance lack rigor, based on their increasingly questionable assumptions about the appropriateness of indices and the way that *detection probability* for different types of sign is treated. Individuals also have variable detectability because of their use of shelter which completely conceals them, their cryptic coloration combined with sessile or slow behaviors, the reduced visibility of small individuals, brief periods of time when they are predictably above-ground, and any of these factors combined with incomplete surveys of study areas. Because detection probability also varies by survey-specific factors such as local and seasonal vegetation cover and tortoise activity, topography, and observer characteristics, application of a universal correction factor to counts of individuals cannot provide a reliable measure of tortoise abundance. Acceptable techniques will estimate detection probabilities as an intermediate step to estimating abundance. Any intermediate estimation steps contribute to the variance (uncertainty) in the estimate of abundance, so some measure of this variance must also be incorporated in order to compare and critique results (Buckland et al. 2001, Kazmaier et al. 2001d, Nomani et al. 2008). Techniques such as the delta method incorporate variance from intermediate estimates into the overall variance of the final abundance estimate (Powell 2007).

Surveys that focus on counts of live animals and estimation of detection probability can falter if sample sizes for both purposes are not adequate. The number of individuals detected for a specified amount of effort (distance traveled on transects or the amount of time spent searching) is the *encounter rate.* Where tortoises occur at low densities, the encounter rate is generally lower, and the number of kilometers surveyed must be correspondingly higher to have an adequate count at the end of the survey.

MARK-RECAPTURE ANALYSIS

Mark-recapture analysis adjusts counts of tortoises by marking individuals detected in one survey of an area and evaluating their proportional representation in future surveys. These approaches have been applied to North American tortoises for years, usually using the Lincoln-Peterson estimator (Begon 1979), which assumes no population losses to death or emigration, or gains from birth or immigration between two survey periods. These assumptions reflect demographic closure. If closure is reasonable to assume, then the proportion of marked animals in a capture survey can be equated with the same proportion in the overall population:

$$\frac{m_2}{n_2} = \frac{n_1}{N}.$$

N represents true abundance of individuals in the population, n_1 is the number of individuals captured and marked in the first session, and m_2 is the number of these marked individuals that were recaptured in the second session of n_2 captured individuals total. As an example, if 20% of the animals caught in the second survey were marked, then it follows that 20% of all individuals (N) were captured and marked (n_1) in the first survey. The ratio of m_2/n_2 is one type of detection probability, and is applied in this example to the count of individuals in the first session (n_1) to arrive at an estimate of abundance (N).

In addition to (1) the demographic closure assumption, this model is valid if (2) all animals are equally likely to be captured in each session, and (3) marks are not lost or overlooked by observers. Starting in the 1970s, mark-recapture methods were used extensively throughout the range of *G. agassizii* to estimate abundance in areas thought to represent best, undisturbed habitat (Turner and Berry 1984). The conventional survey is conducted on one mi² plots encompassing two surveys over 45 calendar days. Schneider (1980) and Turner and Berry (1984) acknowledged that assumption 2 was not met, which seriously compromised initial abundance estimates. If some of the n_1 individuals are juveniles and less likely to be detected in any capture session, then this portion of the population will be less likely to be among the m_2 individuals, depressing this count and biasing high the estimate of N. This problem can be alleviated using various approaches, including stratified Lincoln-Peterson estimation (Overton 1971) to treat each size class separately. For *Gopherus agassizii* individuals larger than 180 mm midline carapace length (MCL), Schneider (1980) estimated 20.1 tortoises/km² for the Chemehuevi long-term study plot in California (in this common scenario, there were insufficient captures to estimate density of smaller tortoises). For analyses of only the largest size classes, "birth" into the size class is obviously not possible; however closure does not exist if many smaller animals grow into the focal size class during the series of surveys. Still, over a 45-day period, low mortality and growth rates of tortoises make demographic closure assumption for one survey season reasonable.

Plots are not demographically closed when the mark-and-recapture sessions are in different years. Jolly-Seber was the first widely used approach to estimate abundance of open populations using marked individuals (Jolly 1965, Seber 1965). This class of models relies on multiple recapture sessions with the number of marked animals in the largest size classes accumulating over this period. Because animals can immigrate or emigrate, can be recruited to the population or size class of interest through birth or growth, or can die, true abundance can vary between years, so marked-to-unmarked proportions have different interpretation from their use in the Lincoln-Peterson method. If individuals can emigrate or die, then additional steps are necessary to estimate the proportion of marked individuals that were not found during a particular session but are captured later: they were likely present but overlooked during the earlier survey. These adjustments require knowledge of future recaptures, so abundance estimates are not possible for either the first or the last survey in a series. The original *G. agassizii* and later *G. morafkai* mark-recapture studies were designed for Lincoln-Peterson estimates requiring two sessions (Schneider 1980, Turner and Berry 1984, Averill-Murray 2000); however, there are also robust models that allow analysis of several years of such data at one site, accounting for the lack of demographic closure between years. These models have not yet been used with tortoise mark-recapture data, but would be appropriate for reanalysis of existing datasets such as Freilich et al. (2000), which applied a closed population model to between-year recaptures during a six-year period, arriving at a single density estimate of 67 adults/km^2 for a plot in what is now Joshua Tree National Park.

This brief description of mark-recapture techniques does not cover all aspects of designing a valid study, but does illustrate why closed population abundance estimates for tortoises in open population situations are likely to be overestimates of abundance of large individuals. Overestimates also occur regularly under the closed population scenario, because even if all individuals in the plot are local residents, some home ranges extend outside the surveyed plot; this makes it difficult to associate the abundance estimate with a particular area (the geographic closure assumption). Royle and Young (2008) have proposed a hierarchical approach, using the same mark-recapture data to estimate home ranges inside the surveyed area as well, but this approach has not been used in tortoise studies. Averill-Murray (2000) suggested using the correction factor of Wilson and Anderson (1985), based on home-range sizes to estimate the larger effective trapping area for individuals captured on the plot.

Averill-Murray's (2000) suggestion has been incorporated in the current monitoring protocol for *G. flavomarginatus* in Mexico (G. Aguirre, personal communication). The protocol has also been loosely implemented for *G. morafkai* in Arizona since 2001, with the additional complication that surveyors have created heterogeneous capture probabilities by expending more search time in areas where they had already found tortoises and especially burrows. Before 2001, each year's plot data were treated as a single session, with recapture possible in future years (Averill-Murray 2000); these data are still amenable to analysis under open-population models, although few years of data are available for each plot.

The first abundance estimates for *G. berlandieri* were for one of the two major environments they inhabit, clay dunes or "lomas," which are small raised features in coastal salt marsh flats. Based on topography, Auffenberg and Weaver (1969) argued that the lomas are closed to immigration and emigration, and uniquely marked 193 individuals on one 0.0855 km^2 loma, a density of 2257 individuals/km^2. Without further explaining their methods or reporting a variance for their estimate, they then contrasted this high density on a loma with an estimated 23.8 tortoises/km^2 at a site in thornscrub vegetation. Long-term open-population mark-recapture studies have been conducted based on individuals captured by grid searches (Rose and Judd 1982, Rose et al. 2011). For one of the larger lomas, Judd and Rose (1983) reported densities of 1500–1600 individuals/km^2, comparable to the density reported by Auffenberg and Weaver (1969). Because their 3.3 ha grid was open to immigration and emigration, and the surveys were repeated over five years, Judd and Rose (1983) compared Lincoln-Peterson and Jolly-Seber estimates, acknowledging evidence that the closed population results increased each year, consistent with expectations from using a closed model for an open population, but also expressing frustration that the Jolly-Seber method appropriate to this situation required so many interim estimates that the apparent precision was quite low.

Individuals of *G. berlandieri* spend most of their time above ground, but thornscrub vegetation conceals them, and tortoises fitted with radio transmitters were observed to conceal themselves further from approaching searchers (R. Kazmaier, personal communication). The ability of individuals to remain hidden in thornscrub may explain why road-cruising yielded an encounter rate of 0.36 individuals per hour, but on-foot searches through thornscrub in the same area encountered only 0.013 individuals per hour (Kazmeier et al. 2001d; also see Rose and Judd 1982, Rose et al. 2011). Such low encounter rates prevent sufficient captures and recaptures for effective abundance estimation. Selective searching along roads has subsequently been used to survey for *G. berlandieri* inside larger plots (Kazmaier et al. 2001d), where researchers were also able to demonstrate equal detectability (assumption 2, above). Kazmaier et al. (2001d) applied open population models (Jolly-Seber) for individuals larger than 120 mm straight-line carapace length to arrive at a density estimate of 26 individuals/km^2 for the ten-year study. They considered the study area to be effectively geographically closed, as it was much larger than the average individual home range, bounded in part by agricultural land, and none of 46 individuals with transmitters emigrated temporarily or permanently. The open model allowed for deaths and recruitment, and they applied the abundance estimate to the 7617 ha area of the surveyed plot plus a buffer representing half of the diameter of a typical home range. A Lincoln-Peterson estimate of density assuming demographic closure over four of the ten years at the same site was 26–28 individuals/km^2 for the 6150 ha survey plot only (Hellgren et al. 2000).

Mark-recapture methods currently are used to monitor *G. flavomarginatus* and *G. berlandieri*, as well as *G. agassizii* and *G. marafkai*. Thus far, results seem to have been reliable only rarely, largely reflecting difficulties with model assumptions. In addition, although a substantial body of data has been collected from the original 31 *G. agassizii* and *G. morafkai* long-term study plots, several factors limit the general utility of the data. Robust population models would be appropriate, but because the plots have been resurveyed a handful of times at most, the results may not be very precise and have not yet been attempted. Single-year closed population abundance estimates for these study plots have not been reported for more than a decade, and estimates that exist from previous years likely underestimate the effective area represented. This result is equivalent to overestimating the density of reproductive-sized animals (Tracy et al. 2004).

For many types of research, study plots where tortoises occur at high densities are therefore preferred, so that sufficient study subjects can be encountered at a lower level of effort. When these plots are used for abundance estimation, however, they are nonrepresentative of unsurveyed (usually low-density) areas, and do not contribute to accurate regional abundance estimates. Interpretation of abundance estimates is therefore also highly dependent on study design.

DISTANCE SAMPLING

Mark-recapture abundance estimates on a few dozen nonrandomly placed plots cannot be readily generalized to unsurveyed areas or to broader regions; they will not provide species abundance estimates for organisms like *Gopherus agassizii* that occur on landscape scales of hundreds to thousands of square kilometers. Distance sampling for tortoises (Anderson et al. 2001) is conducted by walking line transects to sample areas of any size, and measuring the perpendicular distance between the line and each individual seen. This method provides an encounter rate (counts per kilometer walked), which is adjusted to compensate for the decreasing detection probability of tortoises farther from the transect line. This type of detection probability reflects the cryptic appearance and behavior of tortoises when they are not in burrows. The estimate of this detection probability requires many more observations of individuals than does the estimate of encounter rate. Turner and Berry (1984) speculated about using distance sampling for density estimates at regional scales but concluded that the encounter rate would be too low to be effective. Many years later, it is still clear that the number of encounters required for distance sampling estimation constrains use to areas of high tortoise density, or to more intensive sampling in areas of low density, where the encounter also will be low.

Calculation of annual densities starts with estimates of encounter rates in each monitored area:

$$D = \frac{n}{2wLP_aG_0},$$

where L is the total length of kilometers walked in each stratum, and w is the width of the transect, so $2wL$ is the area surveyed. P_a is the detection probability within w meters of the transect centerline. For tortoises, it is also necessary to adjust for G_0, the proportion of available individuals, estimated by behavioral observation. The indirect estimate of D requires direct estimation of n/L, P_a, and G_0, so the variance of D depends on the variance of these quantities as well. Use of Program DISTANCE (Thomas et al. 2010) can simplify calculation and diagnostics associated with the encounter rate and detection probabilities.

A long-term monitoring program for *G. agassizii* based on distance sampling was initiated between 1999 to 2001 (Anderson and Burnham 1996, McLuckie et al. 2002, USFWS 2006). Table 14.1 reports parameter estimates from 2011. Only 32–53% of individuals within 16–20 m of the centerline were observed. If all of the individuals were detected within that distance, the encounter rate and the number of tortoises used to build the density estimate would be two or three times higher. As it is, density estimates in some monitored regions are based on few encounters. For instance, density in Pinto Mountains (encompassing 751 km^2) was estimated based on detection of six tortoises after observers walked 118 km. Any precision in the density estimate comes from this high survey effort, which also translates the count into an encounter rate. The variability in encounter rates between transects is relatively large, reflecting the inherently nonrandom distribution of individuals on the landscape, and accounting for about 70% of the total variance in the density estimate. Coefficients of variation for monitored area density estimates in 2011 were 12–42%.

Pooling observations for large areas of the desert (up to 6500 km^2) is necessary to model detection probability for *G. agassizii*. This area also encompasses considerable variation in local plant phenology and other seasonal effects thought to influence the probability of tortoises being above-ground, or at least visible at the front of burrows. Availability of tortoises is therefore estimated at finer scales (table 14.1) and used to correct detection probabilities to account for individuals that are present but not visible (not available to be detected). Although *G. agassizii* and *G. morafkai* burrows are not on average as deep as *G. polyphemus* burrows, scoping has not been widely used in these species. Instead, radio-telemetry is used to report the availability of a focal set of tortoises several times daily during the period transects are surveyed in the same region. Attempts to model tortoise availability based on temperature or other local conditions have not been successful (Nussear and Tracy 2007, Inman et al. 2009).

Most assumptions of distance sampling are easily met during tortoise surveys. Although individual tortoises may be detected by observers on more than one transect, this is unlikely to be due to movement of the tortoise. Also, the relatively short detection distance from the transect centerline makes these distances straightforward to measure with accuracy; tortoises are also unlikely to have moved far in response to observers before they are detected. Distance sampling also is predicated on detection of all individuals on the centerline, as long as they are available to be detected. Tests conducted with polystyrene tortoise models (Anderson et al. 2001), followed by three years of multiple-pass sampling on transects (USFWS 2006), have demonstrated convincingly that even highly experienced observers can fail to detect individuals that are above ground and within a one to two m of the transect centerline.

Table 14.1. Density of *Gopherus agassizii* individuals larger than 180 mm MCL in each monitored area, 2011

Input and/or estimate	Tortoises seen (n)	Kilometers walked (L)	Tortoise encounter rate (n/L)	Proportion that are visible, G_0	Transect half-width (m), w	Detection rate, P_a	Density (D, tortoises per km^2)	Std(D)
Method to develop estimate	*Line distance sampling*			*Behavioral observation*	*Line distance sampling*		*Compilation*	
Estimate applies to	*Monitored area*			*Regional G0 sites*	*Field team*		*Monitored area*	
Fremont-Kramer	15	264	0.057	0.95			3.5	1.1
Ord-Rodman	9	174	0.052				3.2	1.2
Superior-Cronese	45	820	0.055		20	0.53	3.4	0.8
Chuckwalla	17	280	0.061	0.90			3.9	1.4
Joshua Tree NP	8	147	0.054				3.5	1.3
Pinto Mtns	6	118	0.051				3.3	1.4
Chemehuevi	15	354	0.042	0.84			4.0	1.5
Eldorado Valley	10	331	0.030				2.8	1.1
Fenner	13	179	0.073		16	0.32	6.8	2.8
Ivanpah	20	416	0.048				4.5	1.7
Piute Valley	17	239	0.071				6.6	2.6
Gold Butte–Pakoon	18	1039	0.017	0.65			1.6	0.6
Beaver Dam Slope	23	751	0.031	0.55	20	0.43	3.3	1.2
Mormon Mesa	73	1227	0.059				6.3	2.1
Coyote Springs	52	967	0.054	0.79			4.0	0.9
Upper Virgin River	113	310	0.365	0.74	20	0.64	18.2	2.2

Source: Data for Upper Virgin River are taken from McLuckie et al. 2012; remaining data from Allison, unpublished.

Note: Density derived from the applicable encounter rate, proportion visible, and detection probability. Columns indicate the scale at which each component was estimated. Component variances were combined using the delta method (Powell 2007) to arrive at the variance for density.

Currently, a multiobserver technique is used to test (White et al. 1982, Nichols et al. 2000) or to ensure that all individuals (*G. agassizii* and *G. morafkai*) or burrows (*G. polyphemus*) on the transect centerline are detected. Application of distance sampling to *G. polyphemus* burrows fits the assumptions in a similar way. We delay briefly discussion of how distance sampling estimates of abundance of *G. polyphemus* burrows are converted to abundance of tortoises, because the correction factor is related to occupancy analysis.

OCCUPANCY ANALYSIS

Occupancy analysis was developed for monitoring species that are so widespread that traditional abundance estimation is untenable (Mackenzie et al. 2006). At sufficient densities, occupancy, the proportion of area occupied (PAO), will be correlated with the number of individuals present (Royle and Nichols 2003); however, PAO is of interest in its own right as a descriptor of species distribution or extent, and is more efficiently estimated than abundance under many circumstances (MacKenzie and Royle 2005, Zylstra et al. 2010). Occupancy can also be evaluated for trends, and because of the connection between abundance and occupancy, monitoring programs can consider whether to incorporate estimation of occupancy rather than—or in addition to—abundance.

Occupancy could be expressed simply as the proportion of plots where evidence of at least one tortoise is found. PAO (ψ) also incorporates multiple visits to each site to estimate and account for detection probability, however, similar to the use of repeat surveys in mark-recapture analysis. The history of visits at each site is summarized to indicate whether any tortoises were detected. An encounter history of 01011, for example, indicates the site was visited five times and is occupied, but detected only on the second, fourth, and fifth visits. Likewise, the encounter history 00000 indicates the site was visited five times without the species being detected, which may arise if the species actually is absent from the site or if it is present but was undetected through all five visits. These interpretations depend on assuming closure, so that each site is either occupied across the period of all five visits, or it is unoccupied, although elaborations are possible to relax the closure assumption between one set of visits and another, for instance between years. Detection probabilities can be modeled to vary with time of year and also to fit the scenario that tortoises are equally detectable on each visit, in which case the probability of encounter history 01011 would be

$$\psi(1-p)p(1-p)pp\ .$$

and of encounter history 00000 would be

$$[\![\psi(1-p)]\!]\ {}^\wedge 5+(1-\psi),$$

where the first term is the probability for this history if the site is occupied, and the second term is the probability for this history if the site is not occupied. Estimates are developed for ψ and for p based on these probabilities across all sites. The final estimate of PAO is akin to the proportion of all sites where tortoises were detected, corrected for the probability that the tortoise was present but never encountered across all visits.

Among North American tortoises, estimates of PAO have been generated only for portions of the range of *Gopherus morafkai* (Zylstra et al. 2010; Grandmaison, unpublished). These surveys also served as pilot studies to develop monitoring schemes appropriate to *G. morafkai* in Arizona, where individuals may occur at high densities only in core areas. Zylstra et al. (2010) estimated that 72% of all possible three ha plots in two core areas were occupied, and that for those that were occupied, there was a 0.43 probability of detecting at least one individual during a single survey of a site. In large, flat expanses of desert, tortoises selectively occupy linear washes that provide caliche cave shelters, so that Grandmaison (unpublished) estimated that only 19% of the 175 three ha plots in his sampling frame were occupied, with a 0.33 probability of detecting at least one individual during a single survey of an occupied site. Both studies concluded that to optimize precision of PAO estimates, sites should be surveyed five times in a season (MacKenzie and Royle 2005).

The tables of MacKenzie and Royle (2005) can also be used to identify the number of sites to survey to optimize the PAO estimate. The number of sites identified by Zylstra et al. (2010) would represent 7% and 4%, respectively of the potential habitat at each of their core areas, whereas Grandmaison (unpublished) identified survey sites representing 46% of the potential tortoise habitat at his study area. Meanwhile, because only approximately one-fifth of the latter sites are expected to be occupied; the remaining four-fifths of the sites will nonetheless be surveyed five times, without the expectation of encountering tortoises. These differences illustrate one type of inefficiency that arises when occupancy analysis is applied in areas where tortoises are distributed more sparsely or in localized patches.

A different use of PAO has been implemented for surveying *G. polyphemus.* In this case, occupancy is not evaluated on plots or areas, but for a particular set of burrows. Because use of a scope to examine burrows is often inconclusive, Nomani et al. (2008) proposed a two-stage estimation process requiring detection of burrows (in conjunction with distance sampling or area searches), then three visits to each burrow over a short period of time so that the closure assumption can be invoked; if a burrow is ever found to be occupied, then it was occupied on all visits, even if no tortoise was reported or if part of the burrow could not be examined with the scope. Although scopes are generally accurate, burrows cannot always be examined thoroughly, and tortoise detection probability in occupied burrows was 0.92 (Nomani et al 2008). The estimate of the proportion of burrows occupied can be applied to unsurveyed burrows (in conjunction with distance sampling for burrows, for instance, when it is assumed that some burrows were not detected/visited).

Stober and Smith (2010) considered options for addressing incomplete detection and inaccurate classification of burrows, arguing for various reasons against using PAO but in favor of estimating detection probability. They applied distance sampling techniques only to burrows that were confirmed occupied, arriving at a final abundance estimate with a variance that did not reflect uncertainty about classifying burrow occupancy. However, they provided sufficient data that we can illustrate how to arrive at an abundance estimate that accounts for uncertain occupancy classification. Based on a single scoping event in each *G. polyphemus* burrow, instead of PAO we can calculate an uncorrected average burrow occupancy rate of 0.237 over two seasons in 2007 (CV = 15.7% using the binomial distribution), and 0.134 of burrows could not be classified to occupancy status. To use estimates for both detection and classification of burrows, start with their distance estimate of 151 occupied and unoccupied burrows (CV = 11.6%), then apply the above empirical occupancy estimate, then correct for survey effort to arrive at a live tortoise abundance of 35.8 tortoises on 0.5 km^2 (CV = 19.5%). This number is similar to the estimate of 33 tortoises (CV = 20.7) reported by Stober and Smith (2010), but directly addresses both crypticity and a certain amount of classification uncertainty. If PAO had been used to estimate occupancy rather than relying on a single scoping check per burrow, the occupancy estimate would be the same or probably higher, if repeated visits ever confirmed a tortoise in some burrows that were previously not considered occupied.

TRENDS AND CURRENT TORTOISE ABUNDANCES

Observing relatively rapid increases in numbers of individuals is highly unlikely based on the life history of most tortoises (i.e., delayed reproductive maturity, low reproductive rates, and relatively high mortality early in life). Although declines in contrast can be precipitous, most trends will probably describe long, slow patterns. The preferred way to proceed with comparisons of abundance, PAO, or other measures over time (trends) is to use estimates collected using the same techniques, assuming that any consistent biases will not mask an underlying trend. For example, McCoy et al. (2006) created "most conservative" scenarios to adjust for acknowledged shortcomings in assumptions to assess decade-long declines in *Gopherus polyphemus* on lands that have been managed to protect them. It is more typical, however, for trend assessment to start with a series of estimates that have been developed in tandem with expanding knowledge about the distribution and biology of each species, improved understanding of relative sources of bias and imprecision, and improved survey methods.

Berry (1984) used anecdotal reports and interviews of long-time desert residents to portray the spatial distribution and relative numbers of *G. agassizii* in California during the period from 1920 to 1960. While acknowledging that historical numbers of individuals are speculative, she concluded that densities had decreased in some areas and that the species was now locally extinct in others. Since then, the collective data for *G. agassizii* show appreciable declines in many areas (Berry 1984, Luke et al. 1991, Berry 2003a, Tracy et al. 2004). Tracy et al. (2004) concluded that the apparent downward population trend in the western portion of the Mojave Desert that was identified at the time of listing is valid and ongoing. Results for *G. agassizii* in other portions range were inconclusive; but recent surveys of some areas found too few tortoises to produce population estimates, suggesting that declines have occurred more broadly. Although they were able to make qualitative conclusions about population trends in larger areas, Tracy et al. (2004) also concluded that estimating long-term trends in abundance, habitat, and/or threats across the range of *G. agassizii* had not been feasible based on the combined suite of existing data and analyses.

To estimate current abundance of *G. agassizii* range-wide, here we extrapolate densities estimated using distance sampling to areas that have not been surveyed. To estimate the total area that might support the species north and west of the Colorado River, we used a model that matched known locations of individuals to a suite of environmental variables (Nussear et al. 2009). The model indicates that 83,123 km^2 have characteristics that could provide habitat to support the species (habitat potential of at least 0.5 on a scale of 0 to 1; Liu et al. 2005). Densities reported in table 14.1 indicate that slightly more than 95,000 adults inhabit the surveyed areas; and, extrapolation to the additional areas identified as potential habitat suggests that 200,000 more adults live outside the surveyed areas. Because density estimates often have been made under the best prevailing land management for *G. agassizii,* unsurveyed areas are not likely to support similar densities, however. Moreover, Fry et al. (2011) have identified 14,621 km^2 of the apparent habitat as "impervious surface" arising from human activities such as urbanization and certain agricultural practices, however; so, the actual number of individuals over 180 mm MCL certainly is fewer than 295,000.

The highest reported densities for *G. morafkai* are on rocky slopes in Arizona, where they are found at 23 to 56 individuals/km^2 (Averill-Murray et al. 2002b). No density or abundance estimates are available from the southernmost part of the range, where the species' preferred habitat is Sinaloan thornscrub and deciduous forest (Germano et al. 1994; G. Aguirre, personal communication). As is true of other species of North American tortoises, it is probable that *G. morafkai* occurs at lower densities in areas surrounding the preferred habitat (e.g., Averill-Murray and Averill-Murray 2005, Riedle et al. 2008, Grandmaison et al. 2010). Lack of detectable genetic differentiation among sky islands separated by 25–30 km^2 (Edwards et al. 2004a) supports the idea that these low-density areas are large enough to interconnect the core populations. Therefore, although there is little knowledge of the extent inhabited by tortoises at lower densities, to derive a rough estimate of the number of adults, we used the area within "core" habitat (Germano et al. 1994; 90,061 km^2) multiplied by a density of 40 individuals/km^2 plus the area within a ten km wide "peripheral" zone around each core habitat area (140,724 km^2) multiplied by three individuals/km^2 (Averill-Murray and Averill-Murray 2005). Depending on the accuracy of these assumptions and figures, the current abundance of *G. morafkai* is about 4,000,000 adults.

Gopherus berlandieri generally is perceived as under threat and requiring protection from collection and local sources of mortality. The species' range clearly has contracted from its prehistorical and historical extent. In the U.S., it currently occurs in relatively isolated patches of chaparral within extensive tracts of ranchland, or on unique clay inland dunes (lomas). Occupied areas also occur around the edges of the Rio Grande Valley and otherwise undeveloped parts of its former range, reflecting habitat loss from the center of that range. Recent severe droughts and burgeoning oil development have severely reduced abundances in areas such as the Rio Grande Plains (R. Kazmaier, personal communication). In Mexico, individuals are used for food, kept as pets, and killed on roads; agriculture and urbanization have reduced available habitat (G. Aguirre, personal communication). The most recent field surveys and interviews of human inhabitants in six Mexican states where *G. berlandieri* is known to occur (Niño-Ramírez et al. 1999) showed the species to be more common in the states of Tamaulipas and Nuevo León than Coahuila, Zacatecas, San Luis Potosi, or Veracruz. The survey data also suggest a westward expansion and a southern contraction of the species' range (cf. Smith and Smith 1979, Coghill et al. 2011). Maximum reported population density was 720 individuals/km^2 at one site. The distribution of the species is so poorly known that current range-wide abundance cannot be estimated reasonably (R. Kazmaier, personal communication).

Similar to the range of *G. berlandieri,* the range of *G. flavomarginatus* also has contracted, no longer stretching to the grasslands of western Texas and into New Mexico. Bury et al. (1988) concluded that six remaining populations existed on slopes surrounding the set of basins in the Bolsón de Mapimi, surrounded by burrows at much lower densities. They estimated 7,000–10,000 adults overall. Density of individuals currently is estimated at 84.25/km^2, where the species occurs (Aguirre, unpublished). This figure is much lower than the previous estimate of Morafka (1988). Potential area occupied is based on the modeling work of Martinez-Cardenas (2006), which showed habitat on only 45–51% of the species' range at the Mapimi Biosphere Reserve. A spatial model of detrimental anthropogenic impacts overlaid on the habitat model further reduced the potential area occupied to 23–26%. Extending these estimates to the entire range of the species yields a total abundance estimate of 6200–7100 adults, although this may be an overestimate, as the area outside the Reserve is more disturbed (G. Aguirre, personal communication).

Gopherus polyphemus apparently is restricted to areas with adequately drained soils, but also vegetation successional stages that support food plants. For this reason, it is thought that they occurred originally over large areas of the southeastern U.S., but in colony-like groups of various sizes (chapter 12). Longleaf pine-oak forests support relatively low densities of individuals (Auffenberg and Franz 1982), but this vegetation type is more widespread and connected over larger areas than other available habitat types. The distribution of individuals has been restricted as habitat has been converted for commercial timber harvest or human settlements, in particular. Auffenberg and Franz (1982) expressed concern that many groups of individuals were too small to be ecologically viable. Their attempt to estimate range-wide abundance was complicated by the occurrence of individuals in small habitat patches, requiring the number, extent, and abundance of these groups to be determined first. Where they are present, individuals occur at densities on the order of 10–20/km^2, and have been observed to increase in density in many habitat patches, even as they become more isolated. Recruitment in patches can decline, however, especially when they are particularly small (McCoy and Mushinsky 2007). Even in ostensibly protected areas, declines in *G. polyphemus* abundance on the order of 10% per decade have occurred (McCoy et al. 2006). In Florida, total decline in abundance likely amounts to 70% since the arrival of settlers (McCoy, unpublished). We estimate current abundance range-wide from a variety of reports, our own calculations, and the help of experts (J Berish, C Guyer, J Jensen, T Mann, L Smith, T Tuberville), to be about 1,000,000 adults. The core of the distribution is Florida, where perhaps 75–80% of the remaining individuals occur, with lower numbers in Alabama and Georgia, and very low numbers at the periphery of the species' range in Mississippi, Louisiana, and South Carolina.

CONCLUSIONS

Straightforward approaches exist to refine our understanding of local tortoise abundance after accounting for cryptic and hidden animals, and most of these methods are applicable across the species of North American tortoises. However, local descriptions of tortoise abundance, often in high-density portions of each species' range, are inadequate for conveying abundance and status of such wide-ranging species. Our best efforts to tally tortoise numbers across their entire range tell us that these species differ by orders of magnitude in their absolute abundance. Nonetheless, range-wide abundance estimates also leave us with an incomplete picture of the status of each species. The ability of any species to increase abundance and to maintain the number of populations is also a function of the distribution of individuals on the landscape. Some species currently are at such low local or regional densities that their ability to form stable reproducing units is questionable; and other species occur in increasingly isolated population units that by any measure of population dynamics are vulnerable to stochastic effects. In addition to examples of population isolation due to habitat loss, constructed linear features and constricted areas of habitable land also threaten to isolate populations. For example, *Gopherus morafkai* occurs at highest densities in local mountain ranges and their associated foothills (Zylstra and Steidl 2009). Before fragmentation of the desert by roadways, individuals and genes were exchanged between population centers in mountains through surrounding bajadas (Edwards et al. 2004a, Averill-Murray and Averill-Murray 2005).

All of the species of North American tortoises likely are following the declining-species paradigm, under which reduced population abundance and range extent are associated with factors originating outside the population (Caughley 1994), so it is meaningful to relate declining population size with the magnitude of external threats to population growth. It is probable, however, that some or all populations in each of these species have been reduced sufficiently in size to also be impacted by the suite of genetic and demographic factors that increase a small population's threat of extinction, and include demographic stochasticity, environmental stochasticity, genetic drift, and inbreeding, but also inverse density-dependence, or the "Allee effect" (Stevens et al. 1999). Because adult tortoises are so long-lived, and because these small-population effects are rarely associated with obvious sources of mortality, small-population effects will not be apparent to researchers looking for continuing impacts of external factors. It is important that we look beyond trends in population size as a measure of population risk; reduced population size itself and associated changes in population structure may have become a more immediate source of risk.

Acknowledgements

We gratefully acknowledge D. Grandmaison and R. Kazmaier for providing unpublished data and G. Aguirre, J. Berish, C. Guyer, J. Jensen, R. Kazmaier, T. Mann, L. Smith, and T. Tuberville for providing regional expertise to help estimate abundance of *Gopherus* species. R. Averill-Murray and R. Steidl provided thoughtful comments that greatly improved this chapter.

The findings and conclusions in this article are those of the authors and do not necessarily represent the views of the U.S. Fish and Wildlife Service.

15 Population and Conservation Genetics of North American Tortoises

TAYLOR EDWARDS
J. SCOTT HARRISON

One of the principal goals of conservation is to preserve the diversity of species. Integral to achieving this goal is the conservation of genetic, ecological, and morphological variation within a species and among populations. The field of conservation genetics provides tools and principles for preserving genetic diversity, which improves a species' ability to cope with environmental change and decreases its susceptibility to extinction. In addition, genetic data can be used to estimate effective population size, migration rates, and to assess the role of landscape features and environmental conditions that contribute to the evolutionary history of a species. This information can be directly applied to conservation and management.

Genetic studies contribute to our knowledge of North American tortoises (*Gopherus*) in a wide variety of ways, including: systematics (Lamb et al. 1989, Lamb and Lydeard 1994, Morafka et al. 1994, Spinks et al. 2004, Le et al. 2006, Thompson and Shaffer 2010), taxonomy (Murphy et al. 2011), phylogeography (Lamb et al. 1989, Ostentoski and Lamb 1995, Edwards et al. 2012, Ennen et al. 2012), population structure (Jennings 1985, Rainboth et al. 1989, Glenn et al. 1990, McLuckie et al. 1999, Edwards et al. 2004a, Schwartz and Karl 2005, Ennen et al. 2010, Fujii and Forstner 2010, Sinclair et al. 2010, Latch et al. 2011, Richter et al. 2011, Urena-Aranda and de los Monteros 2012, Clostio et al. 2012), management (Britten et al. 1997, Murphy et al. 2007, Hagerty and Tracy 2010, Ennen et al. 2011, Edwards and Berry 2013), ecological genetics (Hagerty et al. 2011), hybridization (Edwards et al. 2010), paternity (Moon et al. 2006, Davy et al. 2011), and forensics (Schwartz and Karl 2008). In this chapter, we first introduce some of the fundamental principles of population genetics as observed in *Gopherus* and show how they contribute to our understanding of tortoise ecology. We then review how genetic studies help to inform management and conservation efforts.

POPULATION GENETICS, GENETIC DIVERSITY, AND POPULATION GENETIC STRUCTURE

Population genetics evaluates differences in genetic variation among populations or individuals. Differences in the distribution of genetic variation can arise from a combination of processes, including natural selection, isolation, dispersal, demographic processes (e.g., changes in population size) or stochastic processes (genetic drift). Patterns in the distribution of genetic variation at different spatial and / or temporal scales can be used to test hypotheses relating to past or ongoing processes of evolutionary and ecological importance.

Quantifying the amount of genetic variation within and among populations is essential for understanding several aspects of a species ecology and conservation (Frankham 1995). For example, this information can be used to determine if a species constitutes a single, continuous population or multiple, distinct populations. Informed practices for preserving genetic variation within a species are critical for effective conservation.

When genetic variation of a species is partitioned nonrandomly across its geographic range, it is called population structure. In the absence of selection, the genetic structure of a population is influenced primarily by the amount of gene flow (exchange of genetic material via immigration of individuals) among populations. The absence of structure is called panmixia, where mating between individuals is completely random. Patterns of genetic variation can be assessed using a variety of methods and inferences are dependent on the type of genetic marker being measured (table 15.1).

Mitochondrial DNA (mtDNA) is a commonly used genetic marker for population studies. Unlike nuclear genes, mtDNA is maternally inherited and haploid. Because of these characteristics, mtDNA may be affected by factors such as population size reductions (bottlenecks) and isolation at a faster rate

Table 15.1. Descriptive statistics from within-species genetic studies of the species of *Gopherus*

Species	Study Locality	Fst	Divergence (%)	Heterozygosity	Allelic Richness	Marker Type	Source
Gopherus polyphemus							
	Eastern Mississippi			0.50	1.7	STR	Ennan et al. 2010
	Western Mississippi			0.20	2.6	STR	Ennan et al. 2010
	Camp Shelby, Mississippi	0.03		0.21	1.8	STR	Richter et al. 2011
	Kennedy Space Center, Florida	0.03		0.35–0.47	3.3–4.1	STR	Sinclair et al. 2010
	Georgia and Florida	0.24		0.42–0.44	3.0–3.4	STR	Schwartz and Karl 2005
	Range-wide	0.01–0.54		0.23–0.67	13.4	STR	Clostio et al. 2012
	Florida		1.1			mtDNA (RFLP)	Osteneski and Lamb 1995
	Range-wide		2.3			MtDNA	Clostio et al. 2012
	Range-wide	0.57–0.81	1.5			MtDNA	Ennen et al. 2012
Gopherus agassizii							
	Range-wide	0.06				STR	Murphy et al. 2007
	Range-wide	0.01–0.16		0.64–0.80	8.35	STR	Hagerty and Tracy 2010
	Fort Irwin, California	0.005		0.74	14.4	STR	Latch et al. 2011
	Western Mojave RU	0.04	0.1–0.2			MtDNA	Edwards 2003
	San Bernardino County, California	0.04				Allozymes	Rainboth et al. 1989
Gopherus morafkai							
	Southern Arizona	0.04				STR	Edwards et al. 2004a
	Southern Arizona			0.62	11.1	STR	Fujii and Forstner 2010
	Southern Arizona	0.06	0.1–0.2			MtDNA	Edwards 2003
Gopherus berlandieri							
	North Texas			0.47	2.0–6.7	STR	Fujii and Forstner 2010
	Southern Texas			0.53	2.0–9.5	STR	Fujii and Forstner 2010
	Texas	0.08		0.56	5.6	STR	Fujii and Forstner 2010
	Texas		0.08			MtDNA	Edwards unpublished data
Gopherus flavomarginatus							
	Chihuahua, Mexico			0.5		STR	Edwards unpublished data
	Chihuahua, Mexico		0.24			MtDNA	Urena-Aranda and de los Monteros 2012
	Chihuahua, Mexico	0.69		0.02–0.08		Allozymes	Morafka et al. 1994

than some diploid nuclear markers. In addition, the lack of recombination in mtDNA makes it useful for deciphering the evolutionary relationships among populations relative to their geographic distribution (phylogeography). Mitochondrial markers were some of the first used in the study of genetic diversity of *Gopherus* species. More recently, the inclusion of nuclear genetic markers called microsatellites or short (or simple) tandem repeats (STRs) to population genetic studies of *Gopherus* species has greatly improved our understanding of tortoises on a population level. Since STRs have a faster rate of evolution than mtDNA, are inherited from both parents and provide a broader sampling of the genome, STR data may be used to estimate more recent levels of gene flow and demographic history.

The diversity of mtDNA when compared among different *Gopherus* species illustrates their different evolutionary histories (table 15.1). A modest amount of mtDNA genetic diversity exists in *G. polyphemus* just within Florida: nucleotide sequence divergence among sampled populations averages 1.1% and is attributed to episodes of Pleistocene marine inundation of approximately 1.2 million years ago (Ostenoski and Lamb 1995). Across the entire range of the species, Clostio et al. (2012) observed a large amount of mtDNA sequence divergence, 2.3%, with a pronounced phylogenetic break across the Apalachicola River drainage. This genetic barrier was also supported by Ennan et al. (2012), although they observed a broad zone of geographic overlap of mtDNA haplotypes on both sides of the Apalachicola River. Thus, barriers to gene flow over the evolutionary history of *G. polyphemus* seem to have influenced the distribution of mtDNA variation. This pattern is hardly surprising in this wide-ranging species, occupying habitats in a landscape with heterogeneous climate, vegetation, topography, and other environmental factors. In contrast to the entire species' range, variation of mtDNA within local populations of *G. polyphemus* is low, a pattern which seems to be characteristic of all *Gopherus* species (Ostenoski and Lamb 1995).

Other *Gopherus* species exhibit less divergence between populations than are observed in *Gopherus polyphemus*. Edwards (2003) compared the mtDNA sequence divergence of *G. agassizii* within the western Mojave Desert to that of *G. morafkai* in the Sonoran Desert of southern Arizona. Within similar sized geographic areas, the divergence within populations of both species ranged from 0.1% to 0.2%. Edwards (2003) speculated that the low mtDNA genetic variation observed within populations of both desert tortoise species may indicate a prehistoric reduction of population size in both the Mojave and Sonoran Deserts, coinciding with the most recent major glacial-maximum, approximately 22,000 years ago.

The mtDNA of *G. berlandieri* also exhibits very little within population variability. Using the samples of Fujii and Forstner (2010), Edwards (unpublished data) estimated sequence divergence at 0.08% (n = 58, ND2 region, 509 bp). *Gopherus flavomarginatus* exhibits the lowest genetic diversity: Edwards (unpublished data) observed no mtDNA polymorphisms in an assessment of 78 individuals (ND2 region, 1109 bp). Urena-Aranda and de los Monteros (2012) sampled 76 wild individuals of *G. flavomarginatus* representing the entire distribution of the species in the Bolsón de Mapimí in the Chihuahuan Desert, Mexico, and observed that 74 individuals exhibited a single haplotype of only two that were identified (D-loop, 842 bp). The reduced mtDNA genetic variation in *G. flavomarginatus* could have resulted from the species' undergoing an extreme population bottleneck, caused by range reduction; however a lack of mtDNA diversity would also be expected if the species has simply maintained a small population size over its evolutionary history.

With the exception of perhaps *G. agassizii* and *G. polyphemus,* the full extent of genetic variation across the entire range of the different North American *Gopherus* species has not yet been fully described. For example, the *G. berlandieri* samples analyzed by Fujii and Forstner (2010) were limited to Texas; no samples were collected within the species' range in Mexico. *Gopherus morafkai* exemplifies another species with yet-to-be described diversity; Lamb et al. (1989) suggested a distinct mtDNA lineage of the species in Mexico. Tortoises genotyped in southern Sonora and Sinaloa, Mexico, exhibit mtDNA and STR marker divergence from the rest of *G. morafkai* equivalent to that between the two recognized desert tortoise species (Edwards et al. 2012). This pattern could suggest a possible trichotomy of desert tortoise lineages evolving from a single ancestral type five to six million years ago. From a conservation perspective, the current abundance and range of *G. morafkai* would be reduced dramatically by defining a distinct population or species in Mexico.

How genetic variation is partitioned within and among populations can be assessed with greater resolution using highly polymorphic genetic markers, such as STRs. This method can provide an estimate of gene flow and thus the historic and/or current connectivity of habitat across a landscape. In a range-wide study of *G. agassizii* using STRs, Murphy et al. (2007) estimated that 6.1% of the variation was found among different sample locations. This relatively low level of differentiation among locations suggests that gene flow occurs throughout the sampling area. In an STR study of *G. morafkai,* Edwards et al. (2004a) also detected weak population genetic structure, with 3.7% differentiation among tortoise localities, but the authors sampled from a smaller study area, not representative of the entire range of the species. In *G. berlandieri,* Fujii and Forstner (2010) estimated population differentiation of 8.3% between the north and south of the species' range in Texas and suggested that continued gene flow has prevented strong differentiation. Within a fairly small areas such as the Kennedy Space Center, Florida (Sinclair et al. 2010), or the Camp Shelby Joint Forces Training Facility, Mississippi (Richter et al. 2011), as much as 3.0% genetic differentiation is observed among geographically isolated colonies of *G. polyphemus.*

All of the above estimates of population differentiation using STR markers are consistent with models of population structure in which gene flow occurs among sampling locations, but the scale at which these studies were conducted varied significantly. For example, although the estimates of population differentiation within *G. polyphemus* and *G. morafkai* may appear quite comparable, it is important to note that samples collected by Sinclair et al. (2010) and Richter et al. (2011) each covered a maximum distance of only 45 km, whereas the study by Edwards et al. (2004a) spanned 186 km. In a different study of *G. polyphemus,* Schwartz and Karl (2005) observed extremely high genetic differentiation (using STRs) across a much larger study area, spanning Florida and southern Georgia, with an average of 24% of the genetic variation partitioned among sampling localities. Similarly, in a range-wide study, Clostio et al. (2012) observed as much as 54% genetic differentiation of *G. polyphemus* between disparate populations separated by both the Mobile and Apalachicola Rivers.

The difference between *G. polyphemus* and the desert tortoises may be accounted for by the mobility of the different species. Radio telemetry studies of *G. morafkai* confirm that individuals are capable of making long distance movements (32 km; Edwards et al. 2004b). *Gopherus polyphemus* does not appear to disperse as readily, however (McRae et al. 1981, Diemer 1992b, Eubanks et al. 2003). This difference may account for the much greater genetic differentiation observed among localities throughout the species' range, relative to the other *Gopherus* species. Dispersal is an important characteristic to consider for a species whose habitat has been highly fragmented over the last one hundred years. Other factors that could influence gene flow via dispersal ability are sex-biased dispersal or territoriality, but these have not been determined to have an influence on genetic structure in species of *Gopherus* (Hagerty et al. 2011).

For most *Gopherus* species, genetic structure is primarily a function of the distance between populations acting as a barrier to gene flow, called isolation by distance (IBD). Edwards et al. (2004a: 493) stated that "the desert tortoise is perhaps the ideal organism for the IBD model; one that is distributed across the landscape in patches and for which the difficulty of dispersal is a function of geography." In *G. morafkai,* Edwards et al. (2004a) observed that the geographic distance among populations accounts for approximately 30% of the genetic distance measured among sampled populations. For *G. agassizii,* geographic distance explains approximately 68% of the genetic differentiation observed (Murphy et al. 2007; Hagerty et al. 2011). Similarly, in *G. berlandieri,* 50.4% of the observed genetic differentiation among populations can be attributed to geographic distance (Fujii and Forstner 2010). IBD is observed across the range of *G. polyphemus,* with 89% of the genetic differentiation explained by geographic distance (Clostio et al. 2012). Interestingly, the relationship between geographic distance and genetic distance in *Gopherus* appears to break down when observed at finer, geographic scales (<50 km: Sinclair et al. 2010, Latch et al. 2011, Richter et al. 2011). It is also important to note that several recent studies in other taxa have suggested that "ecological distance" also can have a significant impact on the distribution of genetic variation (see Manel et al. 2003, Hokit et al. 2010). Ecological distance is a measure of the least-cost path distance between populations in relation to landscape factors. Ecological distance may be important to consider when deciphering historical evolutionary processes vs. the potential consequences of recent anthropogenic fragmentation. This area of research remains to be explored in *Gopherus.*

The maintenance of genetic diversity across the range of a species requires careful management. Local adaptation of gene complexes may be vital for survival in some habitats. Consequently, mixing among populations via uninformed translocations and reintroductions may result in reduced survival and lower the viability of the population (called outbreeding depression). Of special concern for long-lived species is the fact that the consequences of outbreeding depression are often not expressed until the second generation (Sheffer et al. 1999). In addition, a loss of a population may represent a loss of unique genetic variants represented nowhere else, and thus fail the objectives of a comprehensive conservation plan. Knowledge of the genetic variability of a species across its range can make an important contribution to implementing successful conservation actions.

POPULATION SIZE

Genetic diversity is generated in populations through gene flow (immigration) and mutation, but its preservation is largely dependent on population size. Changes in population size affect the genetic variation of a population. A recent reduction in population size (population bottleneck) may be detected through genetic studies. Schwartz and Karl (2005) detected recent reductions in population size in several Florida populations of *Gopherus polyphemus.* A significant population bottleneck was also observed in the wild population of *G. flavomarginatus* (Urena-Aranda and de los Monteros 2012; Edwards et al. unpublished data). Unfortunately, the genetic signature of a population bottleneck is difficult to identify in tortoises because the loss of variation proceeds more slowly in long-lived organisms (Tessier et al. 2005). In *G. agassizii,* the genetic signature of a population bottleneck was only detected in the northern part of its range, although population declines throughout its distribution have been well documented (Murphy et al. 2007). Advances in the algorithms used to detect genetic bottlenecks are taking steps to address this limitation (see Kuo and Janzen 2003).

A reduction in population size can result in loss of genetic variation as a result of drift and reduced gene flow (Templeton et al. 1990). Small populations also increase the potential for inbreeding, which can reduce individual fitness and population viability, resulting in an even smaller population. This negative cycle is called an extinction vortex (Gilpin and Soulé 1986). Ennen et al. (2010) found western populations of *G. polyphemus* to have low genetic diversity, which they attributed to population declines in this portion of the species' range. The authors suggested that low genetic diversity may be a contributing factor to reduced hatchling success and that reproductive problems in these populations of *G. polyphemus* may be a consequence of increased inbreeding within small populations (inbreeding depression). Harrison et al. (unpublished data) found lower levels of heterozygosity and allelic diversity in small populations relative to large populations of *G. polyphemus* in Georgia. This finding exemplifies the effects that changes in population size can have on the genetic make-up of a population.

An important application of conservation genetics is to predict a population's evolutionary potential. Evolutionary potential describes the ability of a population to evolve in the presence of environmental change. In the context of species conservation, it is not possible for us to determine which individuals contribute most to the evolutionary potential of the species, or more importantly, which adaptive traits will be most critical in the face of environmental change. For a species to persist in a changing environment, genetic diversity provides the foundation for adaptation. In the face of stochastic processes, such as climate change, the prudent approach to species conservation is to preserve the entirety of a species' genetic diversity.

APPLICATIONS OF GENETIC INFORMATION TO CONSERVATION

Knowing the population structure of a species and the distribution of genetic variation allows us to implement conservation efforts that maintain natural variation and preserve a species' evolutionary potential. We now present some examples of how knowledge gained by genetic studies has been applied to conservation efforts of *Gopherus* species.

For *Gopherus agassizii,* genetic information was considered in the designation of new recovery units (RU's) in the revised recovery plan (USFWS 2011b). Two independent studies (Murphy et al. 2007, Hagerty and Tracy 2010) have assessed population genetic structure across the range of the species, in the context of the RU's of the original 1994 recovery plan (USFWS 1994a). Both studies combined suggest that there are as many as nine genetic subdivisions across the range of the species (fig. 15.1). The revised recovery plan reduced the original number of RU's from six to five, however. The preparers of the revised recovery plan based the new RU's primarily on geographic discontinuities that coincide with observed genetic variation and the plan considered "demographic, ecological, and behavioral considerations to be of

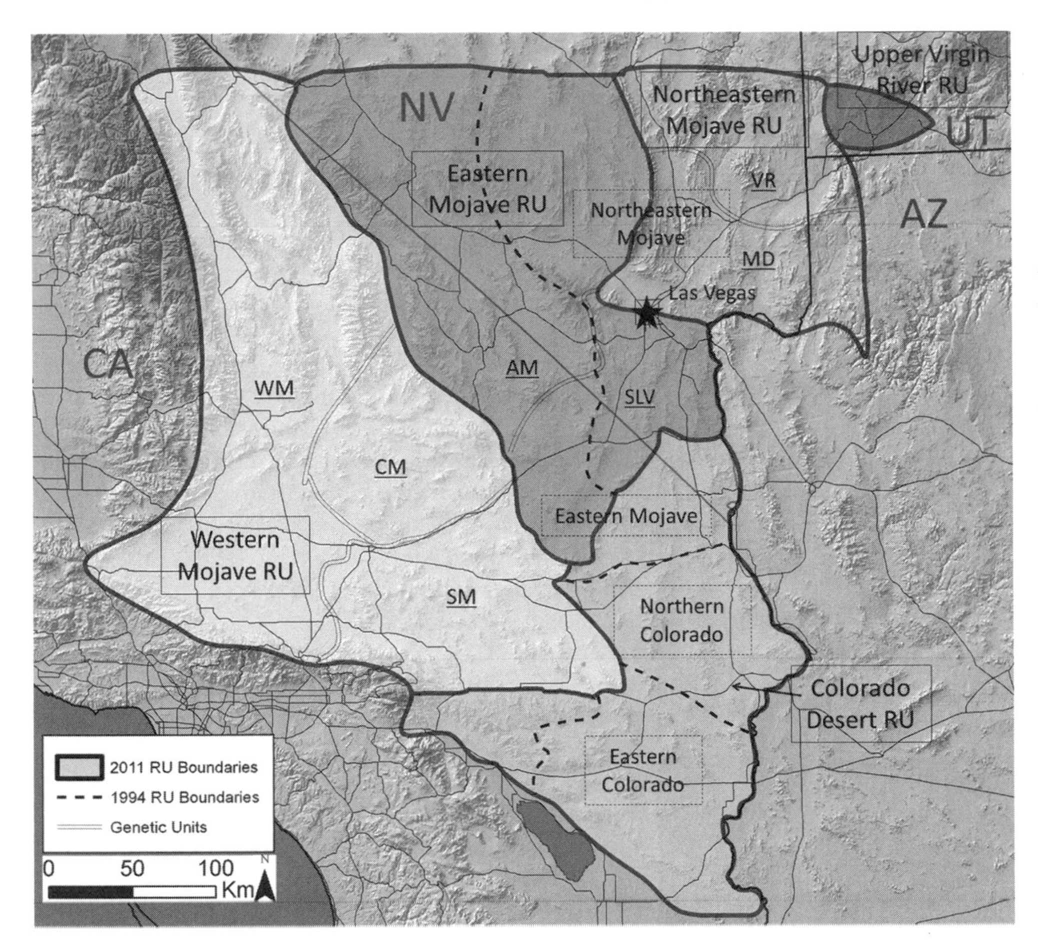

Fig. 15.1. Information from multiple genetic studies contributing to the delineation of Recovery Units (RUs) in the revised recovery plan for *Gopherus agassizii* (USFWS 2011b). Solid line outlines the original six recovery units delimited by the USFWS 1994a Recovery Plan. Dashed line indicates redefinition of the Western Mojave RU by Murphy et al. (2007) into independent Western (WM), Southern (SM), and Central (CM) units. Double line delineates new RU boundaries recommended by Hagerty and Tracy (2010) for Virgin River (VR), Muddy Mountains (MD), Amargosa Desert (AM), South Las Vegas (SLV) and Northern Colorado (NC). Image modified from Murphy et al. (2007).

greater importance than genetic issues alone" (USFWS 2011b: 45). While the advancement of molecular techniques makes it possible to delineate differences between population subunits, an assessment of neutral genetic variability alone does not necessarily mean that those differences are biologically significant (Hedrick 1999). Decisions should not necessarily be made solely on genetics, but should also incorporate relevant ecological and natural history information in defining management units for conservation (Berry et al. 2002a). Ultimately, it is up to resource managers to determine what constitutes a reasonable amount of variation in a population that lends itself to constructive management.

The maintenance of genetic diversity is best achieved by ensuring that there is no reduction in population size and natural patterns of gene flow within and among populations are maintained (Frankham 1995). Artificial habitat fragmentation as a result of anthropogenic landscape change can result in rapid genetic differentiation between remnant patches of habitat (Templeton et al. 1990, McCraney et al. 2010). This rapid differentiation is a consequence of reduction of core habitat, an increase of edge effects, increased rates of drift, and the prevention of normal patterns of gene flow (immigration). BenDor et al. (2009) simulated habitat fragmentation effects on *G. polyphemus* and found that human development that fragments the landscape with as little as 10% habitat loss could reduce the rate of dispersal among subpopulations by as much as 31%. Schwartz and Karl (2005) attribute some of the genetic differentiation they observed among populations of *G. polyphemus* to reduced gene flow caused by anthropogenic habitat fragmentation. In a fine-scale analysis of *G. agassizii,* Latch et al. (2011) observed recent population subdivision, and suggested that roads may have influenced gene flow on a very local level (within a ~25 km radius). Because of the long life span of *Gopherus* species, however, it is generally assumed that current genetic studies can only assess historic rates of gene flow, since landscape changes have been relatively recent in relation to the generation time of tortoises (Edwards et al. 2004a, Sinclair et al. 2010). In a study of recently fragmented populations of *G. polyphemus,* Ennen et al. (2011) observed a greater rate of change in demographic traits than in genetic traits, which they suggest is typical of long-lived species with long generation times. Hagerty and Tracy (2010: 1805) warn that despite the inability to detect current reductions in interpopulation movement via genetic data, habitat fragmentation throughout the range of *G. agasizzi* has "likely removed all possible paths among previously connected populations."

The potential effects of habitat fragmentation in *G. morafkai* were assessed by Edwards et al. (2004a). In the Sonoran Desert, tortoises occur in seemingly isolated patches of upland habitat separated by low, desert valleys (Barrett 1990). Historic connectivity of these patches, resulting from gene flow, was identified in a genetic study by Edwards et al. (2004a), and the authors suggest that movements between populations have been an important part of the species' evolutionary history. Thus, despite the low density of tortoises that occur in valley bottoms separating core populations, these individuals are critical in maintaining gene flow (Averill-Murray and Averill-Murray 2005). Currently, landscape changes, such as roads, urban development, agriculture, and canals, likely make such movements impossible (USFWS 2010a). The fact that each of these foothill habitats contains only a small population of tortoises, and considering the life history traits of tortoises (long-lived and slow to mature), recovery of a small population after a reduction in population size (from drought, disease, etc.) may require immigration from neighboring populations. Thus, persistence of tortoise populations distributed in small patches throughout the Sonoran Desert likely relies on gene flow among populations to ensure long-term viability.

When natural patterns of gene flow can no longer take place, or when a small population is suffering from the effects of reduced genetic diversity, translocation may be a viable management option. Without careful management, however, the introduction of genetically differentiated individuals into wild populations of tortoises could be detrimental. In addition to concerns about health and survivorship of translocated individuals (Tuberville et al. 2005), biologists and managers should consider the potential of outbreeding depression caused by introduction of genetically differentiated individuals (Edwards and Berry 2013). Several studies of *Gopherus* have detected evidence of translocated tortoises in their genetic datasets from samples collected in the wild (Schwartz and Karl 2005, Murphy et al. 2007, Fujii and Forstner 2010, Clostio et al. 2012). These undirected translocations often occur in large numbers (Glenn et al. 1990, Enge et al. 2002, Murphy et al. 2007), resulting from relocations during development or as escaped or released captives. The introduction of "pet" tortoises into wild populations is of particular concern, considering the abundance of hybrid tortoises observed in captivity. In Arizona, Edwards et al. (2010) found that 33% of captive tortoises from the Phoenix, Arizona, area were not the local *G. morafkai,* but instead either *G. agassizii, G. morafkai* × *G. agassizii* hybrids, or even *G. morafkai* × *G. berlandieri* hybrids.

Another caveat of developing translocation and repatriation strategies or addressing the problems associated with escaped or released pet tortoises is polyandry and sperm storage (Moon et al. 2006, Davy et al. 2011). Davy et al. (2011) give a conservative estimate of 64% of female *G. agassizii* being polyandrous, with a resulting 57% of clutches observed that were sired by multiple males. Murphy et al. (2007) report a captive female *G. agassizii* that continued to produce viable clutches of eggs for 15 years following isolation from males. Polyandry and long-term sperm storage allow a single mature female to represent a source of genetic material from multiple individuals. This information should make adult females an important target of any managed translocation or repatriation effort, since they have the potential to successfully introduce additional genetic diversity to the population beyond just their own. A female tortoise of unknown history (such as from captivity), even if genotypically suitable for repatriation, could potentially introduce admixed or hybrid progeny into the population after release, however (Edwards and Berry 2013).

The combination of tortoises' being popular pets and also protected by law means that legal conflicts of ownership are inevitable. Genetic testing has been used to inform identity and placement of confiscated tortoises. In particular, the species or population identity of tortoises has important legal implications, since some species and populations are federally protected and even when held legally as pets, individuals are not allowed to be taken across state lines without permits. Schwartz and Karl (2008) used DNA assignment testing to assess the population of origin for confiscated *G. polyphemus* in Florida. Desert tortoises confiscated by the Arizona Game and Fish Department and of suspicious origins are genotyped so that the appropriate legal action can be taken and the tortoise can be properly placed in a suitable captive facility (Edwards 2008). Unfortunately, genotyping does not entirely solve the problem for desert tortoises. Because the geographic boundary of the federally protected portion of the range of the *G. agassizii* is defined by the Colorado River, individuals of the species that naturally occur in the Black Mountains of Northwestern Arizona (McLuckie et al. 1999, Edwards et al. 2010) are not afforded the same legal protection as the rest of the species under the current recovery plan (USFWS 2011b). Thus, law enforcement officers cannot always rule out that a confiscated *G. agassizii* in Arizona may have originated legally in the state.

The only captive breeding effort specifically for conservation of a *Gopherus* species has been for *G. flavomarginatus,* as part of the repatriation effort into the United States (Truett and Phillips 2009). A population of captive tortoises is being maintained by the Turner Endangered Species Fund in a seminatural setting in part of their historic range in the Chihuahuan Desert in New Mexico. Genetic information is being used to inform mate pairing, to maximize genetic diversity in the captive population and to reduce the potential for inbreeding (Edwards et al. unpublished data). Captive breeding and repatriation can be an effective conservation strategy, but requires special consideration when applied to long-lived species like tortoises (Williams and Osentoski 2007). Small populations are more susceptible to inbreeding, which can decrease heterozygosity of individuals and lead to a reduction of fitness. Inbreeding can also lead to the expression of recessive alleles, resulting in a decrease in population viability. One way to improve genetic diversity of *G. flavomarginatus* populations is to introduce new individuals into the breeding program from zoological institutions or private collections. Genetic screening of potential new animals has uncovered multiple *G. flavomarginatus* × *G. polyphemus* hybrids in captivity, which are ineligible for inclusion in the repatriation effort (Edwards et al. unpublished data).

CONCLUSIONS

Understanding the population ecology and genetics of a species is fundamental to implementing effective strategies for conservation. For a broad overview of genetic applications to turtle conservation, see Shaffer et al. (2007). As other chapters of this book have exemplified, there are several life history traits such as dispersal ability, survivorship of adults, longevity, and delayed sexual maturity that are shared among species of the genus *Gopherus*. The ability to "share" information from one species to inform management decisions about another is sometimes necessary in making timely conservation actions. Conservation biologists and resource managers are often forced to move forward based on the best information available. As is apparent in this book, most of what we know about the genus *Gopherus* has come from studies of *G. polyphemus* and *G. agassizii*, despite other species' perhaps having greater immediate threats of extinction. Application of the tools and principles of population genetics is imperative for successful management and conservation of all species in the genus.

Acknowledgements

We thank R. W. Murphy, K. H. Berry, M. E. Kaplan, C. Dolan, and an anonymous reviewer for their helpful comments on this chapter.

16 Demography of North American Tortoises

EARL D. MCCOY
GUSTAVO AGUIRRE L.
RICHARD T. KAZMAIER
C. RICHARD TRACY

Life-history traits include those characters ultimately influencing the growth and persistence of populations in the face of environmental change. In general, the traits of males, as well as the sex ratios of populations, may not be as important as the traits of females to population growth in North American tortoises, because females of most tortoise species are known to store sperm. The overriding importance of female traits may be especially strong when sperm can be stored for multiple years with little loss of ability to fertilize eggs, which has been demonstrated for *Gopherus agassizii* (Palmer et al. 1998). Females should be able to produce eggs even when they mate infrequently, so traits of females may be key to predicting population growth. The important female traits include age of first reproduction, clutch size, number of clutches produced per year, and age-specific survivorship. Environmental factors influence all of these traits.

Here, we present an outline of the demographic data available for females of the species of North American tortoises. We also present comparative data for males. The type and extent of data vary among species, often reflecting the biases of the researchers gathering the data. Geographical and environmental variation in the data are illustrated, but the actual extent of such variation probably is greater than reported. As demography depends on many rates, such as birth rate and growth rate, information from other chapters (e.g., chapters 7 and 13) can supplement the information presented in this chapter.

A Note on Tortoise Size

"Size" of a tortoise typically is indicated by one or more shell measurements. In most cases, the principal measurement is midline carapace length (MCL), which is the straight line measurement from the anterior edge of the nuchal scute to the caudal edge of the most posterior scute (supracaudal or 12th marginal). For most species, MCL and straight carapace length (SCL) are synonymous; but, for some species, such as *G. berlandieri,* they are not. The supracaudal scute curls under in males of *G. berlandieri* so much that MCL can be shorter than SCL. When the carapace is hinged and movable, other measurements, such as curved carapace length (CCL) or midline plastron length (MPL, from the gular notch to the anal notch) must be taken; but, this is not an issue for North American tortoises. Height typically is the measurement from the bottom of the plastron at its midpoint (junction of the abdominal and femoral scutes) to the top of the carapace and width is the measurement of the carapace at its widest point.

A Note on Tortoise Age

Because tortoises grow at different rates throughout their lifespans, and the rates at which individuals grow are further modified by the productivity of the environment, aging them is a difficult process. Individuals of the same size in two different locations may be quite different in age. One way that aging may be accomplished is through the use of plastral scute annuli. These are annual growth rings, similar to those in trees. Scute annuli tend to become indistinguishable as individual growth slows with age, however, and annuli often are obscured because of abrasion with sandy substrates. Individuals also may lay down more than one growth ring per year. Thus, one must be cautious about interpreting age from growth rings (Wilson et al. 2003). The use of scute annuli for aging is confined to relatively young individuals, for example, up to 12–15-year-olds in *G. polyphemus* (Mushinsky et al. 1994, Aresco and Guyer 1999b) and up to 18-year-olds in *G. berlandieri* (Hellgren et al. 2000).

GOPHERUS FLAVOMARGINATUS

Gopherus flavomarginatus females reportedly become sexually mature at a mean age of 13.9 years, based upon Richards's growth model and measures of scute annuli of live and preserved animals (Germano 1994). The mean carapace length of adult females from the Mapimi Biosphere Reserve (MBR) is 352 mm (SD = 15.1, range = 315–385, n = 57), and the minimum size at maturity for females with evident eggs in oviduct through inguinal palpation is 315 mm and 326 mm CL. This minimum size corresponds to age 18–20 years, fitted by a logistic growth model to age estimations derived from interval growth equations (Frazer and Ehrhart 1985). The only publication on female gonad condition for *G. flavomarginatus* reports the appearance of large follicles (14 to 16 mm in diameter) and corpora lutea in a female 326 mm CL (Legler and Webb (1961). Two young wild-caught females were tracked for nine years, and when they reached 242 and 287 mm CL, they were approximately 15 and 17 years old (as estimated from growth rings), and at those sizes, they showed no evidence of follicles or eggs as determined by inguinal palpation. Their progesterone, estrogen, and testosterone levels were one order of magnitude lower than those of reproductive females larger than 300 mm CL (Gonzalez et al. 2000). The 15- and 17-year-old females did show a small increase in progesterone during the summer, however, suggesting that they were already undergoing follicular growth and atresia and yolk reabsorption, which are phenomena related to the proximity of sexual maturity (Kuchling 1999).

Gopherus flavomarginatus males reach sexual maturity earlier, and at a smaller size, than females. Mean carapace length of adult males from the MBR showing reproductive activity and behavior is 302 mm (SD = 4.11, range = 256–348, n = 29). Male asymptotic size is also smaller in comparison to female asymptotic size (317.5 mm vs. 362.6 mm). Estimated age at maturity is 12.8 years. The substantially larger size of females may be related to the body volume necessary to accommodate eggs. Average egg size and weight for 318 field-collected eggs in the MBR was 47.6 (SD = 2.9, range = 40.3–57.5) × 44.1 mm (SD = 2.1, range = 38.0–49.0) and 54.3 g (SD = 7.6, range = 38.0–72.0) (Aguirre, unpublished data).

Females in the wild produce one or two clutches each year, although captive females fed ad libitum are able to produce as many as three clutches annually. Females may skip reproduction under unfavorable habitat productivity conditions. Clutch size is highly variable, ranging from two to 12 eggs. Mean clutch size is 5.4 eggs (SD = 2.04, n = 60), and the mean total clutch mass is 298 g (SD = 119.0, range = 85–708). No relationship between clutch size and laying date, or between clutch size and female carapace length, width, or height; plastron length; or body mass has been found in *G. flavomarginatus*. No relationship between clutch size and female body size is typical of other species of *Gopherus*, but a positive relationship between them also has been documented (Iverson 1980, Landers et al. 1980, Rose and Judd 1982, Turner et al. 1986, Judd and Rose 1989, Diemer and Moore 1994). How much of clutch size variability is attributable to female body size is difficult to estimate, and primary control of clutch size may be related to female genotype (Bjorndal and Carr 1989). How environmental variation may affect clutch size is not known; the effects of within- and among-year variation in habitat productivity on energy gain remain to be understood.

According to optimal offspring size models, a negative correlation between egg mass and clutch size would indicate a trade-off in these variables. No evidence for such a trade-off in *G. flavomarginatus* has been demonstrated; no correlation between egg mass and clutch size or between egg length or width and clutch size has been shown. It has been suggested that in other reptiles, the relationship can be obscured by high variability in clutch size, related to differences in resource acquisition among individuals (Schwarzkopf 1993). This situation may also obtain for *G. flavomarginatus*.

Survival of all life stages up to sexual maturity is poor. Nest mortality recorded at MBR is 87%, mostly caused by predation. Given low hatching success, primarily because of nest mortality and egg infertility (33%, on average), a female typically can produce only three hatchlings during an eight-year period (Adest et al. 1989). Survival of hatchlings to maturity has been shown to be only about 5%. Radio telemetric monitoring of hatchlings revealed 64% mortality within the first year of life (Tom 1994). Combined, these factors serve to reduce recruitment to extremely-low levels (Aguirre et al. 1997). Although dead immature animals are encountered occasionally in the field, the remains of dead adults rarely are encountered, suggesting low mortality of adults of *G. flavomarginatus*.

The overall adult sex ratio at MBR is 1 male: 1.2 females. Primary sex ratio is not known. The population at MBR is characterized by a seemingly high proportion of large individuals, which may have resulted from enhancement of the survival of small cohorts in the past (Adest et al. 1989). As recruitment declines, the less mortality-prone adults would persist, and if population size is declining, the mortality among hatchlings/juveniles is most likely to be responsible for the observed population structure (Aguirre et al. 1997). A chaos process model is likely to operate on recruitment and dispersal within *G. flavomarginatus* populations inhabiting an environment where precipitation is naturally low, seasonal, or highly variable from one year to the next (Morafka 1994).

Monitoring the population structure of one population in the MBR revealed substantial change over a five-year period. Juveniles decreased by about 17% of the total population and subadults decreased by about 10% (Adest et al. 1989, Aguirre et al. 1997). Maturation accounted for less than 25% of the change; therefore, emigration and mortality are suspected to be the main factors responsible for the change. Overall, the population was comprised of 12.7% juveniles, 6.3% subadults, and 81% adults during these five years. Surveys of the same population 20 years later, showed a continued predominance of adults (76%), but a modest increase of subadults (18%) and a decrease of juveniles (6%). Large proportions of adults may reflect elevated reproductive success in the past, diminished recruitment, emigration of immature, immigration of adults, and great longevity coupled with slow adult growth (Adest et al. 1989). Some of the oldest individuals of both sexes in this population, which are estimated to be about 60 years old, remain reproductive. Had juveniles been sampled adequately, then high juvenile mortality would likely be responsible for the observed population structure, and the population size likely would be declining. On the other hand, if juveniles have not been sampled adequately, because of their secretiveness, then the population size could have been constant or even increasing (Tom 1994, Aguirre et al. 1997).

GOPHERUS BERLANDIERI

The extent of variation in the demography of the *Gopherus berlandieri* across its range is unknown. Although nearly two thirds of the geographic range of the species is in Mexico, demographic data from that part of its range are virtually nonexistent. Within Texas, published research has focused on only two regions, the Lower Rio Grande Valley in Cameron County and Chaparral Wildlife Management Area (WMA) in Dimmit and LaSalle Counties. Chaparral WMA is ~300 km northwest of the more extensively studied Lower Rio Grande Valley populations, and ~200 km southeast of the northwestern extent of the range of this species. Demography of *G. berlandieri* appears to differ dramatically between these two locations (Auffenberg and Weaver 1969, Rose and Judd 1982, Hellgren et al. 2000, Kazmaier et al. 2001a), but explanations for differences are complicated by pronounced differences in climate and habitat. For example, Chaparral WMA is dominated by arid mesquite (*Prosopis glandulosa*)-*Acacia* thornscrub communities, but the Lower Rio Grande Valley is dominated by more mesic tropical thornscrub.

Males of *G. berlandieri* tend to be larger than females, and size and dimorphism can be very different regionally. At least within Texas, body size generally decreases moving north and west through the range of the species (Rose and Judd 1982). Among studied populations, the smallest adult males are found at Chaparral WMA (mean = 150 mm SCL, n = 1022; Kazmaier unpublished data) and the largest males in Cameron County near Port Isabel (mean = 196 mm SCL, n = 7; Auffenberg and Weaver 1969). Similarly, the smallest females also are found at Chaparral WMA (mean = 141 mm SCL, n = 998; Kazmaier, unpublished data) and the largest females at Laguna Atascosa National Wildlife Refuge in the northeastern corner of Cameron County (mean = 170 mm SCL, n = 39; Bury and Smith 1986). Logistic growth curves at Chaparral WMA suggested that males have an asymptotic size of 172 mm SCL and females have an asymptotic size of 154 mm SCL (Kazmaier et al. 2001a; Kazmaier, unpublished data). Unfortunately, although asymptotic size clearly is different in more southern populations, age-structured growth data are not available for any of the other populations to allow its calculation.

Based upon x-ray analyses, females begin egg production at 128 mm SCL at Chaparral WMA (Hellgren et al. 2000; Kazmaier, unpublished data) and 142 mm in the Lower Rio Grande Valley (Judd and Rose 1989). The former value corresponds to an age at maturity of about five years for Chaparral WMA (Hellgren et al. 2000), but age data were unavailable for the later Lower Rio Grande Valley estimate. Based upon growth curves extrapolated from scute rings, the age at maturity is approximately 13 years (Germano 1994), but similarly derived growth curves from an age-validated population (Hellgren et al. 2000) suggest this could be an overestimate. In contrast to this long time to maturity, one report suggests that the species reaches maturity in three to five years (Auffenberg and Weaver 1969). This estimate is based on an assumption that sexual maturity is reached at 105 mm SCL, however, which seems too small given the aforementioned egg production estimates.

Age and size of maturity for males have not been examined, but males begin developing sexually dimorphic characters at ~120 mm SCL or about four years of age at Chaparral WMA (Hellgren et al. 2000). Males in the Lower Rio Grande Valley were diagnosable at 125 mm SCL (Judd and Rose 1983), but another report suggests a lower value of 105 mm SCL (Auffenberg and Weaver 1969). Examination of 1616 male tortoises from Chaparral WMA (1994–2005) revealed only 17 individuals (1%) that were diagnosable as males below 120 mm SCL, the smallest of which was 108 mm SCL (Kazmaier, unpublished data). All 17 of these individuals had higher than expected annuli counts based upon their body size, however, and these instances could represent "stunted" individuals (Kazmaier, unpublished data), or the inadequacy of scute-ring counts in aging *G. berlandieri* (Wilson et al. 2003). Similarly, only six of 304 male tortoises (2%) captured from several sites in the Lower Rio Grande Valley were less than 120 mm SCL, and these tortoises also exhibited higher than expected annuli counts (Kazmaier, unpublished data).

X-rays and ultrasounds of 353 female tortoises at Chaparral WMA revealed 113 clutches with an average clutch size of 1.9 eggs, with a range of 1–4 (Hellgren et al. 2000, Kazmaier, unpublished data). Using x-ray techniques and necropsy in the Lower Rio Grande Valley, 33 clutches were documented with an average clutch size of 3.1, with a range of 1–7 (Rose and Judd 1982, Judd and Rose 1989). Another report found a clutch size of 1.4 (n = 88; range 1–3) in the Lower Rio Grande Valley, but this result was based on nest excavation and could be influenced by females partitioning their clutch into multiple nests (Auffenberg and Weaver 1969). Information on clutch frequency largely is lacking, but one report modeled annual average clutch frequency at 1.3 for Chaparral WMA (Hellgren et al. 2000). Two peaks in nesting (June–July, August–September) appear to occur in the Lower Rio Grande Valley (Auffenberg and Weaver 1969), which also could suggest double clutching. One report suggested that approximately 30% of females do not produce a clutch in any given year (Judd and Rose 1989).

Egg size, as determined by x-ray, averaged 43.9 × 33.2 mm (n = 32) at Chaparral WMA (Kazmaier, unpublished data). Average egg size for 12 field-collected eggs in the Lower Rio Grande Valley was 41.5 × 26.9 mm (Judd and Rose 1989). Egg size at another site in another Lower Rio Grande Valley site averaged 48.5 × 34.0 mm (n = 9; Auffenberg and Weaver 1969). Few data are available on hatchlings for this species. Average size of ten captive-hatched individuals was 39.7 mm SCL and 21.1 g (Judd and McQueen 1980), and one report of hatchling size was 41 mm SCL (Agassiz 1857). From 3343 individuals captured at Chaparral WMA, the smallest was 35.1 mm SCL (Kazmaier, unpublished data).

An age distribution of 2076 individuals (1994–2005) from Chaparral WMA revealed that only 0.05% of individuals were 18 years or older (Kazmaier, unpublished data). Indeed, 90% of all individuals captured were at ten years of age or younger. In reality, these percentages should be skewed to even younger age classes because of a highly underrepresented 0–3-year-old cohort in the sample, resulting from capture bias. Modeling suggested that 50–70% of the total Chaparral WMA population might be in this poorly sampled 0–3-year-old group. Age data from a population in the Lower Rio Grande Valley (n = 89) produced a dramatically different age distribution, with 21% of the population being 18 years of age or older and only 53% ten years of age or younger (fig. 16.1). From the 2076 individuals for which age data has been gathered at Chaparral WMA, the oldest female was 21 years of age and the oldest male was 24 years of age (both individuals were originally captured at younger ages for which reliable annuli counts could be conducted and subsequently recaptured; Kazmaier, unpublished data). Because the scute-ring technique for aging tortoises in this species becomes problematic beyond approximately 18 years of age (Hellgren et al. 2000; Kazmaier, unpublished data), maximum age in the Lower Rio Grande Valley populations is currently unknown, but ages in excess of 50 years have been documented for captive individuals of *G. berlandieri* (Judd and McQueen 1982).

Age-structured regression of 646 male and 760 female *G. berlandieri* at Chaparral WMA produced annual survival rates of 71.3% for females and 75.9% for males (Hellgren et al. 2000; Kazmaier, unpublished data). In the same population, Kaplan-Meier survival estimates based upon a radiotelemetry study conducted over four years produced annual survival rates of 77.4% for females and 83.4% for males (Hellgren et al. 2000). Differences in the age range of individuals used in these two analyses (5–18 years of age for age-structured regression; 8–11 years of age for radiotelemetry) could explain these differences. A mark-recapture analysis of ten years of data from Chaparral WMA (3132 captures of 2128 individuals; 1990–1999) produced an adult annual survival rate of 79% (Kazmaier et al. 2001b), and a population growth rate of 0.98 (95% confidence limits: 0–945–1.019), suggesting a stable to slightly declining population. In contrast to these results, a mark-recapture analysis from two populations in the Lower Rio Grande Valley produced annual survival rates of 76.0% (females) and 86.2% (males) for at one site and 92.0% (females) and 94.0% (males) at the other (Rose et al. 2011). Rose et al. (2011) suggested the differences in survival were a result of differences in prickly pear (*Opuntia* spp.) densities between the two sites, with the low cactus density site producing survival estimates similar to those at Chaparral WMA. Additionally, the annual survival for juveniles was estimated to be 70.6% for the site with high prickly pear density and 49.6% for the other site (Rose et al. 2011).

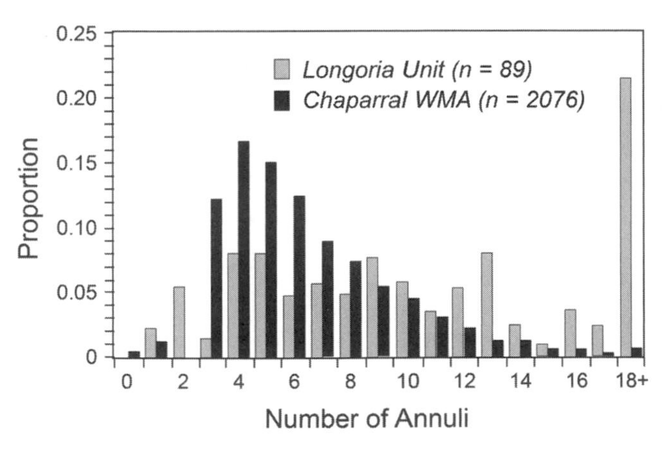

Fig. 16.1. Proportional age distributions of *Gopherus berlandieri* from Chaparral Wildlife Management Area, Dimmit and La Salle counties (1994–2005), and the Longoria Unit of Los Palomas Wildlife Management Area, Cameron County (1999–2002), Texas. Annuli counts have been validated to be accurate ± 1 yr on Chaparral WMA up to 18 years of age (Hellgren et al. 2000; Kazmaier unpublished).

More than 20 years have passed since Rose and Judd (1989) acknowledged that *G. berlandieri* was demographically plastic, but we are little closer to explaining the extent of this plasticity or explaining the mechanisms behind it. More recent research from the more centrally located Chaparral WMA (Hellgren et al. 2000; Kazmaier et al. 2001a, b) has clarified some aspects of the demography of this species. This research also has provided even stronger evidence of pronounced regional variation in this species, however. The extent to which this variation is clinal cannot be addressed until additional populations in the western Rio Grande Plains and Mexico are characterized. Despite that lack of regional data, various studies in the Lower Rio Grande Valley within Cameron County (Auffenberg and Weaver 1969; Bury and Smith 1986; Judd and Rose 1983, 1989, etc.) provide evidence that substantial differences in demography of populations can occur even within this relatively small area (~2,300 km^2).

DESERT TORTOISES (*GOPHERUS AGASSIZII* AND *G. MORAFKAI*)

Recently, a paper (Murphy et al. 2011) has asserted that the genetic differences seen first by Lamb et al. (1989) warrant designating two species of desert tortoise (one found largely in the Sonoran Desert, and the other in the Mojave Desert, mostly north and west of the Colorado River). The literature has heretofore not recognized the two proposed species as different, and the life-history information presented here largely amalgamates data and analyses from desert tortoises in all of North America.

From an analysis of a dataset assembled from the literature (Turner et al. 1986, Henen 1997, Wallis et al. 1999), it is clear that the number of eggs produced by desert tortoises (fig. 16.2) depends both upon the size of females ($F_{1,188} = 148$; $p < 0.001$) and upon the year in which eggs were produced ($F_{6,188}$; $p < 0.001$), with a multiyear mean of 5.8 eggs per female. The "year" variable largely reflects the year-to-year variability in the amount of food and water resources available to females

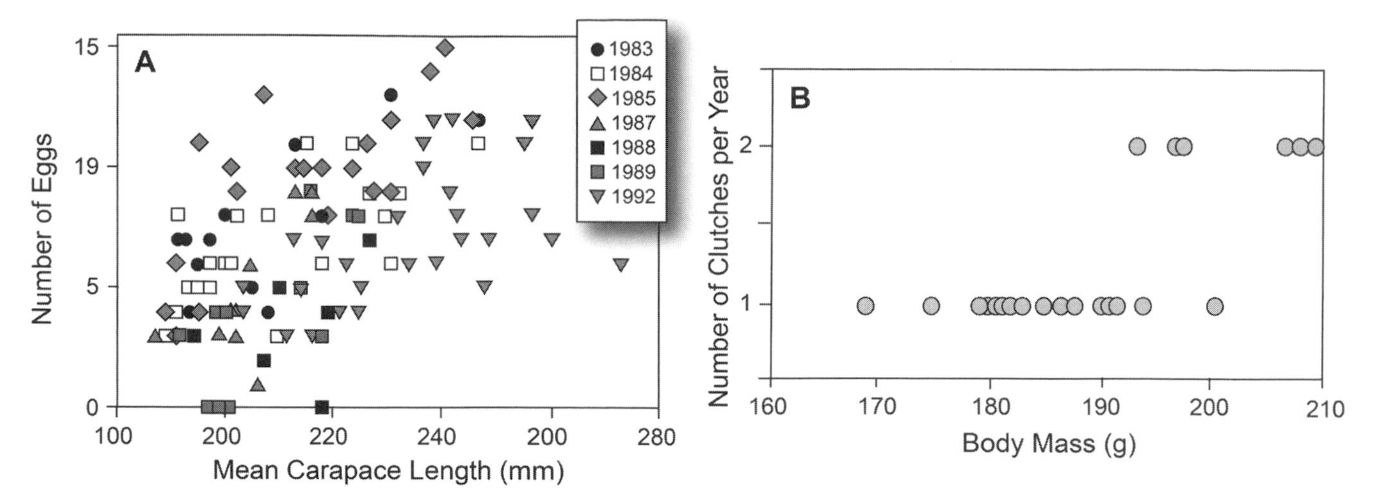

Fig. 16.2. *A*, number of eggs produced by marked tortoises in the eastern Mojave Desert from 1983 to 1992 (Turner et al. 1986, Henen 1997, Wallis et al. 1999); *B*, number of clutches produced by marked tortoises in the eastern Mojave Desert from 1987 to 1989 (Henen 1997).

that can be converted into eggs. Thus, egg production differs in years with adequate rainfall compared to years of drought, yet most females produce some number of eggs in nearly all years (Henen 1997). When females become larger than 185 mm, they can, but not always do, produce a second clutch (fig. 16.2). The number of eggs produced by a female tends to increase with body size, and this increase may be expressed either as production of more eggs in a single clutch or more eggs spread over two clutches (Henen 1997).

The pattern of egg production by females is related to another life-history trait, size at first reproduction. Generally, females begin producing eggs when they are approximately 185 mm in midline carapace length (MCL). It is important to point out that it is size of females that determines egg production, not age of females, as females reach 185 mm MCL at different ages depending upon growth rates. The size of males is largely irrelevant for several reasons. First, females can, and often do, mate with more than one male for each clutch, and second, females can store sperm for at least two years (and maybe as long as five years). Thus, for females, the issue is not finding a mature male, but finding sperm, and finding sperm is generally not a problem for females over a period of several years.

Because female desert tortoises begin producing eggs when they reach a particular body size, as apposed to a particular age, the growth rate of individual females ultimately determines the time required to reach the size at which a female begins producing eggs. Growth rate of females depend on two key attributes of the environment: quality of food available and duration of the season in which food is available. In most years, the habitats in which desert tortoises are found will have grass and forb forage available for one to three months, after which much of the forage in habitats dries and blows away. The conspicuous exception to this pattern is found in the eastern Sonoran Desert, where monsoonal rains create a second season during which forage can grow. Forbs are higher-quality forage than are grasses (and certainly are shrubs), as they generally contain more protein and less fiber than do grasses, and growth rates of tortoises are related to forage quality (fig. 16.3; Tracy et al., unpublished). In combination, the quality of forage and the duration of the season during which food can be obtained ultimately determine the time required for females to reach the size of maturity (fig 16.3; Tracy et al., unpublished). Under laboratory conditions, with very high-quality food available, females can reach adult size in three to five years. In nature, however, mean time to first reproduction for females has been determined for only two populations, and the lower age of reproduction is approximately 13 years on the Beaver Dam Slope at the Nevada / Utah border (Hohman and Ohmart 1980). This area has relatively higher rainfall and longer activity seasons; and because the summers there are relatively mild, individuals typically do not become dormant during the middle summer months as they do in warmer parts of the geographic range. The longer time to the age of first reproduction is from the Nevada Test Site near Las Vegas, where females reached 185 mm in approximately 16 years. This area is at a lower elevation area with hot summer seasons and where individuals become dormant during the mid-summer months. A simulation blending the effects of food quality and duration of the season during which food is available suggests that mean time to first reproduction in females could vary from about 13 years to as many as 20 years (fig. 16.3).

When a flush of annual forage occurs (generally responding to rainfall in winter for *Gopherus agassizii*) (Tracy unpublished), eggs will be produced, often with little difference between egg production in years with a great flush of forage and years with just a modest flush of forage (fig. 16.4). In years with little rainfall, when nearly no forage production occurs, egg production can be greatly reduced, but usually not

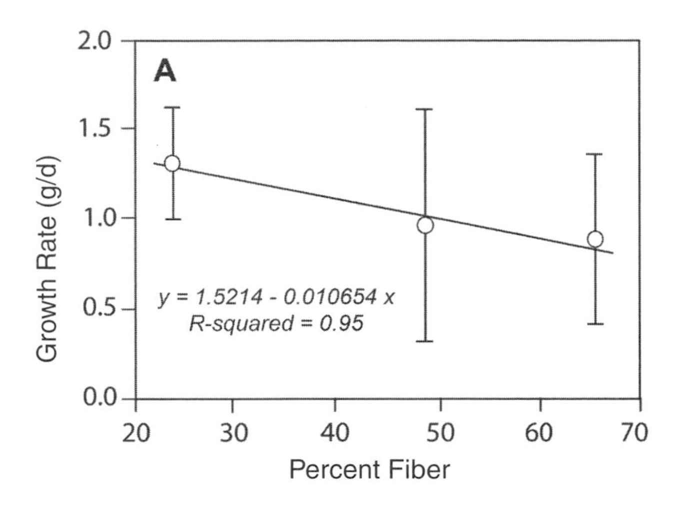

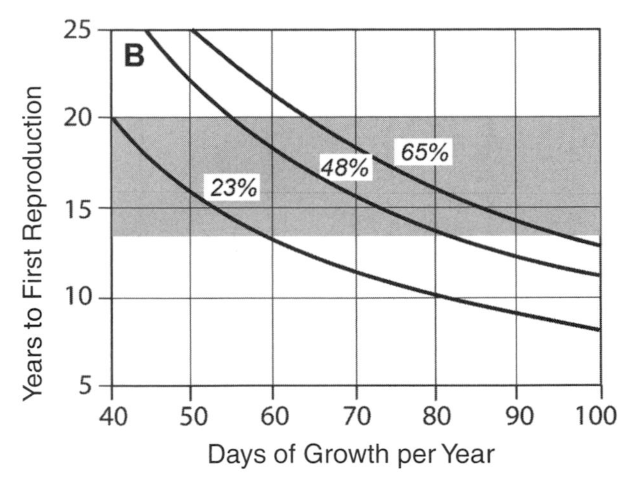

Fig. 16.3. *A,* mean daily body growth rate of immature *Gopherus agassizii* fed experimentally manipulated diets differing in percentage of neutral detergent fiber (Tracy unpublished). *B,* calculated growth rates based on the data from *A.* Diet fiber percentage and number of days of food availability were used to calculate the times for tortoises to reach a body size of 185 mm MCL.

entirely ablated (unless several seasons in a row with no forage occur). The reproductive component of fitness for long lived tortoise species probably should be assessed as an average of egg production over several years. Thus, it may be ecologically more important to assess the frequency of bad years as a predictor of a change of reproductive output of individuals at a particular site rather than the egg production of a sample of individuals in any particular year. Data from range scientists in the U.S. Bureau of Land Management from three sites (Tracy unpublished) illustrates that bad years (years of lower egg production) occur as frequently as 50% of the time at the hotter, drier sites, and as infrequently as 10%, or even less, of the time at cooler, wetter sites (fig. 16.4).

The data on age-specific mortality are scattered. The synthesis in the original recovery plan for *G. agassizii* reported likely mortality rates of preadults (from eggs to individuals of 185 mm carapace length) to be 99%, and annual mortality rates of adults to be about 2% (USFWS 1994a). One study reported mortality of marked adults in a longitudinal study at two sites over a period of seven years, and found a mean mortality rate at one site of 16%, with the highest mortality rate in any one year being 40% (Longshore et al. 2003). At a more equitable site, this study found a mean mortality rate of only 1%, with the highest rate of mortality in any one year being 6% (Longshore et al. 2003). Another study reported mean mortality rates of adults at nine sites in three successive years of 2%, 10%, and 18%; with the highest rate in any one single year being slightly more than 40% (Esque et al. 2010b). Clearly, mortality rates of adults can be substantially different, depending on year (and likely food and water resources in any particular year), place (reflecting the severity of local climate and resulting food and water resources), and presence and diversity of predators and parasites. Preadult mortality rates are extremely difficult to assess. One study assessed mortality rates of eggs, emerging neonates, and dispersing neonates, and found cumulative mortality rates of 22% and 50% in subsequent years (Bjurlin and Bissonette 2004). These results suggest highly variable mortality rates of preadults, depending on environmental conditions including predation, as well as abiotic differences among years (Bjurlin and Bissonette 2004). Mortality rates between the neonate stage and adulthood essentially are unstudied.

GOPHERUS POLYPHEMUS

Although the demography of *Gopherus polyphemus* fits the general mold of a long-lived, slowly reproducing species, this generalization belies the substantial variation in demographic traits that the species displays over its range. Demographic variation seems to be associated with both latitude and longitude, perhaps reflecting genetic isolation (Schwartz and Karl 2008, Ennen et al. 2012) to some degree; but, more likely, this variation reflects mostly ecological variation (KR Smith et al. 1997). For instance, age at sexual maturity depends upon how long individuals take to reach minimum size for maturity, and the length of time can vary greatly, depending upon factors such as nutrition (Germano 1994, Mushinsky et al. 1994, Aresco and Guyer 1999b). Therefore, in a review such as this one, it seems appropriate to specify, as much as possible, a range of values, identified by location.

Sexual size dimorphism is pronounced in *G. polyphemus,* with females being larger than males. Asymptotic size calculated from von Bertallanfy or logistic curves yielded ratios of females to males of 1.08 (301 mm CL / 278 mm CL) in central Florida (Mushinsky et al. 1994) and 1.19 (323 mm CL / 271 mm CL) in Alabama (Aresco and Guyer 1999b). The smaller asymptotic size of females in the first case may reflect the relatively young group of females included in the study.

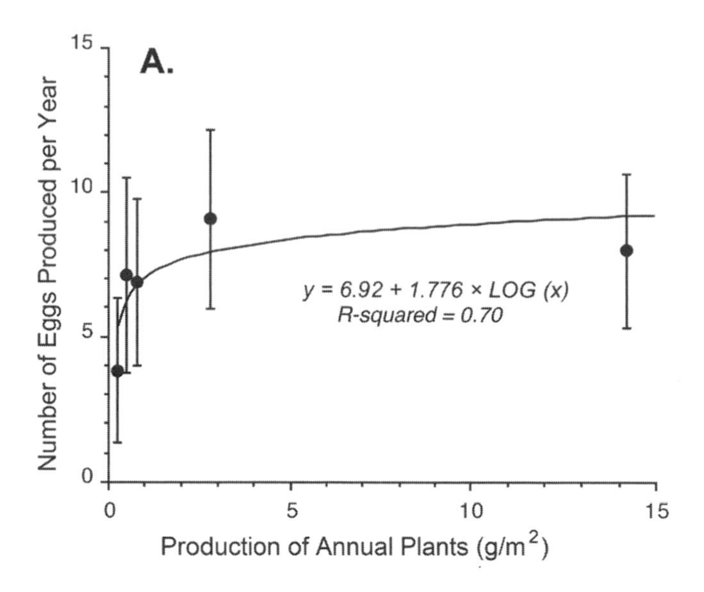

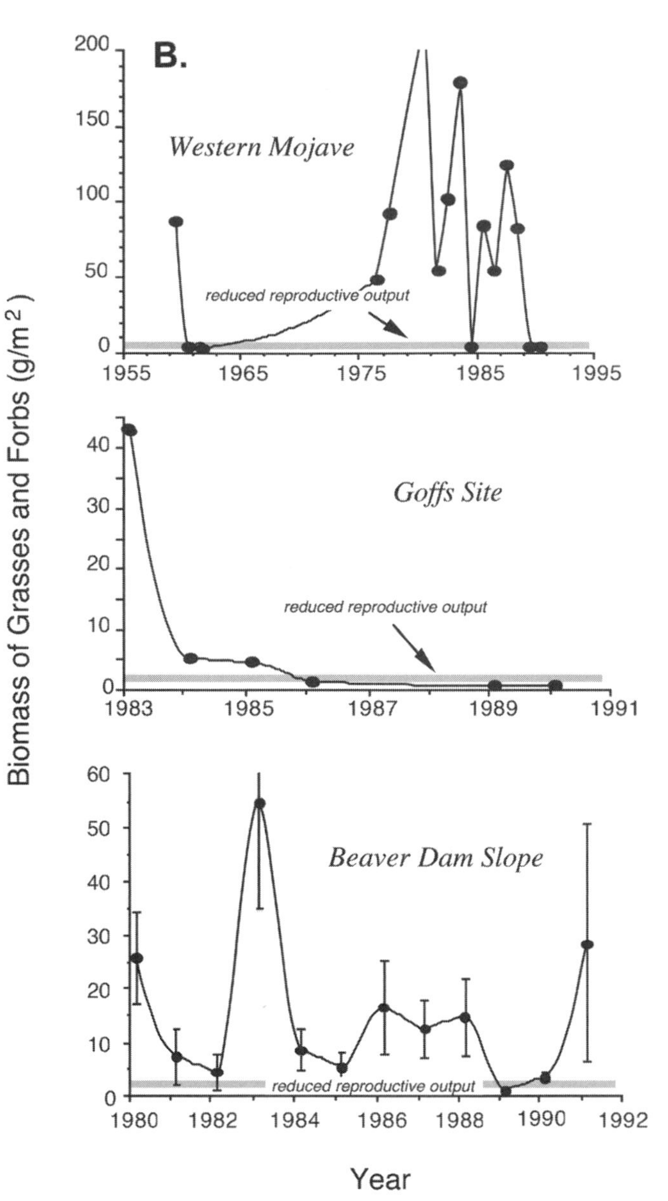

Fig. 16.4. A, number of eggs produced by marked *Gopherus agassizii* in the eastern Mojave Desert as a function of the primary productivity of the study sites (Turner et al. 1986, Henen 1997). *B,* biomass of grasses and forbs on Bureau of Land Management study plots at three sites in the Mojave Desert between 1957 and 1991. The gray line at 2 grams of plant production per square meter is seen as a cutoff below which egg production is reduced.

Mean size of gravid females, a less-reliable indicator of female body size, varied between 255 mm and 308 mm CL across the species' range (Ashton et al. 2007). Maximum attainable size is suggested to be 387 mm CL (Timmerman and Roberts 1994). Female body size at sexual maturity typically is 230–240 mm CL, although smaller gravid females have been observed (chapter 13). High nutrition diets allow females to reproduce at sizes somewhat smaller than 230 mm CL (Small and Macdonald 2001). Males may reach sexual maturity at a body size as small as 180 mm CL (Diemer and Moore 1994, Mushinsky et al. 1994).

Age at sexual maturity depends upon how quickly reproductive size is reached (chapter 7). Females may take 20 years or more to reach sexual maturity in Alabama (Aresco and Guyer 1999b) and Georgia (Landers et al. 1982), 14–18 years in north-central Florida (Diemer and Moore 1994), and as few as 9–11 years in central Florida (Mushinsky et al. 1994). The latter value was obtained in prime habitat; in a near-by location with less-suitable forage, sexual maturity was not reached for 14–16 years (Godley 1989). Males likely reach sexual maturity at slightly younger ages than females in the same location (Landers et al. 1982, Diemer and Moore 1994, Mushinsky et al. 1994).

Gopherus polyphemus produces a single clutch of eggs annually. More than 80% of females in a population may breed annually (KR Smith et al. 1997, Rostal and Jones 2002). Typical clutch sizes vary from five to nine (Diemer and Moore 1994, Butler and Hull 1996; see chapter 13). Clutch size often is a function of body size (CL) (Landers et al. 1980, Diemer and Moore 1994). Clutch size also appears to be a function of nutrition. Individuals translocated to areas with nutrient-enriched soils of reclaimed phosphate mines increased mean clutch sizes by about 50% in four years (Macdonald 1996). Other individuals translocated to reclaimed phosphate mines produced clutch sizes as high as 25 (Godley 1989). Infertility and/or inviability of eggs can be a problem. Hatching success rates can be 33% or less, even for clutches protected from predation (Epperson and Heise 2003), or under laboratory incubation (Moon et al. 2006). Nevertheless, hatching success rates seem often to be relatively high (i.e., 70–90+%; Landers et al. 1980, Hurley 1993, Smith 1995, Butler and Hull 1996). Low hatching success rates may be more prevalent in the western portion of the range, perhaps as a consequence of differences in soils (KR Smith et al. 1997, Epperson and Heise 2003). Nests are prone to high rates of predation. Because of severe prehatching losses, only about six hatchlings are thought to survive per female per decade, even in the best conditions (Landers et al. 1980, Ernst et al. 1994).

Posthatching survival is relatively poor for the first six to seven years (Douglas 1978, Wilson 1991). Mortality appears to be particularly high during the first year posthatching; typically exceeding 90% (Alford 1980, Witz et al. 1992, Epperson and Heise 2003, Pike and Seigel 2006). Data for mortality rates in subsequent ages are sparse, but the data indicate a steady decline in mortality, culminating in annual mortality rates of about 1–2% for sexually-mature individuals (Tuberville et al. 2009, and included references). Long term mark-recapture studies are needed to confirm the estimates of mortality rates currently available (but see Ashton and Burke 2007, Tuberville et al. 2008 for data on translocated individuals).

Because of the rarity and crypticity of small individuals, estimates of population structure often may be questionable. The ratio of juveniles to subadults to adults in two populations in north Florida were reported to be about 27.5% : 20.5% : 52.0% and 45% : 14.5% : 40.5%, respectively (Diemer 1992b). These distributions clearly are not stable, and may indicate that pulses of recruitment are the norm. It has been estimated that a stable-size distribution, if it did exist, would contain about equal percentages of size categories <15.6 cm CL, 15.6–25.4 cm CL, and >25.4 cm CL (Alford 1980). Substantial variation in the size distribution of individuals in populations have been reported, some of which could be related to factors such as the rate of decline in habitable area at a site (McCoy and Mushinsky 1995). These authors suggested that absence of recruitment for only four years, not an unreasonable proposition, would shift about 20% of the distribution from the two smallest size categories to the four larger categories, and that size distributions at relatively small sites may reflect not only poor recruitment, but also emigration of older individuals. It appears that a variety of stochastic factors affect size distributions of *G. polyphemus* (also see Smith 1995), making any generalization difficult.

Several population viability estimates have been produced (Cox et al. 1987, Seigel and Dodd 2000, Miller 2001, Tuberville et al. 2009). The most recent of these estimates were produced using VORTEX (Lacy et al. 2005). As with all such modeling efforts, incomplete data, data averaging, and similar problems affected the resulting models to some unknown degree. The conclusion of the latest modeling effort, that hatchling survivorship is the most critical aspect of viability, contrasts with previous modeling conclusions for *G. polyphemus* and *G. agassizii*, that juvenile and subadult survival are the most critical aspect. This difference is in need of explanation.

CONCLUSIONS

Most of the North American tortoise species share similar life histories, characterized by relatively long time to sexual maturity, low annual reproductive output, and high adult survival. These species tend to follow a demographic pattern common to tortoises (fig. 16.5). The exception to this generalization is *Gopherus berlandieri* which displays a substantially higher rate of turnover of individuals (fig. 16.5).

Growth patterns are similar among North American tortoise species, all growing at a relatively high rate for approximately the same length of time, typically 18–22 years, and growth rates decreasing thereafter (Adest et al. 1989, Germano 1994, Mushinsky et al. 1994). *Gopherus flaavomarginatus*, the largest species, has the highest growth rate, and *G. berlandieri*, the smallest species, the lowest growth rate. Growth rates can vary temporally and spatially, however.

The information presented in this chapter clearly illustrates the need to continue to collect reliable demographic data for all of the North American tortoise species. At a minimum, these data should include size at hatching, maturity, and the growth asymptote; age at maturity; instantaneous adult mortality; and clutch size and number of eggs per clutch. Age at hatching may be important for subsequent calculation of instantaneous preadult (egg to maturity) mortality, but overwintering in the nest complicates its use. Instantaneous preadult mortality can be calculated from these data assuming a stable age distribution. It also could be estimated, potentially, from mark-recapture data, but these data are not easily obtained. Using Leslie or Lefkowitz Matrix modeling

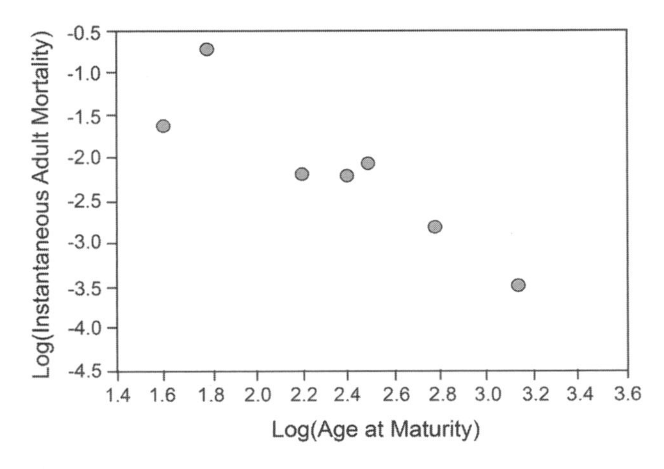

Fig. 16.5. Instantaneous adult mortality (log scale) versus age at maturity (log scale) for seven species of tortoise (McCoy and Mushinsky unpublished). The left-most point represents *Gopherus berlandieri*.

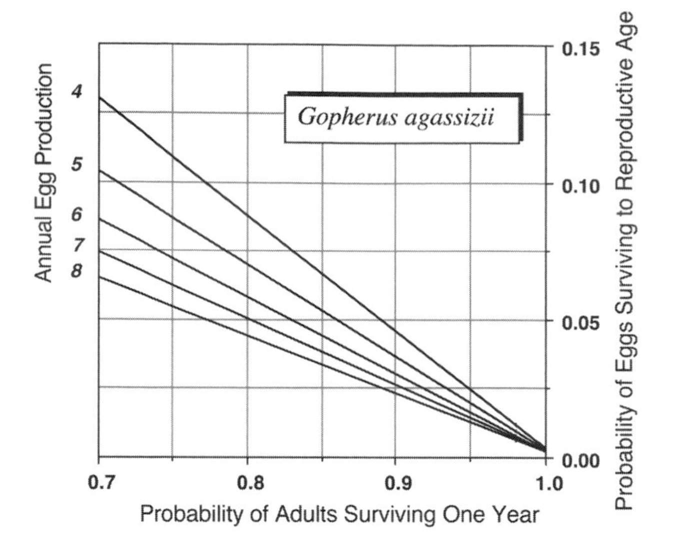

Fig. 16.6. Simulation of the effects of combinations of adult and juvenile survival rates on annual egg production needed to produce a population that is neither growing nor declining (stable population) of *Gopherus agassizii*. The simulation assumes that reproductive age is 15 years. Illustration from R. Tracy, USFWS (1994a), after Mertz (1971).

(Mertz 1971) to estimate the required preadult survivorship to produce a stable population typically yields an unrealistic result. For example, modeling *G. agassizii* (fig. 16.6), and assuming a long-term average egg production of 5.8 eggs per year, a mean age at maturity of 15 years, and an adult annual mortality rate of 10%, indicates that the preadult mortality rate could not exceed approximately 3%. Clearly, populations are not stable in the short term, even though they may be in the long term. For most tortoise species, the difference between short term and long term involves decades, perhaps even centuries. In the face of global climate change, data on the potential effects of highly-skewed sex ratios—because of unusual nest temperature regimes—probably also should be gathered (chapter 6).

Several studies have suggested a relationship between habitat productivity and growth rates in North American tortoises (e.g., Medica et al. 1975, Auffenberg and Iverson 1979, Germano 1988). Yet, other studies (Germano 1994) could not demonstrate a relationship between several measures of growth and precipitation, which is a general indicator of productivity, across all of the species (but, see chapter 7). Although precipitation may not be a good indicator of relative growth potential for all of the species (e.g., *G. polyphemus;* McCoy et al. 2011), comparisons of age and size at maturity in populations of *G. polyphemus* have shown that females from southern populations reach maturity between six and ten years earlier than females from northern populations (Landers et al. 1982, Mushinsky et al. 1994), which could be related to differences in habitat quality. The difference in growth rates also could be related to seasonal effects on opportunities for thermoregulation and growth, or to some other factor(s), however. So, in addition to the need to collect reliable demographic data, we also need to collect reliable data on the environmental variables that potentially cause the large degree of demographic variation that tortoises display, and to find ways to determine unambiguously the cause-and-effect relationships between environmental variation and demographic variation.

17 History of Human Interaction with North American Tortoises

MICHAEL TUMA
CRAIG B. STANFORD

North American *Gopherus* tortoises have a 35 million-year history on the continent (Auffenberg 1974; chapter 2). They have lived alongside human populations for a tiny fraction of their evolutionary history. But an estimated 11,000-year coexistence with human populations has driven the genus *Gopherus* well toward extinction. In this chapter, we review the history of human-tortoise interactions on the North American continent, culling information from diverse sources, including published literature, archaeological site excavation reports, and ethnographic studies. Throughout an 11,000-year history of human-tortoise interactions, humans have primarily used North American tortoises as a source of food.

PALEO-INDIAN AND EARLY ARCHAIC PERIODS

Humans may have arrived in North America using watercraft via the Pacific Coast "Kelp Highway" some 16,000 years ago (Erlandson et al. 2007) or by following herds of megafauna from Siberia across the Bering Land Bridge during the last glacial maximum between 20,000 and 15,000 years ago (Goebel et al. 2008). In either scenario, the first humans to explore the interior of the continent were hunter-gatherers of the Clovis cultural tradition. This culture spread quickly from the north, arriving into areas occupied by gopher tortoises by around 11,000 years ago (Hamilton and Buchanan 2008). During the late Pleistocene, *Gopherus* tortoises ranged as far as Nebraska in the north, Aguascalientes, Mexico, in the south, California's Central Valley in the west, and North Carolina in the east. In addition to the ancestors of modern *Gopherus* species, extinct late Pleistocene tortoises included *G. auffenbergi,* a species from central Mexico, and *G. donlaloi,* a species that ranged from northeastern Mexico to Nebraska and attained sizes of more than half a meter (Reynoso and Montellano-Ballesteros 2004). The first humans to walk into areas supporting tortoise populations likely were amazed to see these burrow dwellers for the first time. Indeed, an adult female *G. donlaloi,* a species that likely shared female-biased sexual size dimorphism with its closest relatives *G. flavomarginatus* and *G. polyphemus,* was no doubt an impressive sight to behold; a colony of these "giants" more impressive still.

Although the Clovis people may have entered areas supporting *Gopherus* tortoises in search of big game, evidence from archaeological sites throughout the region indicates that other smaller game also was targeted for consumption (Cannon and Meltzer 2004). The only known Clovis site definitively to contain tortoise remains in association with cultural materials is the Lehner Ranch site, an encampment in southern Arizona that was occupied by Clovis peoples 11,000 years ago. The remains, likely desert tortoise, were found associated with a roasting pit, along with the remains of extinct Pleistocene mammals, including mammoth, mastodon, bison, jackrabbit, and bear (Haynes and Haury 1982). Thus, since humans first started interacting with North American tortoises, they were eating them. No other Clovis period site contains tortoise remains, though tortoises likely were taken at other sites, at least in small quantities. The age of the sites, the taphonomic processes that contribute to tortoise bone attrition, and the ephemeral nature of Clovis camps outside of cave sites, have likely all contributed to the paucity of tortoise remains recovered from Paleo-Indian archaeological sites. By around 10,800 years before present, the large herds of megafauna had gone extinct, and the Clovis technology had died with them.

North American archaeologists refer to the period following the close of the late Pleistocene and the abandonment of Paleo-Indian life-ways as the beginning of the Archaic Period. The human descendants of the Clovis that entered this period continued practicing their hunter-gatherer subsistence strategies; but, with the climatic upheaval that led to the extinction of the North American megafauna in the late Pleistocene, the development of increasing seasonality through the early Holocene, and increasing human population sizes, early Archaic Period peoples focused their hunting and gathering efforts on exploiting seasonally available resources within smaller territories. Their focus on occupying smaller areas eventually led to the punctuated development of an array of regional cultural expressions throughout North America. Also during the early Holocene, the ancestors of the modern *Gopherus* tortoises—those that had survived the late Pleistocene extinction—experienced a number of environmental challenges that contributed to local extinctions, fragmentation of populations, and development of dispersal barriers (Morafka 1988). The climate of the early Holocene contributed to range contractions of the *Gopherus* species to a configuration more similar to what we would recognize today, with the exception of *G. flavomarginatus,* which persisted over a wider area of the Chihuahua Desert. Early Archaic Period human groups occupying the ranges of the North American tortoises likely captured and consumed them, although archaeological evidence for this consumption has yet to be recovered from the few sites dating to this period.

MIDDLE AND LATE ARCHAIC PERIODS

About 8500 years ago, the onset of the Middle Holocene Climatic Optimum occurred. It was a period characterized by warmer temperatures and the development and expansion of the xeric vegetation communities such as grasslands, oak savannas, and deserts throughout North America. The Climatic Optimum continued until around 4000 years before present, when global temperatures began a steady, gradual decline and the eastern portion of North America became more mesic. Technological and biological evidence for middle Archaic Period peoples is richer than that for Paleo-Indian and early Archaic peoples, because of their larger populations and their more intensive occupation of sites. Accordingly, North American tortoises of all species are commonly recovered from middle and late Archaic Period sites.

The Henwood Site in the Mojave Desert of California was occupied by middle Archaic Period peoples approximately 5000–8500 years ago. One of the most striking features of the site is the amount of desert tortoise remains recovered, far more than any other species that was hunted and deposited there. One locus appears to be a special-use area, where the site's occupants cooked and ate large quantities of desert tortoises. Deposits dating between approximately 7000 and 7500 years ago contained hundreds of charred tortoise shell remains (Schneider and Everson 1989). Site stratigraphy suggests deposition over a long period, perhaps implicating a sustained and targeted tortoise harvest by the hunter-gatherers who occupied this site for centuries.

Archeological sites across portions of Florida point to a range of subsistence economies practiced regionally by middle Archaic Period peoples. Large settlement sites dating to approximately 5600–4000 years before present are known from inland sites located along rivers and lakes, and coastal sites on the mainland and barrier islands. Two inland sites, the West Williams and Enclave Sites, indicate a fairly strong reliance on terrestrial resources, including *Gopherus polyphemus* (Austin et al. 2009). Another inland site, the Outlet Midden at Lake Monroe, provides evidence that *G. polyphemus* was among the terrestrial resources harvested to supplement a primarily freshwater fish, turtle, and shellfish diet (Quitmyer 2001). The nearby Harris Creek Site shows a pattern more similar to the West Williams and Enclave Sites, with a stronger reliance on *G. polyphemus* and other terrestrial resources (Quinn et al. 2008). *Gopherus polyphemus* have also been recovered from barrier island sites such as Horr's Island and Useppa Island, as well as other middle Archaic Period sites across Florida where marine resources formed the major part of the diet of the site's occupants (Franz and Quitmyer 2005).

Sonora Desert tortoises were harvested by middle to late Archaic peoples who were among the earliest farmers in North America. A large farming community, the La Playa site, was recently discovered in northern Sonora, Mexico. The site, which dates between approximately 3200 and 2000 years before present, contains thousands of features, including more than 1300 *hornos* (corn roasting pits) and expansive agricultural fields with irrigation canals used for intensive maize production. Tortoises were among the remains of animals that had been roasted in the hornos (Carpenter et al. 2009). The diminutive *G. berlandieri* also was a favorite food item for Archaic Period peoples. A hearth feature at the Choke Canyon Site, a late Archaic Period hunter-gatherer camp along the Frio River in south-central Texas, provides evidence that *G. berlandieri* was among a variety of freshwater and terrestrial resources procured and eaten by the site's occupants (Hester 2009).

MacWilliams et al. (2006) reported the recovery of desert tortoise remains from a hearth feature at Cueva de los Corrales, a late Archaic Period site in southern Chihuahua. The recovery of desert tortoise is surprising, given the site's location on the east side of the Sierra Madre Mountains, some 150 miles east of the Sonora Basin. Although tortoises may have been transported to the area through trade with groups from Sonora, a more likely possibility is that the tortoise remains are actually misidentified individuals of *G. flavomarginatus.* The present-day range of *G. flavomarginatus* is located approximately 100 miles east of the site. Because *G. flavomarginatus* persisted over a larger range in the Chihuahua Desert during the early Holocene (Morafka 1988), low desert areas within a short distance of the site and easily accessed by the site's occupants could have supported tortoise populations during this period.

Other middle and late Archaic Period sites show a pattern of exploitation—sometimes intensive—of *Gopherus* tortoises by humans. The archaeological record is particularly rich for examples of human exploitation of desert tortoises (Schneider and Everson 1989) and *G. polyphemus* (Franz and Quitmyer 2005). For example, the Stuart Rockshelter, a middle Archaic Period hunter-gather camp in southern Nevada appears to have been the location of intensive harvest and consumption of tortoises. A large amount of desert tortoise remains were recovered at two late Archaic Period hunter-gatherer camps in southern Nevada, the Atlatl Rockshelter and the Turtle Bone

Site. Desert tortoise remains were the most commonly recovered animal from the Archaic Period Roadside Roast Site. *Gopherus polyphemus* is known from many middle and late Archaic Period sites in Florida, including the Tick Island, Palmer-Taylor, Summer Haven, Boca Weir, and Crescent Beach Sites.

LATE PERIOD

By around 2000 years ago the climate had stabilized roughly into its modern condition, and human populations over most of North America had developed more sophisticated technologies for procuring and processing a wider range of animal resources. The hunter-gatherer way of life persisted in many areas occupied by *Gopherus* tortoises, whereas permanent sedentary villages or long-term camps supported by agriculture and supplemented with the hunting and gathering of wild resources were more typical in other areas by the close of this period. Late Period hunter-gatherers occupied the Mojave Desert, the Gulf Coast region of northwestern Mexico and southern Texas, and small portions of other areas occupied by *Gopherus* tortoises. Human groups that practiced agriculture and settled in permanent communities in *Gopherus* territory occupied the Sonora Basin, the Chihuahua Desert, and most of the southeastern United States.

Numerous Late Period sites throughout the Mojave Desert indicate intensive harvest of tortoises by hunter-gatherers, who roasted tortoises in hearths or pits similar to those used for preparing agave. Tortoise remains were recovered in moderate to high quantities at more than 20 Late Period sites and/or loci throughout the Mojave Desert (Schneider and Everson 1989; Dietler et al. 2011), underscoring the importance of desert tortoise as a food source. Moreover, the number of sites and the amount of tortoise remains recovered at the sites indicate that harvesting pressure was much greater during this period than in previous periods.

In the Sonora Basin, Late Period peoples continued the agricultural way of life of their predecessors, but by 1600 years ago had intensified the practice and committed to living in permanent settlements. Desert tortoises are fairly often recovered from Late Period Hohokam sites, though typically in smaller amounts than observed for sites in the Mojave Desert (Schneider and Everson 1989). Several sites near Dove Mountain in the Tortolita Mountains of southern Arizona indicate that desert tortoises were an important source of food (Schwartz 2008).

During the Late Period the Toyah hunter-gatherers inhabited the southern Texas and northeastern Mexico plains and harvested *G. berlandieri*. Excavations at the Hinojosa Site revealed a 600-year-old Toyah base camp where numerous features, including hearths and clusters of well-preserved bones, were exposed and recorded. *Gopherus berlandieri* remains were among those recovered from the features (Black 1986).

In the southeastern United States, prehistoric human populations swelled after the adoption of agriculture and sedentism, and population centers featured spectacular platform mounds topped with chiefly structures. These communities were hubs of far-reaching trade networks, and were situated within a socio-political hierarchy with allied chiefdoms from which forms of tribute were received or paid in turn for protection or by threat. The tribute payments were provisioned among the elite and used in elaborate ceremonies that included feasting on the platform mounds. As food tribute became the norm, the elite class established their status through demanding tribute of the best deer parts, the finest crops, and the rarest foods (Jackson and Scott 2003). *Gopherus polyphemus* remains have yet to be discovered at a major center outside of their range, such as Moundville, Alabama, which might have been expected given their ease of transport, exotic taste, and local rareness. Neither has *G. polyphemus* been recovered from Lake Jackson Mound, a major Mississippian population center in northern Florida, though investigations there have been minimal. One example of *G. polyphemus* being featured in a feast comes from a smaller population center, at Shield Mound in northeastern Florida. *Gopherus polyphemus* remains were recovered from Kinzey's Knoll, a midden that was interpreted as the refuse accumulation from elite feasting activities (Marrinan 2005). *Gopherus polyphemus* was well represented at the site, implicating the possibility that they were collected from rural areas and offered as tribute at this location.

HISTORIC PERIOD

Gopherus polyphemus was harvested and eaten by many immigrants to the New World. During the 1500s, they were an important food item for Spanish colonists in southeastern Georgia and Florida, where they were recovered in substantial numbers at several sites (Reitz 1991). *Gopherus polyphemus* was one of the many wild species that enslaved African Americans hunted and consumed during the Antebellum Period in the American South. Tortoise remains were recovered from the slave quarters areas of several Antebellum plantations, including Kingsley Plantation in northeastern Florida (Walker 1985) and several plantations at Kings Bay in southwestern Georgia (Adams et al. 1987). Cuban merchants regularly made voyages via schooner ships to the southwest Florida coast to harvest *G. polyphemus* by the hundreds until the 1920s (Trowbridge 1952). There is a long history of consumption of *G. polyphemus* by European Americans in the southeastern longleaf pine ecosystem. During the 1700s, Carolinian farmers of Scottish and Irish descent settled the pinewood uplands and reportedly relied heavily on the collection of wild food resources, including tortoises (Longleaf Alliance 2011). A German who traveled through the area in 1793 remarked that these "crackers" dominated the piney woods, preying on *G. polyphemus,* or "cracker chicken," as a supplementary food source (Longleaf Alliance 2011). Thus, the southern European American tradition of eating tortoises, which would continue for the ensuing two centuries, can be traced to these pioneers of the longleaf pine ecosystem. It is well known that *G. polyphemus* was harvested in great numbers during the Depression throughout its range, reportedly almost to the point of extinction (Lohefener and Lohmeier 1984). The "cracker chicken" was then referred to

Fig. 17.1. Gopher (*Gopherus polyphemus*) pulling in Florida, circa 1980. Reproduced from Taylor (1982b) with permission of the Florida Museum of Natural History.

as the "Hoover chicken," and it was hunted and consumed in great numbers. This practice continued through time until entire populations were decimated, colony by colony, as the demand for their meat increased (Puckett and Franz 1991). Taylor (1982b) conducted an ethnographic investigation into the hunting and consumption of *G. polyphemus* by rural African Americans in northern Florida between 1978 and 1980. He reported that "gopher pullers" harvested tortoises for consumption and to make money from sales, as well as for the enjoyment of the cultural tradition. The tortoises were hunted with a six to seven meter wire that featured a hook used to pull the tortoise from the burrow (fig. 17.1). Although prohibited, sale of *G. polyphemus* was widespread among the gopher pullers. From the late 1970s through late 1980s, significant population reductions were noted across Florida from human predation (Diemer 1987). Many rural European Americans in the pinewoods across the Southeast consider hunting and eating *G. polyphemus* a deeply traditional practice. Prior to the closure of tortoise harvesting in 1988, a community in Okaloosa County, Florida, celebrated an annual tortoise cookout (Enge et al. 2006).

Desert tortoises have been harvested for use as both food and pets throughout the Historic Period. Schneider and Everson (1989) found accounts of desert tortoises being eaten by 14 different Historic Period Native American groups in California, Nevada, and Arizona. Apparently, the Paiute and Chemehuevi were particularly well known for their fondness of eating desert tortoises: in the mid-1800s, travelers through the area remarked that these groups were distinctive from others by the amount of discarded tortoise shells that accumulated at the edges of their settlements (Mollhausen 1858, Battye 1934). Archaeological evidence from the San Agustin de Tucson Mission in Tucson, Arizona, indicates that missionized Native Americans of the O'odham tribe harvested desert tortoises in the vicinity of their ranches (Pavao-Zuckerman and LaMotta 2007). Mestizo Mexican traders who used the Old Spanish Trail, a route used to deliver furs, Indian goods, and livestock between Los Angeles, California, and Santa Fe, New Mexico, from the late 1500s through the mid-1850s, were known to have collected desert tortoises while passing through the Mojave Desert. Tortoises were packed live for the trip, and eaten days or weeks later (Pepper 1963). This practice continued during the construction of railroads across the Mojave and Sonora Deserts in the 1880s through the 1920s. Camp (1916) reported that the Native American and mestizo Mexican American rail laborers who lived along the railroad lines being constructed in the vicinity of Needles, California, considered desert tortoises a delicacy, and sought them as a favorite snack. James (1906: 199) reported that a Mestizo laborer working for him "spoke enthusiastically of the wonderful feasts he used to have on 'tortugas.'" Mestizo fishermen and Seri Indians inhabiting Isla Tiberon, Mexico, periodically harvested desert tortoises (Felger et al. 1981). Desert tortoises may have been collected commercially for pets as early as 1883, when the Southern Pacific Railroad became the first transcontinental railway constructed across the Sonora and Mojave Deserts. Luckenbach (1982) reported that desert tortoises were collected by mestizo railway construction laborers and transported to stations to be sold or given away as promotional items to European American tourists from the east. From the 1930s through 1960s, desert tortoises were commonly collected by mestizo children from desert communities along Routes 66 and 95 to sell to European American tourists as pets (Luckenbach 1982). Grant (1936) described how children collected tortoises for sale in small desert towns, including several boys who had collected approximately 600 tortoises within two miles of Hodge, California, and offered them for sale to tourists. Extensive collection of desert tortoises for use as pets in the 1950s and 1960s was likely an important contributor to population declines, as thousands of desert tortoises were removed from the wild and brought into captivity during this period. By the late 1960s, the use of desert tortoises as pets had become so popular across the country that thousands were captured and exported from California in the 1970s (Berry and Nicholson 1984). Desert tortoises also were collected for use as pets by city dwellers in Sonora, Mexico, who collected tortoises from rural locations (Fritts and Jennings 1994; Bury et al. 2002). Baby desert tortoises hatched from captive wild-caught adults were sold in Hermosillo pet shops until their trade was banned by the federal government in 1994 (Bury et al. 2002). In the early 1990s, residents of several rural communities in Sonora reported that the desert tortoises were periodically collected for use as both food and pets (Trevino et al. 1994). Fritts and Jennings (1994) noted that desert tortoise habitats in the vicinity of towns and communities in Sonora were largely devoid of tortoise populations in the mid-1990s, which they attributed to human predation.

Poaching of bolsón tortoises during the 20th century has been a primary factor in the extirpation and fragmentation of local populations throughout the Mapimian subprovince of the Chihuahua Desert, to which the species is restricted (Morafka 1982, Bury et al. 1988, Lieberman and Morafka

1988, Morafka 1988). Tortoises were first harvested by construction laborers on a railroad being constructed through the center of the species' range in the 1940s. Later, railroad crews collected tortoises in adjacent habitats along the rail line and transported them to the Pacific coast, where restaurants and markets featured them as delicacies (Morafka 1982). In the 1950s, Mexican Federal Highway 49 was constructed through the bolsón tortoise's range, which further fragmented populations and provided additional human access to adjacent tortoise habitat. Predation from these two linear features eventually led to the extirpation of tortoise populations as far as five km from the highway and ten km from the railroad by the late 1970s (Morafka 1982). Mestizo farmers and ranchers who colonized the area in the 1940s and 1950s hunted bolsón tortoises intensively, and by the 1970s had extirpated tortoise populations within the vicinity of their *ejidos* (hamlets) (Morafka 1982). Morafka (1982, 1988) conducted interviews with long-time residents of the area, who consistently reported that the bolsón tortoise was formerly much more abundant in the early 20th century. Tenneson (1985) interviewed a 50-year-old man who estimated that tortoise populations had declined by 75–80% of their former size, leading to significant declines, fragmentation of populations, and local extirpations. The same informant reported transporting ox-carts full of tortoises to sell at the local market as a boy. TR VanDevender reported seeing subadult bolsón tortoises for sale in Durango for about $2 each in 1978 (Morafka 1982).

Little is known about the historic poaching of *G. berlandieri,* although it undoubtedly occurred throughout the Historic Period, given the apparently widespread traditional use of *Gopherus* tortoises by European Americans along the Gulf Coast, as well as mestizo Mexicans and Mexican Americans. Rose and Judd (1982) reported that dried and varnished *G. berlandieri* shells were sold in Mexican tourist markets in border cities throughout the tortoise's range. In each case, whether the tortoise shell was used as a ceremonial bowl or a curio, the tortoise likely was killed for consumption prior to conversion of the shell into a novelty.

RECENT PERIOD

Poaching of North American tortoises for consumption or for use as pets has been prohibited in recent times. Most *Gopherus* tortoise species were protected from poaching by the late 1990s in the United States and Mexico (table 17.1), with the exception of *Gopherus berlandieri* which was listed in 1967 following concerns that it was being over-harvested for the pet trade (Rose and Judd 1982). Additionally, *G. flavomarginatus*

Table 17.1. Summary of federal, state, and global status of North American tortoises under laws that protect tortoise populations from wild harvest and international and interstate trade of poached tortoises

Species	Federal Status	State Status	CITES Status	IUCN Red List Status
Gopherus agassizii	Threatened (Mojave Desert population) since 1990	Fully Protected (Arizona) since 1987 Threatened (Mojave population in Arizona) since 1988 Threatened (California) since 1989	CITES Appendix II since 1975	Vulnerable
Gopherus morafkai	Candidate for Threatened (US) since 2010 Threatened (MX) since 1994	Fully Protected (Arizona) since 1987 Species of Special Concern (Arizona) since 1996 none (Sonora and Sinaloa, MX)	CITES Appendix II since 1975	Vulnerable
Gopherus berlandieri	None (US) Threatened (MX) since 2001	Threatened (Texas) since 1967 Protected nongame species (Texas) since 1977 none (Coahuila, MX) none (Nuevo Leon, MX) none (Tamaulipas, MX)	CITES Appendix II since 1975 (in Mexico since 1993 when Mexico signed IUCN charter)	Least Concern
Gopherus flavomarginatus	Endangered (US) since 1979 Federal Mexican Government declared a portion of the Bolsón de Mapimí as a Biosphere Reserve in 1977	none (Coahuila, MX) none (Durango, MX) none (Chihuahua, MX)	CITES Appendix I since 1975 (in Mexico since 1993 when Mexico signed IUCN charter)	Vulnerable
Gopherus polyphemus	Threatened (populations west of the Tombigbee River in Alabama) since 1987	Threatened (west of the Tombigbee River in Alabama) since 1987 Protected nongame species (east of the Tombigbee River in Alabama) since 1987 Fully Protected (Florida) since 1988 Threatened (Florida) since 2008 Threatened (Georgia) since 1987 Threatened (Louisiana) since 1989 Endangered (Mississippi) since 1974 Endangered (South Carolina) since 1976	CITES Appendix II since 1975	Vulnerable

within the Bolsón de Mapimí Biosphere Reserve has been protected from harvest since 1977. The listing of all tortoises under the Convention for the International Trade of Endangered Species (CITES) Appendix II in 1975 protected all species from international transport. Despite these regulations, poaching of tortoises from the wild continues today.

Desert tortoises continue to be targeted for consumption and for use as pets. Grandmaison et al. (2009) reported that 8% of motorists attempted to illegally collect tortoises positioned on roads in a study at 38 sites in Arizona involving 474 human-tortoise interactions, and concluded that a "biologically significant" number of desert tortoises is collected by motorists who take them home as pets. Berry et al. (1996) used law enforcement records, visual observations of suspected poachers, and signs of burrow excavation to determine the extent of desert tortoise poaching in the western Mojave Desert. They found signs of humans having excavated burrows in the study area, and inferred that an influx of Cambodian nationals into the area was at least partially responsible for poaching that led to declines in local tortoise populations. Many cultures from Asian countries traditionally hunt and eat wild tortoises, and immigrants from other countries may contribute to the removal of desert tortoises. In 1979, TR VanDevender discovered a roadside camp in central Sonora that contained the remains of six freshly butchered and consumed desert tortoise carcasses (Bury et al. 2002). Evidently, poaching of desert tortoises by mestizos in central Sonora, Mexico, for use in meat stew is a sporadic occurrence. There, *tortugueros* (tortoise hunters) harvest tortoises when they are active during the late summer monsoonal rain season, and sell them at local markets in small communities (Bury et al. 2002). Continued poaching of gopher tortoises is primarily from the perpetuation of cultural practices in rural areas that have been a mainstay of European American southern culture for generations. Gopher tortoise has been a favorite stew item for decades in the South, and recipes may still be found in print and online. One of the authors (MT) frequently encountered European American persons living in rural southern Mississippi during the mid-1990s who admitted to eating *G. polyphemus*. Rural African Americans likely continue the traditional harvest and consumption of *G. polyphemus* as well. Poaching continues to be a serious issue in local areas throughout the Southeast, as evidenced several recent news reports. One report from Leesburg, Florida, detailed the discovery of more than 200 tortoise carcasses that had been butchered for their meat (Virginia Smith, Daytona News Journal, 6 February 2004). Another news report from Florida told of the discovery of freshly butchered gopher tortoise carcasses along a highway, and the apprehension by wildlife officials of a man who was caught with more than five pounds of butchered tortoise remains in his freezer (Rob Haneisen, MetroWest, 22 March 2006).

Human poaching of *G. flavomarginatus* is likely an ongoing problem because of the cultural preferences for tortoise meat in Mexico, although several ranchers in areas that support tortoise populations prohibit poachers from accessing their properties (Morafka 1982). Still, tortoises are highly sought after for food, and unprotected areas are hit hard by poachers. Human predation was suspected at 93% of *G. flavomarginatus* sites visited by Lieberman and Morafka (1988). Several roads and communities located within the Mapimian subprovince continue to serve as staging areas for locals entering adjacent tortoise habitat on hunting forays. Morafka (1988) reported finding freshly butchered tortoise carcasses at several roadside camp sites had been used by laborers exploring for oil reserves in the area in the early 1980s.

CONCLUSIONS

Human predation has impacted populations of North American tortoise species—both extant and extinct—for more than 11,000 years. Human predation of *Gopherus* tortoises has likely contributed to local population depletions or extirpations for thousands of years, and for one species—*Gopherus flavomarginatus*—has contributed to significant range contraction. Desert tortoise populations have endured a long history of human exploitation, and they continue to be harvested, possibly in great numbers, predominantly for use as pets. *Gopherus polyphemus* populations have been equally exploited over a long period and continue to be harvested in great numbers for use as food. *Gopherus berlandieri* and *G. flavomarginatus* likely are targeted by the continued mestizo practice of eating tortoises. The future of human-tortoise interactions, primarily poaching of wild tortoises, will likely contribute to additional population declines, fragmentation, and extirpations in the years to come. Agencies and conservation groups that protect *Gopherus* tortoises will need to consider that, in addition to protecting expanses of high quality habitat, education of all cultural groups within local human populations should be a high priority for species recovery. Additionally, land managers should strive to ensure that areas that support healthy tortoise populations remain remote and free from human access to the extent possible. Finally, laws protecting North American tortoises from poaching should be enforced more diligently, and policies should be strengthened for those species that are currently under-protected.

18 Threats and Conservation Needs for North American Tortoises

Kristin H. Berry
Matthew J. Aresco

Since the 1500s, at least ten species of tortoises and freshwater turtles have become extinct (Turtle Conservation Coalition 2011), and many more are critically endangered with extinction. All five extant species of *Gopherus* face serious threats to well-being and continued existence. Threats facing populations at local, regional, and range-wide levels are time-sensitive and have the potential to change rapidly in type and severity.

In this chapter, we summarize the degree of endangerment as assessed internationally through the Red List of Threatened Species, nationally by country, and more locally by state or population segment. We report species by species on threats facing populations and habitat, conservation efforts undertaken by governments and nonprofit organizations, and prospects for the future.

THREATENED AND ENDANGERED RANKINGS INTERNATIONALLY, NATIONALLY, AND BY STATE OR POPULATION SEGMENT IN 2012

IUCN and the Red List of Threatened Species

Government and other agencies assign rankings of rarity and endangerment to plant and animal species at international and national levels and regionally within countries by states. The International Union for Conservation of Nature's (IUCN) Species Programme and Species Survival Commission (www.iucn.org/redlist.org) are the best known. The IUCN assesses conservation status of species, subspecies, varieties, and selected subpopulations on a global scale and publishes the findings online (IUCN 2010). Species are categorized on risk of global extinction as critically endangered, endangered, and vulnerable, as well as extinct or extinct in the wild. As of 2012, only one *Gopherus* species, *G. flavomarginatus,* had been assessed recently (van Dijk and Flores-Villela 2007), and it was assigned a vulnerable rating. *Gopherus agassizii* and *G. polyphemus* also have vulnerable ratings, and *G. berlandieri* is categorized as lower risk / least concern. Reevaluation is overdue, and will need to include the newly described *G. morafkai* (Murphy et al. 2011). All *Gopherus* species are likely to be upgraded. In 2011, a coalition of international and national conservation organizations listed *G. flavomarginatus* among the top 40 tortoises and freshwater turtles at very high risk of extinction globally (Turtle Conservation Coalition 2011).

Convention on International Trade in Endangered Species of Wild Fauna and Flora (CITES 2011)

A second international form of protection, CITES (www.cites.org), provides a framework for domestic legislation on import and export of species. Both the United States and Mexico are parties to CITES. Covered species are listed in two appendices. Appendix I (*G. flavomarginatus*) includes species threatened with extinction; trade is permitted only in exceptional circumstances. Appendix II (*G. agassizii, G. polyphemus, G. berlandieri, G. morafkai*) includes species not necessarily threatened with extinction but for which trade must be controlled.

Federal Legislation in the United States and Mexico

All *Gopherus* species described prior to 2011 are protected in whole or in part through federal legislation within the U.S. or Mexico. *Gopherus flavomarginatus* was federally listed as endangered by the U.S under the Endangered Species Act of 1973, as amended, and by Mexico in 2000 under Secretaría de Medio Ambiente y Recursos Naturales (SEMARNAT) (U.S. Dept. of the Interior, Fish and Wildlife Service [USFWS] 1979, SEMARNAT 2000). Listings for *G. polyphemus* and *G. agassizii* have occurred by limited area or population segment, often following years between petitions for listing and final rules. For *G. polyphemus,* populations west of the Tombigbee and Mobile rivers in southwest Alabama, Mississippi, and Louisiana were listed as threatened in 1987, and a recovery plan was approved in 1990, but with no critical habitat designation (USFWS 1987, 1990a). Eastern populations in Alabama, Georgia, Florida, and South Carolina were petitioned for federal listing as threatened in 2006. In July 2011, the USFWS determined that listing was warranted but precluded by higher priority actions (USFWS 2011a). Federal listing protects *G. polyphemus* from both direct and incidental take, which includes destroying tortoise burrows.

For *G. agassizii,* the first population listed as threatened in 1980 was the historic site described by Woodbury and Hardy (1948) on the Beaver Dam Slope of Utah (USFWS 1980). Four years later, a petition was submitted to list populations within the U.S. as threatened; the USFWS determined that federal listing was warranted but precluded because of other, higher priorities (USFWS 1985). In 1990, the Mojave population, defined as occurring west and north of the Colorado River in California, Nevada, Utah, and Arizona in both the Mojave and western Sonoran Deserts, was finally listed as threatened (USFWS 1990b). Critical habitat designations and a recovery plan for the Mojave population were published in 1994 (USFWS 1994a, 1994b). A revised recovery plan was published in 2011 (USFWS 2011b). The desert tortoise population in Mexico was listed as threatened in 1994 (SEMARNAT 1994). In 2010, after receiving a petition to list the Sonoran population of desert tortoise as threatened, the USFWS published a finding of warranted but precluded by other, higher priorities (USFWS 2010a); this finding does not include the Sinaloan population segment. To complicate the situation further, populations occurring east and south of the Colorado River were described as a new species, *G. morafkai,* in 2011, and thus Sonoran and Sinaloan population segments were subsumed under the new species (Murphy et al. 2011). *Gopherus berlandieri,* although not federally listed in the U.S., was listed as threatened by Mexico in 2002 (SEMARNAT 2002).

Protections Provided by State Laws and Regulations

State laws and regulations provide some protection for *G. polyphemus, G. agassizii, G. morafkai,* and *G. berlandieri* within the U.S. These laws or regulations, available on the World Wide Web by state for U.S. tortoise populations, are designed to protect wild populations and to manage captive tortoises. They cover such topics as harm, collecting, possession, registration, propagation, imports and exports, and adoption of captives, but are inconsistent from one state to another for the same species. The primary gap in regulatory protections is the protection of habitat on private land even where a species is federally listed.

THREATS TO WELL-BEING OF SPECIES AND POPULATIONS

The type and severity of threats vary by species and population segment, land ownership or land administrator, and whether habitat has been designated critical habitat (tables 18.1–18.4). Some general patterns exist across species (high mortality rates among adults and/or juveniles, habitat loss and deterioration, inadequate regulations and/or enforcement of existing regulations to protect the species and habitats), but specifics vary. We ranked threats as high, medium, or low as of 2012 (tables 18.1–18.4). Many anthropogenic activities have cumulative and synergistic effects that increase the severity of an activity or combinations of activities on tortoises and their habitats (USFWS 2010a, 2010b, 2011b; Leu et al. 2008). Urban and exurban development, for example, not only results in habitat loss but also in deterioration of adjacent lands from associated infrastructures, dumping of trash, recreational activities, invasion and establishment of exotic plants, predation by subsidized predators (dogs, ravens, coyotes, raccoons, opossums), increased fires, collecting, and vandalism. The most pressing issue for all five species is sufficient protection of viable and representative populations and habitat.

The species with the fewest identified threats, *Gopherus flavomarginatus,* has the smallest geographic range, is remote from major urban areas, and the least studied. Populations are reported to be in a precarious situation (Traphagen et al. 2007; M Traphagen, personal communication; E Goode, personal communication). The greatest threats are lack of recovery or management plans and protected habitats. The species has no advocate, government agency, or nongovernment organization (NGO) actively monitoring and reporting on status of populations and habitat. Other high- to moderate-level threats are from government resettlement programs of agriculturists in ejidos; subsequent clearing of habitat for plantings of corn, cactus, and other crops; deterioration of habitat from livestock grazing; fragmentation of habitat from roads and railroads; and loss of individual tortoises from road kills and take by humans (Aguirre et al. 1997, Trevino et al. 1997, Traphagen et al. 2007, van Dijk and Flores-Villela 2007).

Two themes predominate in threats to *G. polyphemus:* human predation and habitat loss (table 18.1). Human predation throughout most of the 20th century caused significant population declines or extirpation in southern Alabama, southwestern Georgia, and the Panhandle and northern peninsula of Florida (Auffenberg and Franz 1982, Mushinsky et al. 2006). Habitat loss remains the greatest range-wide threat (Enge et al. 2006, Florida Fish and Wildlife Commission [FWC] 2007, USFWS 2009a). For example, in Florida, since the 1960s, an estimated 50 to 80% population decline was inferred from habitat reduction caused by urbanization (residential and commercial development, roads, infrastructure), agriculture (row crops and citrus groves), and phosphate mining (FWC 2001a, 2006; Mushinsky et al. 2006). An estimated 690,000 ha of former habitat is now classified as urban, representing a 15.7% loss (FWC 2006). In Florida and elsewhere, populations are also threatened by habitat loss, degradation, fragmentation, and modification caused by common agricultural and silvicultural practices and lack of management (i.e., prescribed fire, essential for maintaining habitat quality). The potential threat of mycoplasmal upper respiratory tract disease (URTD) to wild populations has been debated extensively in recent years. Although the disease can cause mortality of individual tortoises, little is known of population-level effects and the contribution to overall population declines, however

Table 18.1 Threats to *Gopherus polyphemus*, with estimated level of importance

Threats	Threat Level	References
Direct mortality		
Illegal collection and human predation	Medium	Taylor 1982b; Berish 2001; Enge et al. 2006; Mushinsky et al. 2006; FWC 2007; USFWS 2009a; Aresco, pers. obs.; D. Johnson, FWC Law Enforcement
Gassing of burrows (rattlesnake roundups)	Low	Speake and Mount 1973; Smith et al. 2006
Disease		
upper respiratory tract disease	Medium	Smith et al. 1998; Brown et al 2002; Seigel et al. 2003; McCoy et al. 2005; Mushinsky et al. 2006; Smith et al. 2006; Berish et al. 2010
herpesvirus, ranavirus, iridovirus	Unknown	FWC 2007
Pesticides / herbicides	Unknown	Lohoefener and Lohmeier 1984; Diemer 1986
Subsidized predators		
birds	Low	FWC 2007
dogs	Medium	Causey and Cude 1978; Hawkins and Burke 1989
coyotes	Medium	Mushinsky et al. 2006; Aresco et al. 2010
imported fire ants	Medium	Epperson and Heise 2003; Smith et al. 2006; FWC 2007
Vehicle kills	High	Landers and Buckner 1981; Diemer 1987; Enge et al. 2006; Mushinsky et al. 2006; Aresco, pers. obs.
Illegal relocations	High	Ashton and Ashton 2008
Permitted translocations, improperly conducted	High	Burke 1989b; Berish 2001; Mushinsky et al. 2006; Ashton and Ashton 2008
Entombment under incidental take permits issued before August 2007 (Florida only)	High	FWC 2001b; Fleshler 2005; FWC 2007
Entombment from unpermitted land clearing and development activities	High	Diemer 1989; Aresco pers. obs.
Habitat		
Habitat loss		
urbanization, human population growth, new roads, infrastructure	Very High	Auffenberg and Franz 1982; Diemer 1986; McCoy et al 2002; Enge et al. 2006; FWC 2001a, 2006, 2007; Mushinsky et al. 2006; Smith et al. 2006; USFWS 2009a
rural, exurban	High	Enge et al. 2006; Mushinsky et al. 2006
agriculture	High	Hermann et al 2002; Mushinsky et al. 2006
phosphate mining, sand and clay extraction	High	Diemer 1986; Mushinsky et al. 2006
Habitat modification, conversion, alteration, fragmentation		
silviculture	High	Landers and Buckner 1981; Lohoefener and Lohmeier 1981; Auffenberg and Franz 1982; Kautz 1998; Aresco and Guyer 1999a, b; Conner and Hartsell 2002; Hermann et al. 2002; USFWS 2009a
agriculture	High	Hermann et al. 2002; Mushinsky et al. 2006
fire exclusion / lack of habitat management	High	Landers and Speake 1980; McCoy and Mushinsky 1992b; Mushinsky and McCoy 1994; Aresco and Guyer 1999a; Enge et al. 2006; McCoy et al. 2006
livestock grazing	Low	Auffenberg and Franz 1982; Diemer 1986
invasive exotic plants	Medium	Cogongrass, *Imperata cylindrica* (Shilling et al. 1997; Mushinsky et al. 2006; Smith et al. 2006; FWC 2007)
Adequacy of population and habitat management by government agencies	Varies by state, low to high	Mushinsky et al. 2006; Smith et al. 2006; USFWS 2009a; FWC 2011

(Mushinsky et al. 2006, USFWS 2009a). Mycoplasmal URTD may be a chronic disease with high morbidity but low mortality (Berish et al. 2010).

Gopherus agassizii, despite 22 years of federal protection as a threatened species (USFWS 1990b) and designation of 13 critical habitats (USFWS 1994b), has not recovered. Numerous threats contribute to the status, and severity varies by region (table 18.2). High on the list of threats are rapidly growing human populations, cities and towns, exurban areas and associated infrastructures, all of which result in new habitat loss and deterioration and high mortality rates (USFWS 1994a, 2010b). Within critical habitats, the density of highways and paved roads averages 0.5 km/km^2 (range = 0.14–1.14; USFWS 2010b); few roads have exclusion fencing to protect tortoises from road kills. Utility and transportation corridors affect 8 to 20% of eight critical habitats (USFWS 2010b). Most critical habitats have deteriorated from decades of livestock grazing, and from 33 to 95% of seven critical habitats are still grazed. Unauthorized off-highway (OHV) recreation and military-related activities are major concerns in several critical habitats (U.S. District Court 2009, 2011). Invasive alien plants (*Schismus* spp., *Bromus* spp., and *Brassica tournefortii*) have altered the food supply for tortoises in many areas (e.g., Oftedal 2002, Brooks and Berry 2006). The alien plants are primary

Table 18.2 Estimated level of threats to *Gopherus agassizii* in 13 critical habitats and other noncritical habitats (inclusive of the "Mojave" population occurring north and west of the Colorado River)

Threats	Threat Level	References
Direct Mortality		
Illegal collection and release of captive or exotic tortoises	Medium in Mojave Desert, potentially lower in Colorado Desert	Johnson et al. 2006; summary of releases in Murphy et al. 2007; Berry et al. 2008
Vandalism	Low to Medium, depending on area	Berry 1986b; Berry et al. 2006a, 2008
Disease	High	Jacobson et al. 2009
upper respiratory tract disease	High	Jacobson et al. 1991, 1995; Brown et al. 1999a; Christopher et al. 2003; Jacobson and Berry 2012
herpesvirus	Present but little information on prevalence	Christopher et al. 2003; Johnson et al. 2006; Jacobson et al. 2012
shell disease	Medium to high in Colorado Desert, eastern Mojave Desert, CA	Jacobson et al. 1994; Christopher et al. 2003; Berry et al. 2006a
Subsidized predators	High	Especially high near urban and exurban areas and, for coyotes, also during drought (Fedriani et al. 2001; Esque et al. 2010b)
avian	High	Boarman 1993; Boarman and Berry 1995; Kristan and Boarman 2003
dogs	High	Lenth et al. 2008; Young et al. 2011
coyotes	High	Esque et al. 2010b
Vehicle kills	Low to high, depending on critical habitat and presence of exclusion fencing	USFWS 1994a, 2010b; Von Seckendorff Hoff and Marlow 2002; Boarman and Sazaki 2006
Translocation	High	See text
Habitat		
Habitat loss		
human population growth, urbanization, and infrastructure	High	USFWS 1994a, 2010b; Berry et al. 2006b; Esque et al. 2010b
rural, exurban	Medium	USFWS 1994a, 2010b
agriculture	New direct losses are low in terms of km^2	USFWS 1994a, 2010b; impacts include subsidies for predators, invasive species, and fugitive dust
military maneuvers	High in CA; low elsewhere	Esque et al. 2005; Berry et al. 2006a; USFWS 2010b; 13–26% of 3 critical habitats are in military installations
solar energy	High	USFWS 2010b; L. LaPré pers. comm.
Habitat fragmentation, creation of barriers	High	
paved roads, canals, railroads, power lines, utility corridors	High; most roads have no exclusion fencing to protect tortoises	Boarman and Sazaki 2006; Brooks and Lair 2009; USFWS 2010b
unpaved roads	Low to medium	See above
routes, trails	High	Routes and trails from OHVs, often unauthorized USFWS 1994a, 2010b; USDC 2009, 2011
Habitat modification, conversion, deterioration	Medium to high	
ranching, livestock grazing	Medium to Low, depending on area	Brooks et al. 2006; USFWS 2010b
mining	Low to medium	Chaffee and Berry 2006
recreation	From low to high depending on region; high in the Mojave Desert, CA	Jennings 1997; Bury and Luckenbach 2002; Berry et al. 2008; Brooks 2009; USFWS 2010b. Unauthorized vehicle use is high in several critical habitats (USDC 2009, 2011).
invasive plants	High	Brooks and Pyke 2001; Oftedal 2002; Brooks and Berry 2006; Brooks 2009; USFWS 2010b
fire	UT, AZ, and NV: very high; CA: medium	Brooks and Esque 2002; Brooks and Matchett 2006; Brooks and Minnich 2006; McLuckie et al. 2007; USFWS 2010b
Other factors		
climate change and drought	Ongoing and future threat	Henen et al. 1998; Berry et al. 2002b; Duda et al. 1999; Seager et al. 2007; Hunter 2007; IPCC 2007; McLachlan et al. 2007; Esque et al. 2010b
contaminants	Low now, higher later in more arid conditions	Jacobson et al. 1991; Seltzer and Berry 2005; Chaffee and Berry 2006; Rowe 2008
Adequacy of population and habitat management by government agencies	High	
human immigration and interdiction	Low	Locally, impacts from border patrol and immigrants have been observed in Chuckwalla critical habitat

drivers for wildfires that severely damaged from 13 to 26% of three critical habitats between 2005 and 2010 (Brooks and Matchett 2006, McLuckie et al. 2007, USFWS 2010b). Solar energy development, a new threat, will consume substantial habitat bordering critical and other protected habitats, will cause further losses with new transmission corridors, and degrade habitats in natural areas, parks, and wilderness (Desert Renewable Energy Conservation Plan, Independent Science Panel 2012; LaPré 2010). Sources of direct mortality to the tortoises include subsidized predators (coyotes, ravens, and dogs), vehicle kills on and off paved highways, illegal collections, and translocations. Diseases, particularly URTD caused by *Mycoplasma agassizii,* and the shell disease cutaneous dyskeratosis have contributed to declines of some populations (Jacobson et al. 1991, 1994; Brown et al. 1999a; Christopher et al. 2003).

Populations of *G. morafkai* in the U.S. face threats similar to those of *G. agassizii:* habitat loss, conversion, fragmentation, and deterioration (table 18.3). Although many positive actions have been taken (USFWS 2010a), deficiencies remain for controlling invasive nonnative plant species, wildfires, vandalism, dog predation, OHV use, releases of captive tortoises and exotic species, military use, and management to counter climate change. The nonnative and invasive buffelgrass (*Pennisetum ciliare*), which disrupts Sonoran ecosystems and contributes to wildfires, is an increasing threat (USFWS 2010a). Threats to *G. morafkai* in Mexico are even more severe. Existing laws to protect tortoises and habitat are not enforced, e.g., an estimated 22% of Sonoran habitat is planted to buffelgrass with an estimated 98% likely to be adversely modified in the near future (Stoleson et al. 2005, USFWS 2010a).

Historically, populations of *G. berlandieri* were impacted by intensive brush clearing that converted natural thornscrub-grassland habitat to agricultural or range land (table 18.4). More recently, establishment of the exotic buffelgrass has contributed to habitat deterioration. Droughts and wildfires have caused significant losses to some tortoise population and habitat, e.g., the tortoise population on the Chaparral Wildlife Management Area (WMA) declined by ~95% following drought and a catastrophic wildfire fueled by buffelgrass in 2008, and the population on Las Palomas WMA declined by >90% from the combined effects of prolonged drought and intense heat (Kazmaier et al. 2012; R Kazmaier, personal communication). Fracking and mineral extraction are important new threats, because mineral rights are not owned by the government on state and federal "protected" lands (R Kazmaier, personal communication). The fracking zone affects nearly 50% of the species' range in Texas with the heaviest impacts in the interior Rio Grande Plains, an area formerly considered secure (R. Kazmaier, personal communication). New roads, increased vehicle traffic, and habitat fragmentation are associated with fracking. The combined threats of fracking, climate change (extreme drought and fires), and the spread of exotic grasses may result in catastrophic declines of this species.

For *Gopherus* species living in arid habitats, climate warming is a threat (Intergovernmental Panel on Climate Change [IPCC] 2007, Seager et al. 2007, McAuliffe and Hamerlynck 2010). Droughts affect almost every aspect of the well-being and activities of *G. agassizii* (e.g., Henen et al. 1998, Christopher et al. 1999, Duda et al. 1999, Berry et al. 2002b, Esque et al. 2010b). Drought has been implicated in deaths of *G. berlandieri* in Texas (Kazmaier et al. 2012). More frequent and prolonged droughts are likely to lead to drying and desertification of habitats, local extinctions, and further fragmentation of remaining populations. Some populations may require rescue and translocation to more mesic sites (Hunter 2007, McLachlan et al. 2007).

CONSERVATION EFFORTS

The Importance of Legal Status as Threatened and Endangered Species

In most cases, actions to protect the *Gopherus* species at international, national (federal), state, and local levels have come slowly and have been insufficient to stem population declines, habitat loss, fragmentation, and deterioration. As of 2012, no federally or state-listed population or species merits de-listing or can be treated as "recovered."

Legal status, especially as a federally threatened or endangered species, has provided some measure of protection. The federal endangered status of *G. flavomarginatus* contributed to the establishment of the Mapimí Biosphere Reserve in Mexico (Comisión Nacional de Áreas Naturales Protegidas 2006); however, development of land has continued to occur in the species' habitat inside Reserve boundaries, and status of populations and habitat within the Rancho Sombretillo are unknown (Traphagen et al. 2007; M. Traphagen, personal communication). Within the U.S., federal listings often require years, and then may only occur after scientists or NGOs petition the USFWS, threaten court action, or take the USFWS to court. The next phases of protection for a federally listed species within the U.S. are designation of critical habitat, preparation of a recovery plan, and implementation of the recovery plan. Of the five species of *Gopherus,* only two, *G. polyphemus* and *G. agassizii,* are federally listed; one, *G. morafkai,* has designated critical habitat. Only *G. agassizii* and western populations of *G. polyphemus* have signed recovery plans (USFWS 1990a, 1994a, 2011b). Many recommendations described in each of these recovery plans have not been implemented. For example, soon after the federal listing of *G. polyphemus* in 1990, federal and state agencies prepared several new land-use plans or revised existing plans, a process essential for changing directions in uses on federal and private land (Berry 1997, USFWS 2010b). That process required 10 to 15 years. In 2011, another new land-use planning process for the species was initiated to speed development of renewable energy on federal lands (Desert Renewable Energy Conservation Plan Independent Science Panel 2012).

In the U.S., *G. morafkai* benefitted from the federal listing of the closely related *G. agassizii* in 1990 (USFWS 1990b). As a preemptive measure, government agencies formed the Arizona Interagency Desert Tortoise Team (AIDTT), and developed a management plan for the Sonoran Desert

Table 18.3. Threats experienced by *Gopherus morafkai* (Sonoran and Sinaloan populations in U.S. and Mexico), with estimated level of importance

Threats	Threat Level	References
Direct Mortality		
Collection for pets or release of captive or exotic tortoises	AZ, MX: medium	AZ: Jarchow et al. 2002; Edwards et al. 2010; Grandmaison and Frary 2012; C. Jones pers. comm. MX: Bury et al. 2002; M. Vaughn pers. comm.
Collection for food	MX: medium (Sonora) to high (Sinaloa)	MX: Bury et al. 2002, M. Vaughn pers. comm.
Vandalism	Probably medium	AZ and MX: documented, associated with proximity to human populations, illegal immigration, food (USFWS 2010a; Bury et al. 2002)
Disease	Probably low in 2011	MX: a risk if captives are released; disease documented on coast at Punta Tepopa (M. Vaughn pers. comm.)
upper respiratory tract disease	Probably low	AZ: Jones 2008; Edwards et al. 2010; USFWS 2010a MX: Dickinson et al. 2005
herpesvirus	Unknown	
shell disease	Probably low	
Predators		
birds	Low to none	USFWS 2010a
dogs	Medium	AZ and MX: USFWS 2010a
Vehicle strikes	Medium	USFWS 2010a; M. Vaughn pers. comm.
Habitat		
Habitat loss	High currently and in the near future	
urbanization, human population growth, and infrastructure	High	AZ: USFWS 2010a MX: Bury et al. 2002; M. Vaughn pers. comm.
rural, exurban	AZ: high in the future because of close proximity to human populations	AZ: USFWS 2010a MX: Bury et al. 2002
agriculture	Very high to high	AZ: USFWS 2010a MX: Bury et al. 2002; USFWS 2010a; Stoleson et al. 2005; planting for food crops or illegal drug crops also high
solar energy	Regionally a threat to specific populations	AZ: proposed project in Black Mountains affects 0.05% of habitat
Habitat fragmentation, creation of barriers	High	AZ: Sun Corridor "Megapolitan"
paved roads, canals, railroads, power lines		AZ: Boarman and Sazaki 2006; Andrews et al. 2008; USFWS 2010a MX: P. Rosen pers. comm.
unpaved roads		Grandmaison and Frary 2012
routes, trails (OHV)	High	USFWS 2010a
Habitat modification, conversion, deterioration	High currently and in the near future	
ranching, livestock grazing	AZ: low MX: high due to ineffective management and continued overgrazing	AZ: Oftedal 2002; Brooks et al. 2006; Grandmaison et al. 2010 MX: designation of 2 mil ha of Sonoran Desert habitat to grasslands for livestock production, loss of 20% of tortoise habitat; stocking rates at 2 to 5 times recommended rates
ironwood and mesquite tree harvest	MX: high	MX: loss of 4% and 2% of Sonoran Desert habitat to mesquite clearing and ironwood harvest (Suzan et al. 1999; USFWS 2010a; P. Rosen pers. comm.)
mining and gas exploration	MX: high AZ: also see contaminants	MX: M. Vaughn pers. comm.
recreation	AZ: high	AZ: OHV use an increasing threat (USFWS 2010a; Brooks and Lair 2009; Grandmaison et al. 2010) (see also collecting)
invasive plants	Very high, most severe threat to mod, curtailment of habitat & range (FWS 2010b)	Summary in USFWS 2010a; Brooks 2009; Bury et al. 2002; Esque et al. 2002, 2003; Thomas and Guertin 2007; Stevens and Fehmi 2009
fire	High potential, coupled with invasion and establishment of invasive species (above)	USFWS 2010a: Bury et al. 2002; Esque et al. 2002; Brooks and Matchett 2006; Brooks and Minnich 2006
Other factors		
climate change and drought	Potential threat in the near future in concert with other factors (USFWS 2010b)	AZ: IPCC 2007; Seager et al 2007; McAuliffe and Hamerlynck 2010 MX: high (P. Rosen pers. comm.); drought probably responsible for die-off of tortoises on Tiburon Island and coast of Sonora (M. Vaughn pers. comm.)
contaminants	Potential threat but not documented	AZ: USFWS 2010a; Seltzer and Berry 2005
Adequacy of population and habitat management by government agencies	High	
human immigration and interdiction	AZ: impacts limited, but severe locally	USFWS 2010a

Table 18.4 Threats experienced by *Gopherus berlandieri*, with estimated level of importance

Threats	Threat Level	References and Personal Communications
Direct Mortality		
Collecting	Medium	For pets: Auffenberg and Weaver 1969; Luckenbach 1982; Judd and Rose 2000. For food: Rose and Judd 1982; Judd and Rose 2000. MX: tourist market curios: Judd and Rose 2000
Release of captive tortoises	Probably medium	Fujii and Forstner 2010; R. Kazmaier pers. comm.
Vandalism	Probably medium	Ernst and Barbour 1972; Judd and Rose 2000
Disease	Probably medium	
upper respiratory tract disease	Probably medium	Documented in captives, Judd and Rose 2000; and in wild tortoises, Lower Rio Grande Valley R. Kazmaier pers. comm.
herpesvirus	Unknown	
shell disease	Low	High incidence but apparently not fatal (Judd and Rose 2000)
Predators		
birds	Low	Rose et al. 2011; R. Kazmaier pers. comm. (Crested caracaras)
dogs	Probably medium	Rose et al 2011
other subsidized predators	Probably medium	Judd and Rose 2000; Rose et al. 2011; raccoons, skunks, opossums, coyotes
introduced fire ants	Probably medium	Eggs and hatchlings, Judd and Rose 2000
wood rats	Medium	Auffenberg and Weaver 1969; Judd and Rose 2000
Texas indigo snakes	Low	R. Kazmaier pers. comm.
Vehicle strikes	High	Bury and Smith 1986, Judd and Rose 2000, Engeman et al. 2004
Habitat		
Habitat loss	High currently and in the near future	
urbanization, human population growth, residential development and infrastructure	High	Rose and Judd 1982; Judd and Rose 2000, especially lower Rio Grande Valley
rural, exurban	High	Judd and Rose 2000
agriculture	High	Rose and Judd 1982; Judd and Rose 2000; Kazmaier 2000; Kazmaier et al. 2001a
Habitat fragmentation, creation of barriers	High	
agriculture	High	Rose and Judd 1982; Judd and Rose 2000; Kazmaier et al. 2002
paved roads	High	Fujii and Forstner 2010
unpaved roads	High	Engeman et al. 2004
one-lane tracks	High	Engeman et al. 2004
Habitat modification, conversion, deterioration	High currently and in the near future	
large-scale clearing of native thornscrub habitat by chaining, chopping, herbicide	High	Historically caused very high mortality (Judd and Rose 2000) and currently continues to create unsuitable habitat (Kazmaier et al. 2001a)
livestock grazing	TX: low to high depending on grazing intensity MX: high due to ineffective management and continued overgrazing	Auffenberg and Weaver 1969; Kazmaier 2000; Kazmaier et al. 2001b
mineral development / fracking	Very high within the Eagle Ford Shale region	R. Kazmaier pers. comm.
recreation	Unknown	
invasive plants	High	MX and TX: large areas now dominated by exotic grasses, particularly buffelgrass (Judd and Rose 2000; R. Kazmaier pers. comm.)
wildfire	Low naturally; high when coupled with invasion and establishment of invasive grasses and drought	Bury and Smith 1986; R. Kazmaier pers. comm.
Other factors		
climate change and drought	High, particularly in concert with other factors	Drought: Rose and Judd 1982; Rose et al. 2011
woven wire fencing	High	Engeman et al. 2004; Judd and Rose 2000
Adequacy of population and habitat management by government agencies	High	Existing government measures to protect populations and habitat are inadequate and enforcement is weak (Judd and Rose 2000)

population within the state (AIDTT 1996, Arizona Game and Fish Department 2011). In Mexico, *G. morafkai* and *G. berlandieri* received protection on paper from take, import, and export, but enforcement and protection of habitat have not been priorities.

Protected Lands, Reserves, Refuges, and Natural Areas for the *Gopherus* Species

Habitats and populations may be protected or partially protected in wilderness, national parks and monuments, national wildlife refuges, national forests, military reservations, areas of critical environmental concern, reserves, preserves, and natural areas. Few sites have been designated specifically to protect and conserve a *Gopherus* species, or the species may receive benefits from being on land designated for other purposes. Highly desirable are substantial blocks of land that support viable and undisturbed populations, are interconnected, and that taken as a whole, preserve genetic, morphologic, physiologic, and ecological diversity. In many cases, the existing protected areas are insufficient to support viable populations and have anthropogenic activities that threaten the integrity within or adjacent to the boundaries. Further, many habitats on protected lands are fragments that support few animals, for example, *G. flavomarginatus, G. polyphemus,* and *G. berlandieri*. In recent years, some sites have been established for release of captives or translocated tortoises. Some examples illustrate these points. *Gopherus polyphemus* occurs within numerous national forests, military reservations, state wildlife management areas, state protected areas, and mitigation parks; few of these sites support viable populations. *Gopherus agassizii,* which has >28,000 km^2 of critical habitat in 13 blocks, as well as a Research Natural Area and habitat within four National Park Service holdings, is a case in point. Critical habitat designation has not ensured protection: the loss of 303.5 km^2 of critical habitat and displacement of ~1000 animals for expansion of the Ft. Irwin military installation in the central Mojave Desert is just one of many examples (USFWS 2010b). Although a recovery plan was published in 1994 (USFWS 1994a), implementation of measures to facilitate recovery has been slow. Deterioration of critical habitat is ongoing from wildfires, livestock grazing, and unauthorized OHV use (e.g., USFWS 2010b). The Mapimí Biosphere Reserve for *G. flavomarginatus* provides another, similar example: biologists observed that land was being cleared for agriculture (Traphagen et al. 2007), cattle grazing is causing habitat deterioration, and tortoises are depleted from take for food in rural areas (Traphagen et al. 2007; M Traphagen, personal communication; E Goode, personal communication).

The presence of a federally- or state-listed species on federal or state lands alone does not ensure protection. For example, within the U.S., 85% of populations and habitat for *G. morafkai* occur on federal and state lands. The federal lands, where 60% of habitat occurs, are on public lands managed by the Bureau of Land Management, three national forests, three Department of Defense facilities, four National Park Service properties, and six National Wildlife Refuges (Jeff Servoss, personal communication). An additional 25% of habitat exists on Arizona State lands, 10% on Indian Tribal lands, and 5% in private holdings. In spite of the high percentage of habitat on government lands, *G. morafkai* still qualifies for threatened status (USFWS 2010a). The objectives for managing the federal and state lands are frequently incompatible with maintaining viable populations of this chelonian. In Mexico, where an estimated 50% of the geographic range of *G. morafkai* occurs, no protected habitats have been established (USFWS 2010a), and data are unavailable on demographic attributes of the tortoises and amounts and condition of habitat.

Most *G. berlandieri* populations occur on private land where they are only protected by state regulations. The only populations in the U.S. on federally protected lands occur on the southeast coast of Texas on three national wildlife refuges (Laguna Atascosa, Lower Rio Grande Valley, and Padre Island National Seashore). Three Texas wildlife management areas (WMA)—Chaparral, James E. Daughtrey, and The Las Palomas—also provide protection; two of these, Chaparral and Las Palomas, have sustained ≥90% losses to populations and habitat from recent fires and drought.

Mitigation Measures

For the *Gopherus* species designated as threatened within the U.S. (*G. polyphemus* and *G. agassizii*), numerous conservation actions have been taken to minimize or mitigate loss of tortoises and habitat to development (e.g., see USFWS 2010b). Among the actions are land acquisition, tortoise exclusion fencing of highways and roads to reduce vehicle strikes and collecting, acquisition of grazing allotments and privileges, support of research projects, and translocation of tortoises. The concepts of mitigation and compensation for threatened and endangered species are relatively new and in transitional phases.

Gopherus polyphemus provides examples of mitigation measures typical for other members of the genus and a lesson that mitigation measures rarely compensate fully for losses to populations and habitat. Mitigation banks that preserve sandhill habitat and serve as translocation sites for tortoises moved from highway and public works projects and development sites have been established. Mitigations have changed since the early 1990s. From 1991 to 2007, developers had three options for dealing with tortoises: on-site relocation, off-site relocation, or an incidental take permit that allowed entombment of tortoises (FWC 2001b). Incidental take permits required developers to pay habitat mitigation fees that were used to purchase habitat elsewhere, but resulted in the loss of an estimated >100,000 tortoises and a net loss of 75–85% tortoise habitat (Fleshler 2005, Mushinsky et al. 2006). In contrast, the new FWC tortoise management plan clearly states that the conservation goal is to restore and maintain secure viable populations in Florida by addressing habitat loss, specifically by increasing the amount of protected habitat (increase total protected habitat to 7912 km^2 by 2022), managing habitat on public lands, restocking tortoise where densities are low on protected lands (60,000 tortoises by 2022), and decreasing tortoise mortality on development sites by ending incidental take permits and requiring translocation (180,000 tortoises by 2022 to protected, managed sites) (FWC 2007, 2011). Severe budget constraints and cuts have prevented Florida from attaining its annual land acquisition goals, however; funding for the Florida Forever land buying program was terminated in 2011 by the Florida Legislature with no new public lands added in 2009–2010 (FWC 2011). Furthermore, mitigation is not required for land conversion to agriculture or silviculture because agricultural interests are exempt from these regulations under the Florida Endangered Species Act. This exemption creates a regulatory loophole whereby converted lands can then be sold for development and tortoises legally removed prior to permits normally required for development activities (Mushinsky et al. 2006). Despite the best plans and intentions of regulatory agencies for *G. polyphemus,* significant challenges exist to achieving conservation goals: a lack of acquisition and management funds and low priority for such funding by most lawmakers.

The Role of Nongovernmental Organizations (NGOs)

Since the 1970s, NGOs and nonprofit organizations have been a driving force in protection, conservation, education, and research for *Gopherus* species. The World Wildlife Fund-U.S. and the Turtle Recovery Program financially support research on *G. flavomarginatus* (Aguirre et al. 1997), and members of the Turtle Conservancy, Behler Chelonian Center, have made efforts to acquire key parcels of habitat in recent years (E. Goode, personal communication). The Desert Tortoise Council in the Southwest and Gopher Tortoise Council in the Southeast are two focused nonprofit corporations dedicated to conservation, education, and preserving representative populations of *G. agassizii, G. morafkai,* and *G. polyphemus* throughout their respective geographic ranges (see www.deserttortoise.org, www.gophertortoisecouncil.org). The Desert Tortoise Council has also supported research and education for *G. flavomarginatus* and *G. berlandieri.* The Gopher Tortoise Conservation Initiative (GTCI) founded by Ray and Patricia Ashton is another focused nonprofit organization with similar objectives for *G. polyphemus* in Florida (www.ashtonbiodiversity.org/gtci.php). The GTCI (along with the Humane Society of the United States, Defenders of Wildlife, and other groups) was instrumental in securing a FWC policy change that ended the issuance of incidental take permits allowing entombment of *G. polyphemus* in 2007. In recent years, Southeastern Partners in Amphibian and Reptile Conservation, a consortium of government agencies, organizations, corporations, private individuals, and academic institutions, has also promoted *G. polyphemus* conservation and habitat management (Bailey et al. 2006).

Numerous NGOs have petitioned for federal listings of *G. polyphemus* and the desert tortoises. Save Our Big Scrub, Inc., and Wild South, two nonprofit environmental organizations, were the petitioners for Federal listing of *G. polyphemus* in the eastern part of its range in 2006. NGOs have lobbied Congress for funds to acquire land, litigated in court to gain protection, and to implement recovery actions for the two desert tortoise species, e.g., the Natural Resources Defense Council, Defenders of Wildlife, California Turtle and Tortoise Club, Center for Biological Diversity, WildEarth Guardians, Western Watersheds Council, and Public Employees for Environmental Responsibility.

On behalf of *G. agassizii,* government agencies and NGOs have acquired private inholdings in critical habitat, Natural Areas and other protected lands; they have also worked together and separately to acquire and retire livestock allotments in critical habitat and to erect vehicle and livestock exclusion fencing to protect tortoises and habitat. The Desert Tortoise Preserve Committee, a nonprofit corporation, is one such NGO with multipurposes: land acquisition and mitigation, preserve management and habitat restoration, and education (www.tortoise-tracks.org). This group has worked with the U.S. Department of the Interior's Bureau of Land Management and California Department of Fish and Wildlife to establish the Desert Tortoise Research Natural Area and other protected lands in California and to acquire a livestock allotment in critical habitat. The Wildlands Conservancy also acquires lands for protection of natural resources.

Neither *G. flavomarginatus* nor *G. berlandieri* has the broad-based support system of advocates and watchdogs enjoyed by the other three species of *Gopherus.* Traphagen (personal communication) noted that the greatest threat to *G. flavomarginatus* was lack of strong advocates within Mexico. Judd and Rose (2000) cited a lack of public awareness of the protected status of *G. berlandieri* and the need for public education as vital to conservation.

Headstarting Programs

Two species, *G. flavomarginatus* and *G. agassizii,* have had or currently have head starting programs. The first in situ head starting program for *G. flavomarginatus* was established at Instituto de Ecologia at Mapimí Biosphere Reserve, Mexico, to learn more about life history attributes and to enhance recruitment into local populations (Aguirre et al. 1997, Morafka et al. 1997). The program subsequently was abandoned. Ariel Appleton maintained an ex situ program for captive individuals at the Appleton Research Ranch, Arizona, for >20 years (Appleton 1978). Subsequently, the Appleton tortoises were transferred to the Living Desert Zoo and Gardens State Park and a ranch managed by the Turner Endangered Species Fund (TESF) in New Mexico (Truett and Phillips 2009). The TESF initiated a new head starting program with an objective of determining whether the species can persist in the wild in New Mexico (Truett and Phillips 2009).

The first head starting program for *G. agassizii,* initiated in 1990 for the Department of the Army, was designed to obtain information about life history attributes and survivorship of juveniles (Morafka et al. 1997). Three similar programs, with additional objectives, were established in 2002 by Edwards Air Force Base, in 2006 by the U.S. Marine Corps, and in 2011 by the Mojave National Preserve (Nagy et al. 2011; Buhlmann et al. 2013; B Henen, personal communication).

Vital rate studies of desert tortoises indicate that head-starting programs may have low relative value in recovery actions compared with eliminating threats to breeding females (Reed et al. 2009).

Translocation

Gopherus species have been moved from one place to another for decades and for numerous reasons. Captives have been released and wild tortoises have been moved short and long distances; some releases and translocations have been authorized by government agencies whereas others have not (see Murphy et al. 2007 for a summary for *G. agassizii*). Few releases and translocations have been studied, and most studies were conducted for only a few years—insufficient to determine long-term success (Berry 1986a, Field et al. 2007, Nussear et al. 2012). Translocation is a controversial technique because of valid concerns over negative effects on resident populations, potential spread of disease, genetic mixing, predation, moving tortoises to unprotected or unmanaged sites, and poor site fidelity (Burke 1989b, Dodd and Seigel 1991, Berish 2001, Mushinsky et al. 2006, Martel et al. 2009). Early releases often failed because of a lack of understanding and knowledge of the proper techniques required for success and, in some cases, tortoises were simply dumped on a tract of land with no follow-up monitoring or concern for fates of the tortoises. Translocation also can be time-consuming and costly when done correctly (Mushinsky et al. 2006, Ashton and Ashton 2008). Recent translocation studies of *G. polyphemus* have shown that soft release methods using long-duration enclosures (6–12 months duration) combined with frequent monitoring, attention to forage conditions, and habitat management significantly increase site fidelity and greatly improve the probability of augmenting or reestablishing populations (Tuberville et al. 2005, 2008, 2011; Ashton and Ashton 2008; Aresco et al. 2010).

Translocation of tortoises from lands that will be destroyed by development or committed to other activities has become an important tool for minimizing losses of individuals, especially for *G. agassizii* and *G. polyphemus.* The protocols for these two species differ, depend on legal protections and regulations in the states where they occur, and the outcomes may differ considerably because of behaviors and habitat types. For *G. polyphemus,* regulatory agencies that prohibit incidental take on private lands (e.g., USFWS where species are federally listed and FWC in Florida) attempt to prevent mortality by requiring translocation of all individuals on a development site to protected, managed land (USFWS: Conservation Banks; FWC: Recipient Sites) where tortoises were either extirpated or at unnaturally low densities (FWC 2009, USFWS 2009b).

For *G. agassizii* translocation projects, the types of mitigation and compensation required can be site-specific and may depend on whether habitat is designated critical habitat, federally managed or privately owned. Testing for URTD caused by *Mycoplasma agassizii* and *M. testudineum,* as well as herpesviruses may be required (USFWS 2011c), and may depend on such circumstances as size of the project and location (USFWS 2011c). No long-term studies have been undertaken on settling and survival of translocated individuals, although short-term studies have been published (e.g., Field et al. 2007, Nussear et al. 2012). Preliminary results of a multiyear translocation project in the central Mojave Desert have not been promising because of high death rates (e.g., Berry et al. 2011).

CONCLUSIONS

With the exception of the newly described *Gopherus morafkai,* the *Gopherus* species are protected in whole or in part by international, federal, and state regulations. Unfortunately, designation of a species as threatened, endangered, or fully protected by government agencies has not resulted in the regulations and enforcement essential to protecting individuals, populations, and habitats. General patterns for all species include high mortality rates among adults and/or juveniles, deterioration and fragmentation of habitat, and habitat loss. The list of threats to each species is numerous and continuing to grow. The status of all species can be considered as declining, with *G. flavomarginatus* at greatest risk. The most pressing issue for all *Gopherus* species is sufficient protection of viable and representative populations and habitat. Species with strong support from advocacy groups and NGOs are likely to fare better than those species without such support.

We anticipate that the threats to populations and habitats, described herein as of 2012, are likely to increase in scope and severity with the growing human populations and increasing socioeconomic pressures for use and development of land. Government agencies are making efforts to minimize and mitigate the negative impacts to federally listed species within the U.S. through such measures as land acquisition, establishment of protected areas, restoration of disturbed lands, limits on livestock grazing, highway exclusion fencing, head starting programs, and translocations of displaced wild and captive tortoises. Some measures are more effective than others. Restoration or natural recovery of disturbed habitats in the Southwest, for example, is likely to require centuries, if not longer, especially with climate change (e.g., Abella et al. 2010). Translocation and head starting of tortoises have yet to be demonstrated as effective tools to augment and enhance depleted populations for the genus, at least for *Gopherus agassizii,* on a long-term basis. In the meantime, we continue to observe declines in populations and losses to habitats. We encourage all concerned NGOs and government agencies to engage in vigorous efforts to tackle the complex and difficult problems facing this taxonomic group.

Acknowledgements

Thanks are due to C. Darst, T. Egan, E. Goode, B. Henen, C. Jones, R. Kazmaier, A. McLuckie, K. Nagy, P. Rosen, J. Servoss, M. Traphagen, and M. Vaughn for advice, reports from the field, and access to data.

REFERENCES

Abella SR. 2010. Disturbance and plant succession in the Mojave and Sonoran deserts of the American Southwest. Int. J. Environ. Res. Public Health 7:1248–1284.

Adams WH, WR Adams, and J Kearney-Williams. 1987. Foodways on the plantations at Kings Bay: hunting, fishing, and raising food. Department of Anthropology Reports of Investigations 5. University of Florida, Gainesville.

Adest GA and LG Aguirre. 1995. Natural and life history of the Bolson tortoise, *Gopherus flavomarginatus*. Publ. Soc. Herpetol. Mex. 2:1–5.

Adest GA, LG Aguirre, DJ Morafka, and JV Jarchow. 1989. Bolson tortoise (*Gopherus flavomarginatus*) conservation. I. Life history. Vida Silv. Neotrop. 2:7–13.

Adest GA, MA Recht, LG Aguirre, and DJ Morafka. 1988. Nocturnal activity in the Bolson tortoise (*Gopherus flavomarginatus*). Herpetol. Rev. 19:75–76.

Agassiz L. 1857. Contributions to the Natural History of the United States of America. 2 Vols. Little Brown, Boston.

Aguirre LG. 1995. Conservation of the Bolson tortoise, *Gopherus flavomarginatus*. Publ. Soc. Herpetol. Mex. 2, 6–9.

Aguirre LG, GA Adest, and DJ Morafka. 1984. Home range and movement patterns of the Bolson tortoise, *Gopherus flavomarginatus*. Acta. Zool. Mex 1:1–28.

Aguirre G, DJ Morafka, and GA Adest. 1997. Conservation strategies for the Bolson tortoise, *Gopherus flavomarginatus*, in the Chihuahuan Desert. IN Proceedings: Conservation, Restoration, and Management of Tortoises and Turtles—An International Conference, ed. J VanAbbema, 333–338. New York Turtle and Tortoise Society.

Albrech P, SJ Gould, GE Oster, and DB Wake. 1979. Size and shape in ontogeny and phylogeny. Paleobiology 5:296–317.

Alberts AC, DC Rostal, and VA Lance. 1994. Studies on the chemistry and social significance of chin gland secretions in the desert tortoise, *Gopherus agassizii*. Herpetol. Monogr. 8:116–123.

Alcock J. 2009. Animal Behavior: An Evolutionary Approach. Sinauer Associates, Sunderland, MA.

Alexy KJ, KJ Brunjes, JW Gassett, and KV Miller. 2003. Continuous remote monitoring of gopher tortoise burrow use. Wildl. Soc. Bull. 31:1240–1243.

Alford RA. 1980. Population structure of *Gopherus polyphemus* in northern Florida. J. Herpetol. 14:177–182.

Alho CJR, TMS Danni, and LFM Padua. 1985. Temperature-dependent sex determination in *Podocnemis expansa* (Testudinata: Pelomedusidae). Biotropica 17:75–78.

Anderson DJ 1982. The home range: a new nonparametric estimation technique. Ecology 63:103–112.

Anderson DR and KP Burnham. 1996. A monitoring program for the desert tortoise. Colorado Cooperative Fish and Wildlife Research Unit, Colorado State University, Fort Collins, CO.

Anderson DR, KP Burnham, BC Lubow, L Thomas, PS Corn, PA Medica, RW Marlow. 2001. Field trials of line transect methods applied to estimation of desert tortoise abundance. J. Wildl. Manage. 65:583–597.

Anderson NJ. 2001. The thermal biology of the gopher tortoise (*Gopherus polyphemus*) and the importance of microhabitat selection. MS thesis, Southeastern Louisiana University, Hammond, LA.

Andrews KM, JW Gibbons, and DM Jochimsen. 2008. Ecological effects of roads on amphibians and reptiles: a literature review. Herpetol. Conserv. 3:121–132.

Andrews R. 1982. Patterns of growth in reptiles. IN Biology of the Reptilia, ed. C. Gans and FH Pough, Vol. 13, Ecophysiology, 273–305. Academic Press, New York.

Appleton AB. 1978. Bolson tortoises (*Gopherus flavomarginatus* Legler) at The Research Ranch. IN Proc. Desert Tortoise Council Symp.. 164–174.

Arata AA. 1958. Notes on the eggs and young of *Gopherus polyphemus* (Daudin). Quar. J. FL. Acad. Sci. 21:274–280.

Aresco MJ and C Guyer. 1998. Efficacy of using scute annuli to determine growth histories and age of *Gopherus polyphemus* in southern Alabama. Copeia 1998:1094–1100.

Aresco MJ and C Guyer. 1999a. Burrow abandonment by gopher tortoises in slash pine plantations of the Conecuh National Forest. J. Wildl. Manage. 63:26–35.

Aresco MJ and C Guyer. 1999b. Growth of the tortoise *Gopherus polyphemus* in slash pine plantations of southcentral Alabama. Herpetologica 55:499–506.

Aresco MJ, RN Walker, and M Greene. 2010. Site fidelity, movements, survival, and minimum area required of gopher tortoises (*Gopherus polyphemus*) translocated from development sites to managed conservation lands. Status Report to the Florida Fish and Wildlife Conservation Commission, Permit WX06116b, Tallahassee, FL. 68 pp.

Arizona Game and Fish Department. 2011. Arizona's State Wildlife Action Plan: 2011–2021, DRAFT. Arizona Game and Fish Department, Phoenix, AZ. 210 pp.

Arizona Interagency Desert Tortoise Team (AIDTT). 1996. Management plan for the Sonoran Desert population of the desert tortoise in Arizona. 60 pp.

Ashton RE Jr and KJ Ashton. 1991. *Gopherus polyphemus,* gopher tortoise: drinking behavior. Herpetol. Rev. 22:55–56.

Ashton RE and PS Ashton. 2008. The Natural History and Management of the Gopher Tortoise, *Gopherus polyphemus* (Daudin). Krieger, Malabar, FL.

Ashton KG and RL Burke. 2007. Long-term retention of a relocated population of gopher tortoises. J. Wildl. Manage. 71:783–787.

Ashton KG, RL Burke, and JN Layne. 2007. Geographic variation in body and clutch size of gopher tortoises. Copeia 2007:355–363.

Ashton KG, BM Engelhardt, and BS Branciforte. 2008. Gopher tortoise (*Gopherus polyphemus*) abundance and distribution after prescribed fire reintroduction to Florida scrub and sandhill at Archbold Biological Station. J. Herpetol. 42:523–529.

Atkins A, E Jacobson, J Hernandez, AB Bolten, L Xiaomin. 2010. Use of a portable point-of-care (Vetscan VS2) biochemical analyzer for measuring plasma biochemical levels in free-living loggerhead sea turtles. J. Zoo. Wildl. Med. 41:585–593.

Auffenberg W. 1962. A redescription of *Testudo hexagonatus* Cope. Herpetologica 18:25–34.

Auffenberg W. 1966a. The carpus of land tortoises (Testudinidae). Bull. Florida State Mus. 10: 59–191.

Auffenberg W. 1966b. On the courtship of *Gopherus polyphemus.* Herpetologica 22:113–117.

Auffenberg W. 1969. Tortoise Behavior and Survival. Rand McNally, Chicago, IL.

Auffenberg W. 1974. Checklist of fossil land tortoises (Testudines). Bull. Florida State Mus. 18:121–251.

Auffenberg W. 1976. Genus *Gopherus.* Pt. 1. Osteology and relationships of extant species. Bull. Florida State Mus. 20:47–110.

Auffenberg W and R Franz. 1978. *Gopherus agassizii.* Catalogue of American Amphibians and Reptiles 212.1–212.2.

Auffenberg W and R Franz. 1982. The status and distribution of the gopher tortoise (*Gopherus polyphemus*). IN North American Tortoises: Conservation and Ecology, ed. RB Bury and DJ Germano, 95–126. US Fish and Wildlife Service Report, 12.

Auffenberg W and JB Iverson. 1979. Demography of terrestrial turtles. IN Turtles: Perspectives and Research, ed. M Harless and H Morlock, 541–569. Wiley-International, New York.

Auffenberg W and WW Milstead. 1965. Reptiles in the Quaternary of North America. IN The Quaternary of the United States, ed. HE Wright and DG Frey, 557–568. Princeton University Press, Princeton, NJ.

Auffenberg W and WG Weaver Jr. 1969. *Gopherus berlandieri* in southeastern Texas. Bull. Florida State Mus. 13:141–203.

Austin RJ, L Carlson, and RW Estabrook. 2009. Archaic Period faunal use in the west-central Florida interior. Southeast. Archaeol. 28:148–160.

Averill-Murray RC. 2000. Survey protocol for Sonoran Desert Tortoise monitoring plots: reviewed and revised. Arizona Interagency Desert Tortoise Team.

Averill-Murray RC. 2002. Reproduction of *Gopherus agassizii* in the Sonoran Desert, Arizona. Chelonian Conserv. Biol. 4:295–301.

Averill-Murray RC and A Averill-Murray. 2005. Regional-scale estimation of density and habitat use of the desert tortoise (*Gopherus agassizii*) in Arizona. J. Herpetol. 39:65–72.

Averill-Murray RC, BE Martin, SJ Bailey, and EB Wirt. 2002a. Activity and behavior of the Sonoran Desert tortoise in Arizona. IN The Sonoran Desert Tortoise: Natural History, Biology and Conservation, ed. TR Van Devender, 135–158. University of Arizona Press Tucson.

Averill-Murray RC, AP Woodman, and JM Howland. 2002b. Population ecology of the Sonoran desert tortoise in Arizona. IN The Sonoran Desert Tortoise: Natural History, Biology, and Conservation, ed. TR Van Devender, 109–134. University of Arizona Press, Tucson.

Avise JC, J Arnold, RM Ball, E Bermingham, T Lamb, JE Neigel, CA Reeb, and NC Saunders. 1987. Intraspecific phylogeography: the mitochondrial DNA bridge between population genetics and systematics. Annu. Rev. Ecol. Syst. 18:489–522.

Avise JC, BW Bowen, T Lamb, AB Meylan, and E Bermingham. 1992. Mitochondrial DNA evolution at a turtle's pace: evidence for low genetic variability and reduced microevolutionary rate in Testudines. Mol. Biol. Evol. 9:457–73.

Bailey MA, JN Holmes, KA Buhlmann, and JC Mitchell. 2006. Habitat Management Guidelines for Amphibians and Reptiles of the Southeastern United States. Partners in Amphibian and Reptile Conservation Technical Publication HMG-2, Montgomery, AL. 88 pp.

Bailey SJ, CR Schwalbe, and CH Lowe. 1995. Hibernaculum use by a population of desert tortoises (*Gopherus agassizii*) in the Sonoran Desert. J. Herpetol. 29:361–369.

Bakken GS. 1981a. A two-dimensional operative temperature model for thermal energy management by animals. J. Therm. Biol. 6:23–30.

Bakken GS. 1981b. How many equivalent black-body temperatures are there? J. Theoret. Biol. 6:59–60.

Bakken GS. 1989. Arboreal perch properties and the operational temperature experienced by small animals. Ecology 70:922–930.

Bakken GS. 1992. Measurement and application of operative and standard operative temperatures in ecology. Am. Zool. 32:194–216.

Bakken GS and DM Gates. 1975. Heat-transfer analysis of animals: some implications for field ecology, physiology, and evolution. In Perspectives in Biophysical Ecology, ed. DM Gates and R Schmerl, 255–290. Springer-Verlag, New York.

Bakken GS, WR Santee, and DJ Erskine. 1985. Operative and standard operative temperature: tools for thermal energetic studies. Am. Zool. 25:933–944.

Barbault R and G Halffter. 1981. A Comparative and Dynamic Approach to the Vertebrate Organization of the Desert of Mapimi (Mexico) in Ecology of the Chihuahuan Desert, ed. R Barbault and G Halffter. Man and the Biosphere Program, UNESCO, Instituto de Ecologia, Mexico.

Barrett SL. 1990. Home range and habitat of the desert tortoise (*Xerobates agassizii*) in the Picacho Mountains of Arizona. Herpetologica 46:202–206.

Barrett SL and JA Humphrey. 1986. Agonistic interactions between *Gopherus agassizii* (Testudinidae) and *Heloderma suspectum* (Helodermatidae). Southwest. Nat. 31:261–263.

Bartholomew B. 1993. Desert tortoise (*Gopherus agassizii*) nesting behavior. Vivarium 4(4):27–29.

Basiotis KA. 2007. The effects of invasive cogongrass *(Imperata cylindrica*) on the threatened gopher tortoise (*Gopherus polyphemus*). MS thesis, University of South Florida, Tampa.

Baskaran LM, VH Dale, RA Efroymson, and W Birkhead. 2006. Habitat modeling within a regional context: an example using gopher tortoise. Am. Midl. Nat. 155:335–351.

Batsch AJGC. 1788. Versuch einer Anleitung, zur Kenntniß und Geschichte der Thiere und Mineralien, für akademische Vorlesungen entworfen, und mit den nöthigsten Abbildungen versehen. Erster Theil. Allgemeine Geschichte der Natur: besondre der Säugthiere, Vögel, Amphibien und Fische. Akademische Buchhandlung, Jena.

Battye C. 1934. An episode of the early days at Needles, California. Santa Fe Magazine 28(12):37–39.

Baxter PC, DS Wilson, and DJ Morafka. 2008. The effects of nest date and placement of eggs in burrows on sex ratios and potential survival of hatchling desert tortoises, *Gopherus agassizii*. Chelonian Conserv. Biol. 7:52–59.

Beaupre SJ, ER Jacobson, HB Lillywhite, and K Zamudio. 2004. Guidelines for use of live amphibians and reptiles in field and laboratory research. 2nd ed. rev. by the Herpetological Animal Care and Use Committee (HACC) of the American Society of Ichthyologists and Herpetologists.

Begon M. 1979. Investigating Animal Abundance: Capture-Recapture for Biologists. University Park Press, Baltimore.

Bell CJ, EL Lundelius, AD Barnosky, RW Graham, EH Lindsey, DR Ruiz, HA Semken, SD Webb, RJ Zakrzewski. 2004. The Blancan, Irvingtonian, and Rancholabrean mammal ages. In Late Cretaceaous and Cenozoic Mammals of North America, ed. MO Woodburne, 232–314. Columbia University Press, NY.

Beltz RE. 1954. Miscellaneous observations on captive Testudininae. Herpetologica 10:45–47.

BenDor T, J Westervelt, JP Aurambout, and W Meyer. 2009. Simulating population variation and movement within fragmented landscapes: an application to the gopher tortoise (*Gopherus polyphemus*). Ecol. Model. 220:867–878.

Berish JE. 2001. Management considerations for gopher tortoises in Florida. Final Report. Florida Fish and Wildlife Conservation Commission, Tallahassee, FL. 44 pp.

Berish JE, LD Wendland, RA Kiltie, ED Garrison, and CA Gates. 2010. Effects of mycoplasmal upper respiratory tract disease on morbidity and mortality of gopher tortoises in northern and central Florida. J. Wildl. Dis. 46:695–705.

Berry KH. 1974. Desert tortoise relocation project: status report for 1972. Division Highways, State of California, Bishop. Contr. F-9353. Sec. 3.

Berry KH. 1984. The distribution and abundance of desert tortoises in California from the 1920's to the 1960's and a comparison with the current situation. IN The Status of the Desert Tortoise (*Gopherus agassizii*) in the United States, ed. KH Berry, chap. 4:118–153. Report to the US Fish and Wildlife Service from the Desert Tortoise Council.

Berry KH. 1986a. Desert tortoise (*Gopherus agassizii*) relocation: implications of social behavior and movements. Herpetologica 42:113–125.

Berry KH. 1986b. Incidence of gunshot deaths in desert tortoise in California. Wildl. Soc. Bull. 14:127–132.

Berry KH. 1989. *Gopherus Agassizi* desert tortoise. In The Conservation Biology of Tortoises, ed. IR Swinglandand and MK Klemens, 5–7. Occasional papers of the IUCN Species Survival Commission (SSC) IUCN, Gland, Switzerland.

Berry KH. 1997. The Desert Tortoise Recovery Plan: an ambitious effort to conserve biodiversity in the Mojave and Colorado deserts of the United States. IN Proceedings: Conservation, Restoration, and Management of Tortoises and Turtles—An International Conference, ed. J Van Abbema, 430–440. New York Turtle and Tortoise Society, New York.

Berry KH. 2003a. Declining trends in desert tortoise populations at long-term study plots in California between 1979 and 2002: multiple issues. Proc. Desert Tortoise Council Symp.

Berry KH. 2003b. Using growth ring counts to age wild juvenile desert tortoises (*Gopherus agassizii*) Chelonian Conserv. Biol. 4:416–424.

Berry KH, TY Bailey, and KM Anderson. 2006a. Attributes of desert tortoise populations at the National Training Center, Central Mojave Desert, California, USA. J. Arid Environ. 67S:165–191.

Berry KH and MM Christopher. 2001. Guidelines for the field evaluation of desert tortoise health and disease. J. Wildl. Dis. 37:427–450.

Berry KH, A Emerson, and T Gowan. 2011. The status of 158 desert tortoises 33 months after translocation from Ft. Irwin. Abstract. 36th Annual Desert Tortoise Council Symposium, Las Vegas, NV, www.deserttortoise.org/symposium/abstract.html.

Berry KH, FG Hoover, and M Walker. 1996. The Effects of Poaching Desert Tortoises in the Western Mojave Desert: Evaluation of Landscape and Local Impacts. Proc. Desert Tortoise Council Symp. Las Vegas.

Berry KH, K Keith, and T Bailey. 2008. Status of the desert tortoise in Red Rock Canyon State Park. Calif. Fish and Game 94:98–118.

Berry KH, J Mack, RW Murphy, and W Quillman. 2006b. Introduction to the special issue on the changing Mojave Desert. J. Arid Environ. 67S:5–10.

Berry KH, DJ Morafka, and RW Murphy. 2002a. Defining the desert tortoise(s): our first priority for a coherent conservation strategy. Chelonian Conserv. Biol. 4:249–262.

Berry KH and LL Nicholson. 1984. The distribution and density of desert tortoises in California in the 1970's. IN The Status of the Desert Tortoise (*Gopherus agassizii*) in the United States, ed. KH Berry118–153. Report to the US Fish and Wildlife Service from the Desert Tortoise Council. Order No. 11310–0083–81.

Berry KH, EK Spangenberg, BL Homer, and ER Jacobson. 2002b. Deaths of desert tortoises following periods of drought and research manipulation. Chelonian. Conserv. Biol. 4:436–448.

Berry KH and FB Turner. 1986. Spring activities and habits of juvenile desert tortoises, *Gopherus agassizii*, in California. Copeia 1010–1012.

Birkhead RD, C Guyer, SM Hermann, and WK Michener. 2005. Patterns of folivory and seed ingestion by gopher tortoises (*Gopherus polyphemus*) in a southeastern pine savanna. Am. Midl. Nat. 154:143–151.

Bjorndal KA. 1980. Nutrition and grazing behavior of the green turtle *Chelonia mydas*. Mar. Biol. 56:147–154.

Bjorndal KA. 1987. Digestive efficiency in a temperate herbivorous reptile *Gopherus polyphemus*. Copeia 1987:714–720.

Bjorndal KA and A Carr. 1989. Variation in clutch size and egg size in the green turtle nesting population at Tortuguero, Costa Rica. Herpetologica 45:181–189.

Bjurlin CD and JA Bissonette. 2004. Survival during early life stages of the desert tortoise (*Gopherus agassizii*) in the south-central Mojave Desert. J. Herpetol. 38:527–535.

Black JH. 1976. Observations on courtship behavior of the desert tortoise. Great Basin Nat. 36:467–470.

Black SL. 1986. The Clemente and Herminia Hinojosa Site, 41JW8: A Toyah Horizon Campsite in Southern Texas. Center for Archaeological Research, the University of Texas at San Antonio, Special Report 18:110–112.

Blanck T and R Hudson. 2011. Turtles in Trouble: The World 25+ Most Endangered Tortoises and Freshwater Turtles–2011. IUCN/SSC Tortoise and Freshwater Turtle Specialist Group, Turtle Conservation Fund, Turtle Survival Alliance, Turtle Conservancy, Chelonian Research Foundation, Conservation International,

Wildlife Conservation Society, and San Diego Zoo Global. Lunenburg, MA.

Blanvillain G, D Owens, and G Kuchling. 2010. Hormones and reproductive cycles in turtles. In Hormones and Reproduction of Vertebrates, Vol. 3, ed. DO Norris and K H Lopez, 277–303. Elsevier, San Diego.

Boarman WI. 1993. When a native predator becomes a pest: a case study. In Conservation and Resource Management, ed. SK Majumdar, EW Miller, DE Baker, EK Brown, JR Pratt, and RF Schmalz, 191–206. Pennsylvania Academy of Sciences, Easton, PA.

Boarman WI. 2003. Managing a subsidized predator population: reducing common raven predation on desert tortoises. Environ. Manag. 32:205–217.

Boarman WI and KH Berry. 1995. Common ravens in the southwestern United States, 1968–92. In Our Living Resources: A Report to the Nation on the Distribution, Abundance, and Health of U.S. Plants, Animals, and Ecosystems, eds. ET LaRoe, GF Farris, CE Puckett, PD Doran, and MJ Mac, 73–75. National Biological Survey, Washington, DC.

Boarman WI and M Sazaki. 2006. A highway's road-effect zone for desert tortoises (*Gopherus agassizii*). J. Arid Environ. 65:94–101.

Boellstorff DE and DH Owings. 1995. Home range, population structure, and spatial organization of California ground squirrels. J. Mammal. 76:551–561.

Boglioli MD, C Guyer, and WK Michener. 2003. Mating opportunities of female gopher tortoises, *Gopherus polyphemus,* in relation to spatial isolation of females and their burrows. Copeia 2003:846–850.

Boglioli MD, WK Michener, and C Guyer. 2000. Habitat selection and modification by the gopher tortoise, *Gopherus polyphemus,* in Georgia longleaf pine forest. Chelonian Conserv. Biol. 3:699–705.

Bonin F, B Devaux, and A Dupre. 2006. Turtles of the World, trans. PCH Pritchard. Johns Hopkins University Press, Baltimore, MD.

Borgatti SP, MG Everett, and LC Freeman. 2002. Ucinet 6 for Windows: Software for Social Network Analysis. Analytical Technologies Inc., Harvard, Cambridge, MA.

Bour R and A Dubois. 1984. *Xerobates agassiz* 1857, synonyme plus ancient de *Scaptochelys* Bramble 1982 (Reptilia, Chelonii, Testudinidae). Bull. de la Societe Linneennee de Lyon 53:30–32.

Bramble DM. 1971. Functional morphology, evolution, and paleoecology of gopher tortoises. PhD diss., University of California, Berkeley.

Bramble DM. 1974. Occurrence and significance of the Os Transiliens in gopher tortoises. Copeia 1974:102–109.

Bramble DM. 1982. *Scaptochelys:* generic revision and evolution of gopher tortoises. Copeia 1982:852–867.

Bramble, DM and DB Wake. 1985. Feeding mechanisms of lower tetrapods. In Functional Vertebrate Morphology, ed. M Hildebrand, DM Bramble, KR Liem, and DB Wake, 230–261. Belknap Press, Cambridge, MA.

Brattstrom B. 1961. Some new fossil tortoises from western North America, with remarks on the zoogeography and paleoecology of tortoises. J. Paleontol. 35:543–560.

Brattstrom BH. 1965. Body temperatures of reptiles. Am. Midl. Nat. 73:376–422.

Breininger DR, PA Schmalzer, and CR Hinkle. 1991. Estimating occupancy of gopher tortoise (*Gopherus polyphemus*) burrows in coastal scrub and slash pine flatwoods. J. Herpetol. 25:317–321.

Breininger DR, PA Schmalzer, and CR Hinkle. 1994. Gopher tortoise (*Gopherus polyphemus*) densities in coastal scrub and slash pine flatwoods in Florida. J. Herpetol. 28:60–65.

Britten HB, BR Riddle, PF Brussard, R Marlow, and TE Lee Jr. 1997. Genetic delineation of management units for the desert tortoise, *Gopherus agassizii,* in northeastern Mojave Desert. Copeia 1997:523–530.

Brooks ML. 1999. Alien annual grasses and fire in the Mojave Desert. Madroño 46:13–19.

Brooks ML. 2009. Spatial and temporal distribution of nonnative plants in upland areas of the Mojave Desert. In The Mojave Desert, Ecosystem Processes and Sustainability, ed. RH Webb, LF Fenstermaker, JS Heaton, DL Hughson, EV McDonald, and DM Miller, 101–124. University of Nevada Press, Reno.

Brooks ML and KH Berry. 2006. Dominance and environmental correlates of alien annual plants in the Mojave Desert, USA. J. Arid Environ. 67S:100–124.

Brooks ML and TC Esque. 2002. Alien plants and fire in desert tortoise (*Gopherus agassizii*) habitat of the Mojave and Colorado deserts. Chelonian Conserv. Biol. 4:330–340.

Brooks ML and BM Lair. 2009. Ecological effects of vehicular routes in a desert ecosystem. In The Mojave Desert, Ecosystem Processes and Sustainability, ed. RH Webb, LF Fenstermaker, JS Heaton, DL Hughson, EV McDonald, and DM Miller, 168–195. University of Nevada Press, Reno.

Brooks ML and JR Matchett. 2006. Spatial and temporal patterns of wildfires in the Mojave Desert, 1980–2004. J. Arid Environ. 67S:148–164.

Brooks ML, JR Matchett, and KH Berry. 2006. Effects of livestock watering sites on alien and native plants in the Mojave Desert, USA. J. Arid Environ. 67S:125–147.

Brooks ML and RA Minnich. 2006. Southeastern deserts bioregion. In Fire in California Ecosystems, ed. NG Sugihara, JW van Wagtendonk, KE Shaffer, J Fites-Kaufman, and AE Thode, 391–414. University of California Press, Berkeley.

Brooks ML and DA Pyke. 2001. Invasive plants and fire in the deserts of North America. In Proceedings of the Invasive Species Workshop: The Role of Fire in the Control and Spread of Invasive Species, Fire Conference 2000; The First National Congress on Fire Ecology, Prevention, and Management, ed. KEM Galley and TP Wilson, 1–14. Miscellaneous Publication No. 11, Tall Timbers Research Station, Tallahassee, FL.

Brown AE. 1908. Generic types of nearctic reptiles and amphibia. Proc. Acad. Nat. Sci. Phila. 60:112–127.

Brown DE. 1994. Sinaloan thornscrub. In Biotic Communities of the American Southwest—United States and Mexico, ed. DE Brown, 101–105. University of Utah Press, Salt Lake City.

Brown DE and RA Minnich. 1986. Fire and changes in creosote bush scrub of the western Sonoran Desert, California. Am. Midl. Nat. 116:411–422.

Brown DR, JL Merritt, ER Jacobson, PA Klein, JG Tully, and M.B. Brown. 2004. *Mycoplasma testudineum* sp. nov., from a desert tortoise (*Gopherus agassizii*) with upper respiratory tract disease. Int. J. System. Evol. Microbiol. 54:1527–1529.

Brown DR, IM Schumacher, GS McLaughlin, MB Brown, PA Klein, and ER Jacobson. 2002. Application of diagnostic tests for mycoplasmal infections of desert and gopher tortoises, with management recommendations. Chelonian. Conserv. Biol. 4:497–507.

Brown MB, KH Berry, IM Schumacher, KA Nagy, MM Christopher, and PA Klein. 1999a. Seroepidemiology of upper respiratory tract disease in the desert tortoise in the western Mojave Desert of California. J. Wildl. Dis. 35:716–727.

Brown MB, GS McLaughlin, PA Klein, BC Crenshaw, IM Schumacher, DR Brown, and ER Jacobson. 1999b. Upper respiratory tract disease in the gopher tortoise is caused by *Mycoplasma agassizii*. J. Clin. Microbiol. 37:2262–2269.

Brown MB, IM Schumacher, PA Klein, RK Harris, T Correll, and ER Jacobson. 1994. *Mycoplasma agassizii* causes upper respiratory tract disease in the desert tortoise. Inf. Immun. 62:4580–4586.

Brown TK, KA Nagy, and DJ Morafka. 2005. Costs of growth in tortoises. J. Herpetol. 39:19–23.

Brudvig LA and EI Damschen. 2011. Land-use history, historical connectivity, and land management interact to determine longleaf pine woodland understory richness and composition. Ecography 34:257–266.

Brussard PF, KH Berry, ME Gilpin, ER Jacobson, GJ Morafka, C Schwalbe, CR Tracy, FC Vasek. 1994. Desert Tortoise (Mojave Population) Recovery Plan. US Fish and Wildlife Service Portland, OR.

Buckland ST, DR Anderson, KP Burnham, JL Laake, DL Borchers, and L Thomas. 2001. Introduction to Distance Sampling: Estimating Abundance of Biological Populations. Oxford University Press, Oxford.

Buhlmann KA, TD Tuberville, MG Nafus, M Peaden, and BD Todd. 2013. Desert tortoise head-starting project in Mojave National Preserve: An update. 38th Annual Desert Tortoise Council Symposium, Las Vegas, NV, www.deserttortoise.org/symposium/abstract.html

Bull JJ and RC Vogt. 1981. Temperature-sensitive periods of sex determination in emydid turtles. J. Exp. Zool. 218:435–440.

Bull JJ and EL Charnov. 1989. Enigmatic reptilian sex ratios. Evolution 43:1561–1566.

Bull JJ, RC Vogt, and CJ McCoy. 1982. Sex determining temperatures in turtles: a geographic comparison. Evolution 326–332.

Bulova SJ. 1994. Patterns of burrow use by desert tortoises: gender differences and seasonal trends. Herpetol. Monogr. 8:133–143.

Bulova SJ. 1997. Conspecific chemical cues influence burrow choice by desert tortoises (*Gopherus agassizii*). Copeia 1997:802–810.

Bulova SJ. 2002. How temperature, humidity, and burrow selection affect evaporative water loss in desert tortoises. J. Thermal Biol. 27:175–189.

Burge BL. 1977. Movements and behavior of the desert tortoise *Gopherus agassizii*. MS thesis, University of Nevada, Las Vegas.

Burge BL. 1978. Physical characteristics and patterns of utilization of cover sites used by *Gopherus agassizii* in southern Nevada. Proc. Desert Tortoise Council Symp. 80–111.

Burge BL and WG Bradley. 1976. Population density, structure and feeding habits of the desert tortoise, *Gopherus agassizii*, in a low desert study area in southern Nevada. Proc. Desert Tortoise Council Symp., 57–74.

Burke RL. 1989a. Burrow-to-tortoise conversion factors: comparison of three gopher tortoise survey techniques. Herpetol. Rev. 20:92–94.

Burke RL. 1989b. Florida gopher tortoise relocation: overview and case study. Biol. Conserv. 48:295–309.

Burke RL, MA Ewert, JB Mclemore, and DR Jackson. 1996. Temperature dependent sex determination and hatching success in the gopher tortoise (*Gopherus polyphemus*). Chelonian Conserv. Biol. 2:86–88

Burt WH. 1943. Territoriality and home range concepts as applied to mammals. J. Mammal. 24:346–352.

Bury RB. 1982. An overview in North American tortoises: conservation and ecology, ed. RB Bury, v-vii. US Fish and Wildlife Service, Wildlife Research Report 12.

Bury RB, TC Esque, LA Defalco, and PA Medica. 1994. Distribution, habitat use, and protection of the desert tortoise in the eastern Mojave Desert. IN Biology of North American Tortoises, ed. Bury, RB and DJ. Germano,57–72. Fish and Wildlife Research 13.

Bury BR and DJ Germano. 1994. Biology of North American tortoises: introduction. IN Biology of North American Tortoises, ed. RB Bury and DJ Germano,1–5. Fish and Wildlife Research13.

Bury RB, DJ Germano, TR Van Devender, and BE Martin. 2002. The desert tortoise in Mexico: distribution, ecology, and conservation. IN The Sonoran Desert Tortoise: Natural History, Biology, and Conservation, ed. TR Van Devender, 86–108. University of Arizona Press, Tucson.

Bury RB and RA Luckenbach. 1977. Censusing desert tortoise populations using a quadrat and grid location system. IN Desert Tortoise Council Symposium Proceedings 1977, 169–178.

Bury RB and RA Luckenbach. 2002. Comparison of desert tortoise (*Gopherus agassizii*) populations in an unused and off-road vehicle area in the Mojave Desert. Chelonian Conserv. Biol. 4:457–463.

Bury RB, RA Luckenbach, and LR Munoz. 1978. Observations on *Gopherus agassizii* from Isla Tiburon, Sonora, Mexico. Proc. Desert Tortoise Council Symp. 69–72.

Bury RB, DJ Morafka, and CJ McCoy. 1988. Distribution, abundance and status of the Bolsón Tortoise. IN The Ecogeography of the Mexican Bolsón Tortoise (*Gopherus flavomarginatus*): Derivation of its Endangered Status and Recommendations for its Conservation, ed. DJ Morafka and CJ McCoy, 5–30. Ann. Carnegie Mus. 57

Bury RB and EL Smith. 1986. Aspects of the ecology and management of the tortoise *Gopherus berlandieri* at Laguna Atascosa, Texas. Southwest. Nat. 31:387–394.

Butler JA, RD Bowman, TW Hull, and S Sowell. 1995. Movements and home range of hatchling and yearling gopher tortoises, *Gopherus polyphemus*. Chelonian Conserv. Biol. 1:173–180.

Butler JA and TW Hull. 1996. Reproduction in the gopher tortoise, *Gopherus polyphemus*, in northeastern Florida. J. Herpetol. 30:14–18.

Cablk ME and JS Heaton. 2006. Accuracy and reliability of dogs in surveying for desert tortoise (*Gopherus agassizii*). Ecol. Appl. 16:1926–1935.

Calhoun JB and JU Casby. 1958. Calculation of home range and density of small mammals. US Public Health Monogr. 55:1–24.

Camp CL. 1916. Notes on the local distribution and habits of the amphibians and reptiles of southeastern California in the vicinity of Turtle Mountains. Univ. Calif. Publ. Zool. 12:503–544.

Campbell TW. 1996. Clinical pathology. IN Reptile Medicine and Surgery, ed. DR Mader, 248–257. WB Saunders, Philadelphia.

Cannon MD and DJ Meltzer. 2004. Early Paleoindian foraging: examining the faunal evidence for large mammal specialization and regional variability in prey choice. Quaternary Sci. Rev. 23:1955–1987.

Carr A. 1952. Handbook of Turtles. Cornell University Press, Ithaca, NY.

Carpenter JP, G Sanchez, and ME Villalpando C. 2009. The Late Archaic / Early Agricultural Period in Sonora, Mexico. IN The Late Archaic Across the Borderlands: From Foraging to Farming, ed. BJ Vierra, 13–40. University of Texas Press, Austin.

Caughley G. 1994. Directions in conservation. J. Anim. Ecol. 63:215–244.

Causey MK and CA Cude. 1978. Feral dog predation of the gopher tortoise, *Gopherus polyphemus*, in southeast Alabama. Herpetol. Rev. 9:94–95.

Cayan DR, EP Maurer, MD Dettinger, M Tyree, and K Hayhoe. 2008. Climate change scenarios for the California region. Climatic Change 87 (Suppl. 1):S21-S42.

Chaffee MA and KH Berry. 2006. Abundance and distribution of selected elements in soils, stream sediments, and selected forage plants from desert tortoise habitats in the Mojave and Colorado deserts, USA. J. Arid Environ. 67S:35–87.

Chaloupka M, N Kamezaki, and C Limpus. 2008. Is climate change affecting the population dynamics of the endangered pacific loggerhead sea turtle? J. Exp. Mar. Biol. Ecol. 356:136–143.

Chevalier J, MH Godfrey, and M Girondot. 1999. Significant difference of temperature-dependent sex determination between French Guiana (Atlantic) and Playa Grande (Costa-Rica, Pacific) Leatherbacks (*Dermochelys coriacea*). Ann. Sci. Nat. Zoo. Bio. Anim. 20:147–152.

Christian KA, CR Tracy, and WP Porter. 1984. Diet, digestion, and food preferences of Galapagos land iguanas. Herpetologica 40:205–212.

Christopher MM, KH Berry, IR Wallis, KA Nagy, BT Henen, and CC Peterson. 1999. Reference intervals and physiologic alterations in hematologic and biochemical values of free-ranging desert tortoises in the Mojave Desert. J. Wildl. Dis. 35:212–238.

Christopher MM, KH Berry, BT Henen, and KA Nagy. 2003. Clinical disease and laboratory abnormalities in free-ranging desert tortoises in California (1990–1995). J. Wildl. Dis. 39:35–56.

Clusella Trullas TS, JH van Wyk, and JR Spotila. 2009. Thermal benefits of melanism in cordylus lizards: a theoretical and field test. Ecology 90:2297–2312.

Clostio RW, AM Martinez, KE LeBlanc, and NM Anthony. 2012. Population genetic structure of a threatened tortoise across the southeastern United States: implications for conservation management. Anim. Conserv. 15:613–625.

Coghill LM, J Chaves-Campo, UOG Vázquez, and A Contreras. 2011. Geographic distribution, *Gopherus berlandieri*. Herpetol. Rev. 42:388.

Coleman RM. 1991. Measuring parental investment in non-spherical eggs. Copeia 1991:1092–1098.

Colwell RK. 1974. Predictability, constancy, and contingency of periodic phenomena. Ecology 55:1148–1153.

Comisión Nacional de Áreas Naturales Protegidas (CONANP). 2006. Programa de Conservación y Manejo Reserva de la Biosfera Mapimí, México. Camino al Ausco No. 200. Col. Jardines en la Montana, Tlalpan. CP. 14210, México, DF.

Conner RC and AJ Hartsell. 2002. Forest area and conditions. IN Southern Forest Resource Assessment. ed. DN Wear and JG Greis, 357–402. Southern Research Station, Technical Report GTR SRS-53, Asheville, NC.

Convention on International Trade in Endangered Species of Wild Fauna and Flora (CITES). 2011. Appendices I, II, and III. www.cites.org.

Coombs EM. 1977. Status of the desert tortoise, *Gopherus agassizii,* in the state of Utah. Proc. Desert Tortoise Council 1977:95–101.

Cooper JG. 1861. New Californian animals. Proc. Calif. Acad. Sci. (ser. 1) 2:118–123.

Cope ED. 1872. Second account of new vertebrata from the Bridger Eocene of Wyoming Territory. Proc. Am. Philos. Soc. 12:466–468.

Cope ED. 1873. Second notice of extinct vertebrata from the Tertiary of the Plains. Paleontol. Bull. 15:1–6.

Cope ED. 1875. Check-list of North American Batrachia and Reptilia. Bull. US Natl. Mus. 1:1–104.

Cope ED. 1885 (1884). The vertebrata of the Tertiary formations of the West. Rept. US Geol. Surv. Terr. 3:1–1009.

Cope ED. 1892. A contribution to a knowledge of the fauna of the Blanco beds of Texas. Proc. Acad. Nat. Sci. Phila. 44:226–227.

Cope ED. 1893. A preliminary report on the vertebrate paleontology of the Llano Estacado. 4th Annu. Rept. Geol. Soc. Texas 1–137.

Costanzo JP, SA Dinkelacker, JB Iverson, and RE Lee Jr. 2004. Physiological ecology of overwintering in the hatchling painted turtle: multiple-scale variation in response to environmental stress. Physiol. Biochem. Zool. 77(1):74–99.

Costanzo JP, JB Iverson, MF Wright, and RE Lee Jr. 1995. Cold hardiness and overwintering strategies of hatchlings in an assemblage of northern turtles. Ecology 1772–1785.

Costanzo JP, JR Lee, E Richardand, GR Ultsch. 2008. Physiological ecology of overwintering in hatchling turtles. J. Exp.l Zool., Part A: Ecol. Genet.Physiol. 309(6):297–379.

Cowles RB and CM Bogert. 1944. A preliminary study of the thermal requirements of desert reptiles. Bull. Am. Mus. Nat. Hist. 83:268–280.

Cox J, D Inkley, and R Kautz. 1987. Ecology and habitat protection needs of gopher tortoise (*Gopherus polyphemus*) populations found on lands slated for large-scale development in Florida. Florida Game and Freshwater Fish Commission Non-Game Wildlife Program Technical Report 4.

Crother BI. 2008. Scientific and standard English names of amphibians and reptiles of North America north of Mexico. SSAR Herpetological Circular 37:1–84.

Crumly CR. 1994. Phylogenetic systematics of North American tortoises (genus *Gopherus*): evidence of their classification. IN Biology of North American Tortoises, ed. RB Bury and DJ Germano, 7–32. US Fish and Wildlife Research 13.

Crumly CR and LL Grismer. 1994. Validity of the tortoise *Xerobates lepidocephalus* Ottley and Velazquez [sic]. IN Biology of North American Tortoises, ed. RB Bury and DJ Germano, 32–36. US Fish and Wildlife Service 13.

Curtin AJ, GR Zug, and JR Spotila. 2009. Longevity and growth strategies of the desert tortoise (*Gopherus agassizii*) in two American deserts. J. Arid Environ. 73:463–471.

D'Antonio C and PM Vitousek. 1992. Biological invasions by exotic grasses, the grass/fire cycle, and global change. Annu. Rev. Ecol. and Syst. 23:63–87.

Daudin FM. 1801–1803. Histoire Naturelle des Reptiles. F. Dufart, Paris, 8 Vols. (see Vol. 2, 1803:256–259).

Davy CM, T Edwards, A Lathrop, M Bratton, M Hagan, KA Nagy, J Stone, LS Hillard, and RW Murphy. 2011. Polyandry and multiple paternities in the threatened desert tortoise, *Gopherus agassizii.* Conserv. Genet. 12:1313–1322.

Dean-Bradley K. 1995. Digestive flexibility in response to diet quality in hatchling desert tortoises, Gopherus agassizii. MS thesis, Colorado State University, Fort Collins.

DeFalco LA. 1995. Influence of cryptobiotic crusts on winter annuals and foraging movements of the desert tortoise. MS thesis, Colorado State University, Fort Collins.

DeGregorio BA, KA Buhlmann, and TD Tuberville. 2012. Overwintering of gopher tortoises (*Gopherus polyphemus*) translocated to the northern limit of their geographic range: temperatures, timing, and survival. Chelonian Conserv. Biol. 11:84–90.

Demuth JP. 2001. The effects of constant and fluctuating incubation temperatures on sex determination, growth, and performance in the tortoise *Gopherus polyphemus.* Can. J. Zool. 79:1609–1620.

Desert Renewable Energy Conservation Plan, Independent Science Panel. 2012. Final Report. Independent Science Review for the California Desert Renewable Energy Conservation Plan. Prepared for the Renewable Energy Action Team. Sacramento, CA. www.drecp.org/documents/#science.

Des Lauriers, JR. 1965. New Miocene tortoise from southern California. Bull. So. California Acad. Sci. 64:1–10.

Dickinson VM, IM Schumacher, JL Jarchow, T Duck, and CR Schwalbe. 2005. Mycoplasmosis in free-ranging desert tortoises in Utah and Arizona. J. Wildl. Dis. 41:839–842.

Diemer JE. 1986. The ecology and management of the gopher tortoise in the southeastern United States. Herpetologica 42:125–133.

Diemer JE. 1987. The status of the gopher tortoise in Florida. IN Proceedings of the Third Southeastern Nongame and Endangered Wildlife Symposium, ed. R Odom, K Riddleberger, and J Osier, 72–83. Georgia Department of Natural Resources, Game and Fish Division, Atlanta.

Diemer JE. 1989. *Gopherus polyphemus.* IN The Conservation Biology of Tortoises, ed. IR Swingland and MW Klemens, 14–19. Occasional Papers IUCN Species Survival Commission 5.

Diemer JE. 1992a. Demography of the tortoise *Gopherus polyphemus* in northern Florida. J. Herpetol. 26:281–289.

Diemer JE. 1992b. Home range and movements of the tortoise *Gopherus polyphemus* in northern Florida. J. Herpetol. 26:158–165.

Diemer JE and CT Moore. 1994. Reproduction of gopher tortoises in north-central Florida. IN Biology of North American Tortoises, ed. RB Bury and DJ Germano, 129–137. US Fish and Wildlife Service 13.

Dietler JD, RS Ramirez, C Backes, and LH Hoffman. 2011. Archeological evaluation of 10 sites within the Acorn and Emerson Lake training areas, Marine Corps Air Ground Combat Center, Twentynine Palms, CA. SWCA Cultural Resources Report Database No. 2010–43.

Dillon ME, G Wang, and RB Huey. 2010. Global metabolic impacts of recent climate warming. Nature 467:704–707.

Dodd CK and RA Seigel. 1991. Relocation, repatriation, and translocation of amphibians and reptiles: are they conservation strategies that work? Herpetologica 47:336–350.

Dorsey RJ. 2012. Earliest delivery of sediment from the Colorado River to the Salton Trough at 5.3 Ma: evidence from Split Mountain Gorge. IN Searching for the Pliocene: Southern Exposures, ed. RE Reynolds, 88–93. Proceedings of the 2012 Desert Symposium, California State University, Fullerton.

Douglass JF. 1978. Refugia of juvenile gopher tortoises, *Gopherus polyphemus* (Reptilia, Testudines, Testudinidae). J. Herpetol. 12:413–414.

Douglass JF. 1986. Patterns of mate-seeking and aggression in a southern Florida population of gopher tortoise, *Gopherus polyphemus*. Proc. Desert Tortoise Council 1986:155–199. Palmdale, CA

Douglass JF and JN Layne. 1978. Activity and thermoregulation of the gopher tortoise (*Gopherus polyphemus*) in southern Florida. Herpetologica 34:359–374.

Douglass JF and CE Winegarner. 1977. Predators of eggs and young of the gopher tortoise, *Gopherus polyphemus* (Reptilia, Testudines, Testudinidae) in southern Florida. J. Herpetol. 11:236–238.

Drake KK, KE Nussear, TC Esque, AM Barber, KM Vittum, PA Medica, CRTracy, and KW Hunter Jr. 2012. Does translocation influence physiological stress in the desert tortoise? Anim. Conserv. 15:560–570.

Drew MB, LK Kirkman, and AK Gholson Jr. 1998. The vascular flora of Ichauway, Baker County, Georgia: a remnant longleaf pine / wiregrass ecosystem. Castanea 63:1–24.

Drummond AJ and A Rambaut. 2007. BEAST: Bayesian evolutionary analysis by sampling trees. BMC Evol. Biol. 7:214.

Duda JJ and AJ Krzysik. 1998. Radiotelemetry study of a desert tortoise population. US Army Corps of Engineers. USACERL. Technical Report 98/39.

Duda JJ, AJ Krzysik, and JE Freilich. 1999. Effects of drought on desert tortoise movement and activity. J. Wildl. Manag. 63:1181–1192.

Duda JJ, AJ Krzysik, and JM Meloche. 2002. Spatial organization of desert tortoises and their burrows at a landscape scale. Chelonian Conserv. Biol. 4:387–397.

Dzialowski EM and MP O'Connor. 2001. Thermal time constant estimation in warming and cooling ectotherms. J. Therm. Biol. 26:231–245.

Eales JG. 1979. Thyroid Functions in Cyclostomes and Fishes in Hormones and Evolution, Vol. 1, ed. EJW Barrington, 341–436. Academic Press, New York.

Eales JG. 1988. The influence of nutritional state on thyroid function in various vertebrates. Am. Zool. 28:351–362.

Edwards T. 2003. Desert tortoise conservation genetics. MS Thesis, University of Arizona, Tucson.

Edwards T. 2008. Turtle trouble in Kingman, Arizona. Sonoran Herpetol. 21:62–65.

Edwards T and KH Berry. 2013. Are captive tortoises a reservoir for conservation? An assessment of genealogical affiliation of captive *Gopherus agassizii* to local, wild populations. Conserv. Genet. 14:649.

Edwards T, CS Goldberg, ME Kaplan, CR Schwalbe, and DE Swann. 2003. PCR primers for microsatellite loci in the desert tortoise (*Gopherus agassizii*, Testudinidae). Molec. Ecol. 3:589–591.

Edwards T, CJ Jarchow, CA Jones, and KE Bonine. 2010. Tracing genetic lineages of captive desert tortoises in Arizona. J. Wildl. Manage. 7:801–807.

Edwards T, A Lathrop, A Ngo, K Choffe, and RW Murphy. 2011. STR/microsatellite primers for the desert tortoise, *Gopherus agassizii*, and its congeners. Conserv. Gen. Resourc. 3:365–368.

Edwards T, CR Schwalbe, DE Swann, and CS Goldberg. 2004a. Implications of anthropogenic landscape change on inter-population movements of the desert tortoise (*Gopherus agassizii*). Conserv. Gen. 5:485–499.

Edwards T, EW Stitt, CR Schwalbe, and DE Swann. 2004b. Natural history notes: *Gopherus agassizii* (Desert Tortoise) movement. Herpetol. Rev. 35(4):381–382.

Edwards T, M Vaughn, CM Torres, AE Karl, PR Rosen, KH Berry, and RW Murphy. 2013. A biogeographic perspective of speciation among desert tortoises in the genus *Gopherus:* a preliminary evaluation. IN Merging Science and Management in a Rapidly Changing World: Biodiversity and Management of the Madrean Archipelago III, comp. GJ Gottfried, PF Ffolliott, BS Gebow, LG Eskew, and LC Collins, 243–247. Proceedings RMRS-P-67, Fort Collins, CO, US Department of Agriculture, Forest Service, Rocky Mountain Research Station.

Eendebak BT. 1995. Incubation period and sex ratio of Hermann's tortoise, *Testudo hermanni boettgeri*. Chelonian Conserv. Biol. 1:227–231.

Eglis A. 1962. Tortoise behavior: a taxonomic adjunct. Herpetologica 18:1–8.

Eich AML. 2009. Evolutionary and conservation implications of sex determination and hatchling depredation in Kemp's ridley sea turtles. PhD. diss., University of Alabama at Birmingham.

Elgar MA and LJ Heaphy. 1989. Covariation between clutch size, egg weight and egg shape: comparative evidence for chelonians. J. Zool., Lond. 219:137–152.

Emlen JM. 1966. The role of time and energy in food preference. Am. Nat. 100:611–617.

Enge KM, JE Berish, R Bolt, A Dziergowski, and HR Mushinsky. 2006. Biological Status Report, Gopher Tortoise. Gopher Tortoise Biological Review Panel, Florida Fish and Wildlife Conservation Commission, Tallahassee, FL. 60.

Enge KM, BA Millsap, TJ Doonan, JA Gore, NJ Douglass, and GL Sprandel. 2002. Conservation plans for biotic regions in Florida containing multiple rare or declining wildlife taxa. Florida Fish and Wildlife Conservation Commission Technical Report 20. 146 pp.

Engeman RM, MJ Pipas, and HT Smith. 2004. *Gopherus berlandieri* (Texas Tortoise). Mortality. Herpetol. Rev. 35:54–55.

Ennen JR, RD Birkhead, BR Keiser, and DL Gaillard. 2011. The effects of isolation on the demography and genetic diversity of long-lived species: implications for conservation and management of the Gopher Tortoise (*Gopherus polyphemus*). Herpetol. Conserv. 6:202–214.

Ennen JR, BR Kreiser, and CP Qualls. 2010. Low genetic diversity in several gopher tortoise (*Gopherus polyphemus*) populations in the Desoto National Forest, Mississippi. Herpetologica 66:31–38.

Ennen JR, BR Kreiser, CP Qualls, D Gaillard, M Aresco, R Birkhead, TD Tuberville, ED McCoy, HR Mushinsky, TW Hentges, and A Schrey. 2012. Mitochondrial DNA assessment of the phylogeography of the Gopher Tortoise. J. Fish Wildl. Manag. 3:110–122.

Epperson DM and CD Heise. 2003. Nesting and hatchling ecology of gopher tortoises (*Gopherus polyphemus*) in southern Mississippi. J. Herpetol. 37:315–324.

Erlandson JM, MH Graham, BJ Bourque, D Corbett, JA Estes, and RS Steneck. 2007. The Kelp Highway hypothesis: marine ecology, the coastal migration theory, and the peopling of the Americas. J. Island Coastal Archaeol. 2:161–174.

Ernst CH, and RW Barbour. 1972. Turtles of the United States. University Press Kentucky, Lexington.

Ernst CH and JE Lovich. 2009. Turtles of the United States and

Canada. 2nd ed. Johns Hopkins University Press, Baltimore, MD. 827 pp.

Ernst CH, Lovich JE, and RW Barbour. 1994. Turtles of the United States and Canada. Smithsonian Institution Press, Washington, DC.

Esque TC. 1994. Diet and diet selection of the desert tortoise (*Gopherus agassizii*) in the northeast Mojave Desert. MS thesis, Colorado State University, Fort Collins.

Esque TC, A Burquez, CR Schwalbe, TR Van Devender, PJ Anning, and MJ Nijhuis. 2002. Fire ecology of the Sonoran desert tortoise. In The Sonoran Desert Tortoise: Natural History, Biology, and Conservation, ed. TR Van Devender, 312–333. University of Arizona Press. Tucson, AZ.

Esque TC, KE Nussear, KK Drake, AD Walde, KH Berry, RC Averill-Murray, AP Woodman, W. I. Boarman, PA Medica, J Mack, and JS Heaton. 2010b. Effects of subsidized predators, resource variability, and human population density on desert tortoise populations in the Mojave Desert, USA. Endang. Sp. Res. 12:167–177.

Esque TC, KE Nussear, and PA Medica. 2005. Desert tortoise translocation plan for Fort Irwin's land expansion program at the US Army National Training Center (NTC) at Fort Irwin. Prepared for US Army National Training Center, Directorate of Public Works. www.fortirwinlandexpansion.com/Documents.htm.

Esque TC and EL Peters. 1994. Ingestion of bones, stones and soil by desert tortoises. IN Biology of North American Tortoises, ed. RB Bury and DJ Germano,105–112. Department of Interior, US Fish and Wildlife Research 13.

Esque TC, CR Schwalbe, LA DeFalco, RB Duncan, and TJ Hughes. 2003. Effects of desert wildfires on desert tortoise (*Gopherus agassizii*) and other small vertebrates. Southwest. Nat. 48:103–111.

Esque TC, JA Young, and CR Tracy. 2010a. Short-term effects of experimental fires on a Mojave Desert seed bank. J. Arid Environ. 74:1302–1308.

Estes R and JH Hutchison. 1980. Eocene lower vertebrates from Ellesmere Island, Canadian Arctic Archipelago. Palaeogeogr. Palaeoclim. Palaeoecol. 30:325–347.

Eubanks JO, JW Hollister, C Guyer, and WK Michener. 2002. Reserve area requirements for gopher tortoises (*Gopherus polyphemus*). Chelonian Conserv. Biol. 4:464–471.

Eubanks JO, WK Michner, and C Guyer. 2003. Patterns of movement and burrow use in a population of gopher tortoises (*Gopherus polyphemus*). Herpetologica 59:311–321.

Evenari M. 1985. The desert environment. IN Ecosystems of the World 12A: Hot Deserts and Shrublands, ed. Evanari, I Noy-Meir, and DW Goodall, 1–22. Elsevier, Amsterdam.

Ewert MA. 1979. The embryo and its egg: development and natural history. IN Turtles, Perspectives and Research, ed. M Harless and M Morelock, 333–413. John Wiley and Sons. New York.

Ewert MA, DR Jackson, and CE Nelson. 1994. Patterns of temperature-dependent sex determination in turtles. J. Exp. Zool. 270:3–15.

Ewert MA and CE Nelson. 1991. Sex determination in turtles: diverse patterns and some possible adaptive values. Copeia 1991:50–69.

Fabens AJ. 1965. Properties and fitting of the von Bertalanffy growth curve. Growth 29:265–289.

Fairbanks B and FS Dobson. 2007. Mechanisms of the group size effect on vigilance in Columbian ground squirrels: dilution versus detection. Anim. Behav. 73:115–123.

Fedriani JM, TK Fuller, and RM Sauvajot. 2001. Does availability of anthropogenic food enhance densities of omnivorous mammals? An example with coyotes in southern Calfiornia. Ecography 24:325–331.

Felger RS, MB Moser, and EW Moser. 1981. The desert tortoise in Seri Indian culture. Proc. Desert Council Symp. 113–120.

Field KJ, CR Tracy, PA Medica, RW Marlow, and PS Corn. 2007. Return to the wild: translocation as a tool in conservation of the desert tortoise (*Gopherus agassizii*). Biol. Conserv. 136:232–245.

Fleshler D. 2005. Facing a slow death: Florida allows developers to bury alive or kill gopher tortoises in return for protection of habitat elsewhere in the state, but officials warn that the species now faces a "very high risk of extinction." 21 August 2005, Sun-Sentinel, Fort Lauderdale, FL.

Florida Fish and Wildlife Conservation Commission (FWC). 2001a. Biological status report: gopher tortoise (*Gopherus polyphemus*). Report prepared by Florida Fish and Wildlife Conservation Commission, Gainesville, FL.

Florida Fish and Wildlife Conservation Commission (FWC). 2001b. Available options to address the presence of gopher tortoise on lands slated for development. Report prepared by Florida Fish and Wildlife Conservation Commission, Tallahassee, FL.

Florida Fish and Wildlife Conservation Commission (FWC). 2006. Biological Status Report: Gopher Tortoise. Report prepared by Florida Fish and Wildlife Conservation Commission, Gainesville, FL.

Florida Fish and Wildlife Conservation Commission (FWC). 2007. Gopher Tortoise Management Plan, *Gopherus polyphemus*. Fish and Wildlife Conservation Commission, Tallahassee, FL. 127 pp.

Florida Fish and Wildlife Conservation Commission (FWC). 2008. Gopher Tortoise Permitting Guidelines, *Gopherus polyphemus*. Fish and Wildlife Conservation Commission, Tallahassee, FL. 73 pp.

Florida Fish and Wildlife Conservation Commission (FWC). 2009. Gopher tortoise permitting guidelines: *Gopherus polyphemus*. April 2008 (rev. April 2009). Florida Fish and Wildlife Conservation Commission, Tallahassee, FL. 57 pp.

Florida Fish and Wildlife Conservation Commission (FWC). 2011. Florida's Endangered and Threatened Species Management and Conservation Plan, FY 2009–10, Progress Report, Florida Fish and Wildlife Conservation Commission, Tallahassee, FL. 113 pp.

Francis J. 1988. A 50-million-year-old fossil forest from Strathcora Fiord, Ellesmere Island, Arctic Canada: evidence for a warm polar climate. Arctic 4:314–318.

Frankham R. 1995. Conservation Genetics. Annu. Rev. Genet. 29:305–327.

Franks BR, HW Avery, and JR Spotilla. 2011. Home range and movement of desert tortoises *Gopherus agassizii* in the Mojave Desert of California, USA. Endangered Species Research 13:191–201.

Franz R and SE Franz. 2009. A new fossil land tortoise in the genus *Chelonoidis* (Testudines: Testudinidae) from the northern Bahamas, with an osteological assessment of other neotropical tortoises. Bull. Florida Mus. Nat. Hist. 49:1–44.

Franz R and IR Quitmyer. 2005. A fossil and zooarchaeological history of the gopher tortoise (*Gopherus polyphemus*) in the southeastern United States. Bull. Florida Mus. Nat. Hist. 45:179–199.

Frazer NB and LM Ehrhart. 1985. Preliminary growth models for green, *Chelonia mydas,* and loggerhead, *Caretta caretta,* turtles in the wild. Copeia 1985:73–79.

Freeman L. 1996, Cliques, Galois lattices, and the structure of human social groups. Soc. Networks 18:173–187.

Freilich JE, KP Burnham, CM Collins, and CA Garry. 2000. Factors affecting population assessments of desert tortoises. Conserv. Biol. 14:1479–1489.

Fritts TH and RD Jennings. 1994. Distribution, habitat use, and status of the desert tortoise in Mexico. IN Biology of North American Tortoises, ed. RB Bury and DJ Germano, 49–56. US Fish and Wildlife Research 13.

Fritz U and P Havaš. 2007. Checklist of chelonians of the world. Vert. Zool. 57:149–368.

Fry J, G Xian, S Jin, J Dewitz, C Homer, L Yang, C Barnes, N Herold, and J Wickham. 2011. National Land Cover Database 2006 Percent

Developed Imperviousness. Raster digital data. MRLC.gov. www.mrlc.gov/nlcd06_data.php.

Fuentes M, J Maynard, M Guinea, I Bell, P Werdell, and M Hamann. 2009. Proxy indicators of sand temperature help project impacts of global warming on sea turtles in northern Australia. Endangered Species Research, 9:33–40.

Fujii A and MRJ Forstner. 2010. Genetic variation and population structure of the Texas Tortoise, *Gopherus berlandieri* (Testudinidae), with implications for conservation. Chelonian Conserv. Biol. 9:61–69.

Gaffney ES. 1979. Comparative cranial morphology of recent and fossil turtles. B. Am. Mus. Nat. Hist. 164:65–375.

Gaffney ES and PA Meylan. 1988. A Phylogeny and Classification of the Tetrapods. Vol. 1. Amphibians, Reptiles, Birds, ed., MJ Benton, 157–219. Clarendon Press, Oxford.

Gans C. 1974. Biomechanics: An Approach to Vertebrate Biology. J.B. Lippincott, Philadelphia.

Garner JA and JL Landers. 1981. Foods and habitat of the gopher tortoise in southwestern Georgia. Proc. Ann. Conf. Southeast Assoc., Fish and Wildl. Agencies 35:120–134.

Georges A, S Doody, K Beggs, J Young. 2004. Thermal models of TSD under laboratory and field conditions. In Temperature Dependent Sex Determination in Vertebrates, ed N. Valenzuela and V. Lance, 79–89. Smithsonian Books, Washington, DC.

Georges A, K Beggs, JE Young, and JS Doody. 2005. Modelling development of reptile embryos under fluctuating temperature regimes. Physiol. Biochem. Zool. 78(1):18–30.

Germano DJ. 1988. Age and growth histories of desert tortoises using scute annuli. Copeia 1988: 914–920.

Germano DJ. 1992. Longevity and age size relationships of populations of desert tortoises. Copeia 1992:367–374.

Germano DJ. 1993. Shell morphology of North American tortoises. Am. Midl. Nat. 129:319–335.

Germano DJ. 1994. Growth and maturity of North American tortoises in relation to regional climates. Can. J. Zool. 72:918–931.

Germano DJ, RB Bury, TC Esque, TH Fritts, and PA Medica. 1994. Range and habitats of the desert tortoise. IN Biology of North American Tortoises, ed. RB Bury and DJ Germano, 73–84. US Fish and Wildlife Research 13.

Gerwien RW and HB John-Alder. 1992. Growth and behavior of thyroid-deficiency lizards, *Sceloporus undulatus*. Gen. Comp. Endocrin. 87:312–324.

Gibbard PL, MJ Head, C Walker, and the Subcommission on Quaternary Stratigraphy. 2010. Formal ratification of the Quaternary system/period and the Pleistocene series/epoch with a base at 2.58 Ma. J. Quaternary Sci. 25:96–102.

Gibbons JW and JE Lovich. 1990. Sexual dimorphism of turtles with emphasis on the slider turtle (*Trachemys scripta*) Herpetol. Monogr. 4:1–29.

Gilmore CW. 1916. The fossil turtles of the Uinta Formation. Mem. Carnegie Mus. 7(2):101–161.

Gilmore CW. 1946. The osteology of the fossil turtle Testudo praeextans Lambe with notes on other species of Testudo from the Oligocene of Wyoming. Proc. U.S. Natl. Mus. 96:293–310.

Gilpin ME and ME Soulé. 1986. Minimum viable populations: processes of species extinction. IN Conservation Biology: The Science of Scarcity and Diversity, ed. ME Soulé, 19–34. Sinauer, Sunderland, MA.

Girondot M, V Delmas, P Rivalan, F Courchamp, A Prévot-Julliard, and MH Godfrey. 2004. Implications of temperature-dependent sex determination for population dynamics. IN Temperature-dependent sex determination in vertebrates, ed. N Valenzuela and V Lance. Smithsonian Books, Washington, DC.

Girvan M and MEJ Newman. 2002. Community structure in social and biological networks. Proc. Nat. Acad. Sci. 99:7821–7826.

Gist DH. 1989. Sperm storage within the oviduct of turtles. J. Morph. 199:379–384.

Glenn JL, RC Straight, and JW Sites Jr. 1990. A plasma protein marker for population genetic studies of the desert tortoise (*Xerobates agassizi*). Great Basin Natur. 50:1–8.

Gmelin JF. 1788. Caroli a Linné systema naturae per regna tria naturae, secundum classes, ordines, genera, species, cum characteribus, differentiis, synonymis, locis. Editio decima tertia, aucta, reformata. Lipsiae. (Beer).

Godley JS. 1989. Comparison of gopher tortoise populations relocated onto reclaimed phosphate mined sites in Florida. IN Gopher Tortoise Relocation Symposium, ed. JE Diemer, DR Jackson, JL Landers, JN Layne, DA Wood, 43–58. Florida Game and Fresh Water Fish Commission, Nongame Wildlife Program Technical Report 5.

Goebel T, MR Waters, and DH O'Rourke. 2008. The Late Pleistocene dispersal of modern humans in the Americas. Science 319:1497–1502.

Goin CJ and CC Goff. 1941. Notes on the growth rate of the gopher turtle, *Gopherus polyphemus,* Herpetologica 2:66–68.

Gonzalez TR. 1995. Reproduction of the Bolson tortoise, *Gopherus flavomarginatus,* Legler, 1959. Pp. 32–36 IN Aguirre, G, ED McCoy, H Mushinsky, M Villagran, R Garcia and G Casas (comp.), Proceedings of the North American Tortoise Conference. Mapimi Biosphere Reserve, Durango, Mexico. Sociedad Herpetologica Mexicana, A.C., Instituto de Ecologia, A.C., University of South Florida. Mexico DF

Gonzalez TR, G Aguirre, and GA Adest. 2000. Sex-steroids associated with the reproductive cycle in male and female Bolson Tortoise, *Gopherus flavomarginatus.* Acta Zool. Mex. n.s. 80:101–117.

Gosnell SJ, G Rivera, and RW Blob. 2009. A phylogenetic analysis of sexual size dimorphism in turtles. Herpetologica 65:70–81.

Gould, SJ. 1977. Ontogeny and Phylogeny. Belnap Press of the Harvard University Press, Cambridge, MA.

Grandmaison DD and VJ Frary. 2012. Estimating the probability of illegal desert tortoise collection in the Sonoran Desert. J.Wildl. Manag. 76:263–268.

Grandmaison D, M Ingraldi, and R Schweinsburg. 2009. Tortoises crossing roads: the science for the solution. Proc. Desert Council Symp.15–16.

Grandmaison, DD, MF Ingraldi, and FR Peck. 2010. Desert tortoise microhabitat selection on the Florence Military Reservation, south-central Arizona. J. Herpetol. 44:581–590.

Grant BW and AE Dunham. 1988. Thermally imposed time constraints on the activity of the desert lizard *Sceloporus merriami.* Ecology 69:167–176.

Grant C. 1936. The southwestern desert tortoise, *Gopherus agassizii.* Zoologica 21:225–229.

Gray JE. 1831. Synopsis Reptilium. Treuttel, Wutrz, London.

Gray JE. 1844. Catalogue of Tortoises, Crocodiles, and Amphisbaenians in the Collection of the British Museum, Edward Newman, London.

Gray JE. 1852. Description of a new genus and some new species of tortoises. Proc. Zool. Soc. London, 133–135.

Gray JE. 1855. Catalogue of Shield Reptiles in the Collection of the British Museum. Part I, Taylor and Francis, London.

Gray JE. 1870. Supplement to the Catalogue of Shield Reptiles in the Collection of the British Museum. Part I, British Museum (Natural History), London.

Griffith MAJ. 1991. Neonatal desert tortoise (*Gopherus agassizii*) biology: analyses of morphology, evaporative water loss and natural egg production followed by neonatal emergence in the Central Mojave Desert. MA thesis, California State University, Dominguez Hills..

Guyer C and SM Hermann. 1997. Patterns of size and longevity of gopher tortoise (*Gopherus polyphemus*) burrows: implications for the longleaf pine ecosystem. Chelonian Conserv. Biol. 2:507–513.

Guyer C, SM Hermann, and VM Johnson. 2012. Effects of population density on patterns of movement and behavior of gopher tortoises (*Gopherus polyphemus*). Herpetol. Monogr. 26:122–134.

Hagerty BE, KE Nussear, TC Esque, CR Tracy. 2011. Making molehills out of mountain landscape genetics of the Mojave Desert tortoise. Landscape Ecol. 2:267–280.

Hagerty BE and CR Tracy. 2010. Defining population structure for the Mojave Desert tortoise. Conserv. Gen. 11:1795–1807.

Hailey A and NS Loumbourdis. 1988. Egg size and shape, clutch dynamics, and reproductive effort in European tortoises. Can. J. Zool 66:1527–1536.

Hailman JP, JN Layne, and R Knapp. 1991. Notes on aggressive behavior of the Gopher Tortoise. Herpetol. Rev. 22:87–88.

Haines RW. 1969. Epiphyses and sesamoids. IN Biology of the Reptilia, 1, ed. C. Gans, AA Bellairs, and TS Parsons, 81–115. Academic Press, London.

Halstead BJ, ED McCoy, TA Stilson, and HR Mushinsky. 2007. Alternative foraging tactics of a central place forager examined using correlated random walk models. Herpetologica 63:472–481.

Hamilton MJ and B Buchanan. 2008. Spatial gradients in Clovis-age radiocarbon dates across North America suggest rapid colonization from the north. Proc. Natl Acad. Sci. 105:11651–11654.

Hampton AM. 1981. Field studies of natality in the desert tortoise, *Gopherus agassizi*. Proc. Desert Tortoise Counc. Symp. 1981:128–138.

Hansen K. 1963. The burrow of the gopher tortoise. J. FL. Acad. Sci. 26:353–360.

Hansen RM, MK Johnson, and TR Van Devender. 1976. Foods of the Desert Tortoise, *Gopherus agassizii,* in Arizona and Utah. Herpetologica 32:247–251.

Hanski IA. 1991. Single-species metapopulation dynamics: concepts, models and observations. Biol. J. Linn. Soc. 42:17–38.

Hanson J, T Wibbels, and RE Martin. 1998. Predicted female bias in hatchling sex ratios of loggerhead sea turtles from a Florida nesting beach. Can. J. Zool. 76:1850–1851.

Harless ML, AD Walde, DK Delaney, LL Pater, and WK Hayes. 2010. Sampling considerations for improving home range estimates of desert tortoises: effects of estimator, sampling regime, and sex. Herpetol. Conserv. Biol. 5:374–387.

Harper PAW, DC Hammond, and WP Heuschele. 1982, A herpesvirus-like agent associated with a pharyngeal abscess in a desert tortoise. J. Wildl. Dis. 18:491–494.

Harris A. 1993. Quaternary vertebrates of New Mexico. New Mexico Mus. Nat. Hist., Bull. 2:179–197.

Hawkes L, A Broderick, M Godfrey, and B Godley 2007. Investigating the potential impacts of climate change on a marine turtle population. Global Change Biol. 13:923–932.

Hawkins R and R Burke. 1989. Of pens, pullers, and pets: problems of tortoise relocation. In Gopher Tortoise Relocation Symposium Proceedings, ed. JE Diemer, DR Jackson, JL Landers, JN Layne, and DA Wood. Florida Game and Fresh Water Fish Commission. Nongame Wildlife Technical Report 5, Tallahassee. 99 pp.

Hay OP. 1902. Descriptions of two species of extinct tortoises, one new. Proc. Acad. Nat. Sci. Phila., 383–388.

Hay OP. 1904. A new gigantic tortoise from the Miocene of Colorado. Science 19:503–504.

Hay OP. 1907. Descriptions of new turtles of the genus *Testudo,* collected from the Miocene by the Carnegie Museum; together with a description of the skull of Stylemys. Ann. Carnegie Mus. 4:15–20.

Hay OP. 1908. The fossil turtles of North America. Carn. Inst. Wash. Publ. 75:1–568.

Hay OP. 1916. Descriptions of some Florida fossil vertebrates, belonging mostly to the Pleistocene. 8th Ann. Rept. Fl. Geol. Surv., 39–76.

Haynes CV Jr. and EW Haury. 1982. Archaeological investigations at the Lehner site, Arizona, 1974–1975. Natl. Geogr. Res. 14:325–334.

Hays GC, AC Broderick, F Glen, and BJ Godley. 2003. Climate change and sea turtles: a 150-year reconstruction of incubation temperatures at a major marine turtle rookery. Global Change Biol. 9:642–646.

Hays GC, S Fossette, KA Katselidis, G Schofield, and MB Ravenor. 2010. Breeding periodicity for male sea turtles, operational sex ratios, and implications in the face of climate change. Conserv. Biol. 24:1636–1643.

Hazard LC, DR Shemanski, and KA Nagy. 2009. Nutritional quality of natural foods of juvenile desert tortoises (*Gopherus agassizii*): energy, nitrogen, and fiber digestibility. J. Herpetol. 43:38–48.

Hazard LC, DR Shemanski, and KA Nagy. 2010. Nutritional quality of natural foods of juvenile and adult desert tortoises (*Gopherus agassizii*): calcium, phosphorus, and magnesium digestibility. J. Herpetol. 44:135–147.

Hebert PDN, A Cywinska, SL Ball, and JR deWaard. 2003. Biological identifications through DNA barcodes. Proc. Royal Soc. London B270:313–321.

Hebert PDN, JR deWaard, and JF Landry. 2009. DNA barcodes for 1/1000 of the animal kingdom. Biol. Lett. doi:10.1098/rsbl.2009.0848.

Hedrick PW. 1999. Perspective: highly variable loci and their interpretation in evolution and conservation. Evolution 53:313–318.

Hellgren EC, RT Kazmaier, DC Ruthven Jr., and DR Synatzske. 2000. Variation in tortoise life history: demography of *Gopherus berlandieri*. Ecology 81:1297–1310.

Henen BT. 1997. Seasonal and annual energy budgets of female desert tortoises (*Gopherus agassizii*). Ecology 78:283–296.

Henen BT. 2002a. Energy and water balance, diet, and reproduction of female desert tortoises (*Gopherus agassizii*). Chelonian Conserv. Biol. 4:319–329.

Henen BT. 2002b. Reproductive effort and reproductive nutrition of female desert tortoises: essential field methods. Integr. Comp. Biol. 42:43–50.

Henen BT, CC Peterson, IR Wallis, KH Berry, and KA Nagy. 1998. Effects of climatic variation on field metabolism and water relations of desert tortoises. Oecologia 117:365–373.

Hermann SM, C Guyer, J Hardin Waddle, and M. Greg Nelms. 2002. Sampling on private property to evaluate population status and effects of land use practices on the gopher tortoise, *Gopherus polyphemus*. Biol. Conserv. 108:289–298.

Hernandez-Divers SM, SJ Hernandez-Divers, and J Wyneken. 2002. Angiographic, anatomic, and clinical technique descriptions of a subcarapacial venipuncture site for chelonians. J. Herp. Med. Surg. 12:32–37.

Hernandez-Divers, SM, SJ Stahl, and R Farrell. 2009. An endoscopic menthod for identifying sex of hatchling Chinese box turtles and comparison of general versus local anesthesia for coelioscopy. J. Amer. Vet. Med. 234:80–84.

Hester TR. 2009. An overview of the Late Archaic in southern Texas. IN The Late Archaic across the Borderlands: From Foraging to Farming, ed. BJ Vierra, 259–278. Univ. Texas Press, Austin.

Hildebrand M. 1985. Digging of quadrupeds. IN Functional Vertebrate Morphology, ed. M Hildebrand, DM Bramble, KR Liem, and DB Wake, 89–109. Belknap Press of the Harvard University Press, Cambridge, MA.

Hillard S. 1996. The importance of the thermal environment to juvenile desert tortoises. MS thesis, Colorado State University, Fort Collins.

Hillis DM. 1987. Molecular versus morphological approaches to systematics. Annu. Rev. Ecol. Syst. 18:23–42.

Ho S. 1987. Endocrinology of vitellogenesis. IN Hormones and Reproduction in Fishes, Amphibians, and Reptiles, ed. D Norris and R Jones, 145–169. Plenum Press, New York.

Hohman J and RD Ohmart. 1980. Ecology of the desert tortoise on the Beaver Dam Slope, Arizona. Arizona State University. Report to Bureau of Land Management.

Hokit DG, M Ascunce, J Ernst, LC Branch, and AM Clark. 2010. Ecological metrics predict connectivity better than geographic distance. Conserv. Genet.11:149–159.

Holbrook JE. 1836–1840. North American Herpetology: Or, a Description of the Reptiles Inhabiting the United States. 4 Vols. J. Dobson, Philadelphia.

Holbrook JE. 1842. North American Herpetology: Or, a Description of the Reptiles Inhabiting the United States. 5 vols., 2nd ed. J. Dobson, Philadelphia.

Homer BL, KH Berry, MB Brown, G Ellis, and ER Jacobson. 1998. Pathology of diseases in wild desert tortoises from California. J. Wildl. Dis. 34:508–523.

Howland JM and JC Rorabaugh. 2002. Conservation and protection of the desert tortoise in Arizona. IN The Sonoran Desert Tortoise: Natural History, Biology, and Conservation, ed. TR Van Devender, 334–354. University of Arizona Press, Tucson, AZ.

Hubbard HG. 1893. The Florida land tortoise-Gopher, *Gopherus polyphemus*. Science 22:57–58.

Huey RB, CR Peterson, SJ Arnold, and WP Porter. 1989. Hot rocks and not so-hot rocks: retreat-site selecton by garter snakes and its thermal consequences. Ecology 70:931–944.

Hulbert AJ and CA Williams. 1988. Thyroid function in a lizard, a tortoise and a crocodile compared with mammals. Comp. Biochem. Physiol. A 90:41–48.

Hulbert RC Jr., ed. 2001. The Fossil Vertebrates of Florida. University Press of Florida, Gainesville, FL. 350 pp.

Humason GL. 1972. Animal Tissue Techniques. W.H. Freeman, San Francisco.

Hunt RM. 2002. New amphicyonid carnivorans (Mammalia, Daphoenine) from the early Miocene of southeastern Wyoming. Amer. Mus. Novitates, 3385. 41 pp.

Hunter ML Jr. 2007. Climate change and moving species: furthering the debate on assisted colonization. Conserv. Biol. 21:1356–1358.

Hurley J. 1993. Reproductive biology of the gopher tortoise *Gopherus polyphemus* in Louisiana. MS thesis, Southeastern Louisiana University, Hammond.

Hutchison JH. 1996. The terrestrial Eocene-Oligocene transition in North America, ed. DR Prothero and RJ Emry, 337–353. Cambridge University Press, Cambridge.

Hutchison JH and DM Bramble. 1981. Homology of the plastral scales of the Kinosternidae and related turtles. Herpetologica 37:73–85.

Hutchison VH, A Vinegar, and RJ Kosh. 1966. Critical thermal maxima in turtles. Herpetologica 22:32–41.

ICZN. 1999. International Code of Zoological Nomenclature, 4th ed. The International Trust for Zoological Nomenclature, c/o The Natural History Museum, London.

Independent Science Advisors, Desert Renewable Energy Conservation Plan. 2010. Recommendations of Independent Science Advisors for the California Desert Renewable Energy Conservation Plan. Conservation Biology Institute, San Diego.

Inman R, K Nussear, and C Tracy. 2009. Detecting trends in desert tortoise population growth: elusive behavior inflates variance in estimates of population density. Endangered Species Research 10:295–304.

Intergovernmental Panel on Climate Change (IPCC). 2007. Climate change 2007: the physical science basis—summary for policymakers. Contribution of Working Group I to the Fourth Assessment Report of the Intergovernmental Panel on Climate Change.

International Union for Conservation of Nature (IUCN). 2001. IUCN Red List Categories and Criteria: Version 3.1. IUCN Species Survival Commission. IUCN, Gland, Switzerland, and Cambridge, UK.

International Union for Conservation of Nature (IUCN). 2010. IUCN Red List of Threatened Species. Version 2010.4. www.iucnredlist.org.

Itsch GR and JF Anderson. 1986. The respiratory microenvironment within the burrows of gopher tortoises (*Gopherus polyphemus*). Copeia 1986:787–795.

Iverson JB. 1980. The reproductive biology of *Gopherus polyphemus* (Chelonia: Testudinide). Am. Midl. Nat. 103:352–359.

Iverson JB. 1992a. Correlates of reproductive output in turtles (Order Testudines). Herpetol. Monogr. 6:25–42.

Iverson JB. 1992b. A Revised Checklist with Distribution Maps of the Turtles of the World. Green Nature Books. 400 pp.

Iverson JB, CP Balgooyen, KK Byrd, and KK Lyddan. 1993. Latitudinal variation in egg and clutch size in turtles. Can. J. Zool. 71:2448–2461.

Jackson CJ Jr., JA Trotter, TH Trotter, and MW Trotter. 1976. Accelerated growth rate and early maturity in *Gopherus agassizii* (Reptilia: Testudines). Herpetologica 32:139–145.

Jackson CJ Jr., TH Trotter, JA Trotter, and MW Trotter. 1978. Further observations of growth and sexual maturity in captive desert tortoises (Reptilia: Testudines). Herpetologica 34: 225–227.

Jackson HE and SL Scott. 2003. Patterns of elite faunal utilization at Moundville, Alabama. Am. Antiquity 68:552–572.

Jackson OF. 1980. Weight and measurement data on tortoises (*Testudo graeca* and *Testudo hermanni*) and their relationship to health. J. Small Anim. Pract. 21:409–416.

Jacobson ER. 1993. Blood collection techniques in reptiles: laboratory investigations. IN Zoo and Wild Animal Medicine: Current Therapy 3, ed. ME Fowler, 144–152. WB Saunders, Philadelphia.

Jacobson ER. 2007a. Viruses and viral diseases of reptiles. IN Infectious Diseases and Pathology of Reptiles: A Color Atlas and Text, ed. ER Jacobson, 395–460. CRC Press, Boca Raton, FL.

Jacobson ER. 2007b. Bacterial Diseases of Reptiles. IN Infectious Diseases and Pathology of Reptiles: A Color Atlas and Text, ed. ER Jacobson, 461–527. CRC Press, Boca Raton, FL.

Jacobson ER. 2007c. Parasites and Parasitic Diseases of Reptiles. IN Infectious Diseases and Pathology of Reptiles: A Color Atlas and Text, ed. ER Jacobson, 571–661. CRC Press, Boca Raton, FL.

Jacobson ER and KH Berry. 2012. *Mycoplasma testudineum* in free-ranging desert tortoises, *Gopherus agassizii*. J. Wildl. Dis. 48:1063–1068.

Jacobson ER, KH Berry, B Stacy, LM Huzella, VF Kalasinsky, ML Fleetwood, MG Mense. 2009. Oalosis in wild desert tortoises, *Gopherus agassizii*. J. Wildl. Dis. 45:982–988.

Jacobson ER, KH Berry, JFX Wellehan Jr., F Origgi, AL Childress, J Braun, M Schrenzel, J Yee, and B Rideout. 2012. Serologic and molecular evidence for Testudinid herpesvirus 2 infection in wild Agassiz's desert tortoises, *Gopherus agassizii*. J. Wildl. Dis. 48:747–757.

Jacobson ER, MB Brown, IM Schumacher, BR Collins, RK Harris, and PA Klein. 1995. Mycoplasmosis and the desert tortoise (*Gopherus agassizii*) in Las Vegas Valley Nevada. Chelonian Conserv. Biol. 1:279–284.

Jacobson ER, JM Gaskin, M Brown, RK Harris, CH Gardiner, JL LaPointe, HP Adams, and C Reggiardo. 1991. Chronic upper respiratory tract disease of free-ranging desert tortoises, *Xerobates agassizii*. J. Wildl. Dis. 27:296–316.

Jacobson ER, J Schumacher, and ME Green. 1992. Field and clinical techniques for sampling and handling blood for hematologic and selected biochemical determinations in the desert tortoise, *Xerobates agassizii*. Copeia:237–241.

Jacobson ER, M Weinstein, K Berry, C Hardenbrook, C Tomlinson, and D Freita. 1993. Problems with using weight vs. length relationships to assess tortoise health. Vet. Rec. 132:222–223.

Jacobson ER, T Wronski, J Schumacher, C Reggiardo, and KH Berry. 1994. Cutaneous dyskeratosis in free ranging desert tortoise, *Gopherus agassizii,* in the Colorado Desert of Southern California. J. Zoo Wildl. Med. 25:68–81.

James GW. 1906. The Wonders of the Colorado Desert (Southern California): Its Rivers and Its Mountains, Its Canyons and Its Springs, Its Life and Its History, Pictured and Described; Including an Account of a Recent Journey Made Down the Overflow of the Colorado River to the Mysterious Salton Sea, Vol. 1. Little, Brown, Boston.

Janzen FJ. 1994. Climate change and temperature-dependent sex determination in reptiles. Proc. Natl. Acad. Sci. U.S.A. 91:7487–7490.

Janzen FJ and JG Krenz. 2004. Phylogenetics: which was first, TSD or GSD? IN Temperature-Dependent Sex Determination in Vertebrates, ed. N Valenzuela and VA Lance, 121–130. Smithsonian Books, Washington, DC.

Janzen F and GL Paukstis. 1991 Environmental sex determination in reptiles: ecology, evolution, and experimental design. Quart. Rev. Biol. 66:149–179.

Jarchow JL, HE Lawler, TR Van Devender, and CS Ivanyi. 2002. Care and diet of captive Sonoran desert tortoises. IN The Sonoran Desert Tortoise: Natural History, Biology, and Conservation, ed. TR Van Devender, 289–311. University of Arizona Press, Tucson, AZ.

Jennings RD. 1985. Biochemical variation of the Desert Tortoise, *Gopherus agassizii*. MS thesis, University of New Mexico, Albuquerque, NM.

Jennings WB. 1993. Foraging ecology of the desert tortoise, *Gopherus agassizii* in the western Mojave Desert. MS thesis, University of Texas, Arlington, TX.

Jennings WB. 1997. Habitat use and food preferences of the desert tortoise, *Gopherus agassizii* in the western Mojave Desert and impacts of off-road vehicles. IN Proceedings: Conservation, Restoration, and Management of Tortoises and Turtles—An International Conference, ed. J Van Abbema, 42–45. New York Turtle and Tortoise Society and WCS Turtle Recovery Program, New York.

Jennings WB. 2002. Diet selection by the desert tortoise in relation to the flowering phenology of ephemeral plants. Chelonian Conserv. Biol. 4:353–358.

Jennrich RI and FB Turner. 1969. Measurement of non-circular home range. J. Theor. Biol. 22:227–237.

Jodice PGR, DM Epperson, and GH Visser. 2006. Daily energy expenditure in free-ranging gopher tortoises (*Gopherus polyphemus*). Copeia 2006:129–136.

Johnson AJ, DJ Morafka, and ER Jacobson. 2006. Seroprevalence of *Mycoplasma agassizii* and tortoise herpesvirus in captive desert tortoises (*Gopherus agassizii*) from the Greater Barstow Area, Mojave Desert, California. J. Arid Environ. 67S:192–201.

Johnson AJ, AP Pessier, JF Wellehan, R Brown, and ER Jacobson. 2005. Identification of a novel herpesvirus from a California desert tortoise (*Gopherus agassizii*). Vet. Micro. 111:107–116.

Johnson AJ, AP Pessier, JF Wellehan, TM Norton, NL Stedman, DC Bloom, W Belzer, VR Titus, R Wagner, J Brooks, J Spratt, and ER Jacobson. 2008. Ranavirus infection of free-ranging and captive box turtles and tortoises in the United States. J. Wildl.. Dis. 44:851–863.

Johnson AJ, L Wendland, TM Norton, and ER Jacobson. 2010. Development and use of an indirect enzyme-linked immunosorbent assay for detection of iridovirus eposure in gopher tortoises (*Gopher polyphemus)* and eastern box turtles (*Terrapene carolina carolina*). Vet. Micro. 142:160–167.

Johnson VM and C Guyer. 2007. Phenology of attempted matings in gopher tortoises. Copeia 2007:490–495.

Johnson VM, C Guyer, SM Hermann, J Eubanks, and WK Michener. 2009. Patterns of dispersion and burrow use support scramble competition polygyny in *Gopherus polyphemus*. Herpetologica 65:214–218.

Johnston CS. 1937. Osteology of *Bysmachelys canyonensis,* a new turtle from the Pliocene of Texas. J. Geol. 45:439–445.

Jolly GM. 1965. Explicit estimates from capture-recapture data with both death and immigration—stochastic model. Biometrika 52:225–247.

Jones CA. 2008. *Mycoplasma agassizii* in the Sonoran population of the desert tortoise in Arizona. MS thesis, University of Arizona, Tucson.

Jones DN Jr. 1996. Population biology of the gopher tortoise, *Gopherus polyphemus,* in southeast Georgia. MS thesis, Georgia Southern University, Statesboro.

Jones J and B Dorr. 2004. Habitat associations of gopher tortoise burrows on industrial timberlands. Wildl. Soc. Bull. 32:456–464.

Jorgensen CB. 1998. Role of urinary and cloacal bladders in chelonian water economy: historical and comparative perspectives. Biol. Rev. Camb. Philos. Soc. 73:347–366.

Judd FW and JC McQueen. 1980. Incubation, hatching, and growth of the tortoise, *Gopherus berlandieri*. J. Herpetol. 14:377–380.

Judd FW and JC McQueen. 1982. Notes on longevity of *Gopherus berlandieri* (Testudinidae). Southwest. Nat. 27:230–232.

Judd FW and FL Rose. 1977. Aspects of the thermal biology of the Texas Tortoise, *Gopherus berlandieri* (Reptilia, Testudines, Testudinidae) J. Herpetol. 11:147–153.

Judd FW and FL Rose. 1983. Population structure, density, and movements of the Texas tortoise *Gopherus berlandieri*. Southwest. Nat. 28:387–398.

Judd FW and FL Rose. 1989. Egg production by the Texas tortoise, *Gopherus berlandieri,* in southern Texas. Copeia 1989:588–596.

Judd FW and FL Rose. 2000. Conservation Status of the Texas Tortoise *Gopherus berlandieri*. Museum of Texas Tech University, Occasional Papers 196:1–11.

Kabigumila J. 2001. Sighting frequency and food habits of the leopard tortoise, *Geochelone pardalis,* in northern Tanzania. East African Wildlife Society, Afr. J. Ecol. 39:276–285.

Karl AE. 1998. Reproductive strategies, growth patterns, and survivorship of a long-lived herbivore inhabiting a temporally variable environment. PhD diss., University. California, Davis. 178 pp.

Kautz RS. 1998. Land use and land cover trends in Florida 1936–1995. FL. Sci. 61:171–187.

Kazmaier RT. 2000. Ecology and demography of the Texas tortoise in managed thornscrub ecosystem. PhD diss., Oklahoma State University, Stillwater.

Kazmaier RT, EC Hellgren, and DC Ruthven III. 2001a. Habitat selection by the Texas tortoise in a managed thornscrub ecosystem. J. Wildl. Manage. 65:653–660.

Kazmaier RT, EC Hellgren, and DC Ruthven III. 2002. Home range and dispersal of Texas tortoises, *Gopherus berlandieri,* in a managed thornscrub ecosystem. Chelonian Conserv. Biol. 4:488–496.

Kazmaier RT, EC Hellgren, DC Ruthven III, and DR Synatzske. 2001b. Effects of grazing on the demography and growth of the Texas tortoise. Conserv. Biol. 15:1091–1101.

Kazmaier RT, EC Hellgren, and DR Synatzske. 2001c. Patterns of behavior in the Texas tortoise, *Gopherus berlandieri:* A multivariate ordination approach. Can. J. Zool. 79:1363–1371.

Kazmaier RT, EC Hellgren, DR Synatzske, and JC Rutledge. 2001d. Mark-recapture analysis of population parameters in a Texas tortoise (*Gopherus berlandieri*) population in southern Texas. J. Herpetol. 35:410–417.

Kazmaier RT, S Patten, and DC Ruthven III. 2012. The tortoises of Texas: the importance of long-term data for understanding a poorly studied species. 10th Annual Symposium on the Conservation and Biology of Tortoises and Freshwater Turtles, Tucson, Arizona. Joint Annual Meeting of the Turtle Survival Alliance and the IUCN Tortoise and Freshwater Turtle Specialist Group. Abstracts.

Kinlaw AE. 2006. Burrows of semi-fossorial vertebrates in upland communities of central Florida: their architecture, dispersion and ecological consequences. PhD diss., University Florida, Gainsville.

Kirkman LK, RJ Mitchell, RC Helton, and MB Drew. 2001. Productivity and species richness across an environmental gradient in a fire-dependent ecosystem. Am. J. Bot. 88:2119–2128.

Koerner HE. 1940. The geology and vertebrate paleontology of the Fort Logan and Deep River formations of Montana. Am. J. Sci. 238(12):837–861.

Kohel KA, DS MacKenzie, DC Rostal, JC Grumbles, and VA Lance. 2001. Seasonality in plasma thyroxine in the desert tortoise, *Gopherus agassizii*. Gen. Comp. Endocrinol. 121:214–222.

Kolbert E. 2006. Field Notes From a Catastrophe: Man, Nature and Climate Change. New York: Bloomsbury.

Kristan WB and WI Boarman. 2003. Spatial pattern of risk of common raven predation on desert tortoises. Ecology 84:2432–2443.

Krzysik AJ. 2002. A landscape sampling protocol for estimating distribution and density patterns of desert tortoises at multiple spatial scales. Chelonian Conserv. Biol. 4:366–379.

Kuchling G. 1999. The Reproductive Biology of the Chelonia. Springer, Berlin.

Kuo CH and FJ Janzen. 2003. BottleSim: a bottleneck simulation program for long-lived species with overlapping generations. Mol. Ecol. Notes, 3, 669–673. DOI: 10.1046/j.1471-8286.2003.00532.x.

Kushlan JA and FJ Mazotti. 1984. Environmental effects on a coastal population of gopher tortoises. Herpetologica 18:231–239.

Lacy RC, M Borbat, JP Pollak. 2005. VORTEX: A Stochastic Simulation of the Extinction Process, Version 9.50, User's Manual. Chicago Zoological Society, Brookfield, IL.

Lamb T, JC Avise, and JE Gibbons. 1989. Phylogeographic patterns in mitochondrial DNA of the desert tortoise (*Xerobates agassizi*) and evolutionary relationships among North American gopher tortoises. Evolution 43:76–87.

Lamb T and C Lydeard. 1994. A molecular phylogeny of the gopher tortoise, with comments on familial relationships within Testudinoidea. Mol. Phylogen. Evol. 3:283–291.

Lamb T and AM McLuckie. 2002. Genetic differences among geographic races of the desert tortoise. In The Sonoran Desert Tortoise: Natural History, Biology, and Conservation, ed. TR Van Devender, 67–85. University of Arizona Press, Tucson, AZ.

Lambe LM. 1913. Descriptions of new species of *Testudo* and a remarkable specimen of *Stylemys nebrascensis* from the Oligocene of Wyoming. Ottawa Nat. 27:57–63.

Lance VA, JS Grumbles, and DC Rostal. 2001. Sex differences in plasma corticosterone in desert tortoises, *Gopherus agassizii*, during the reproductive cycle. J. Exp. Zool. 289: 285–289.

Lance VA and DJ Morafka. 2001. Post natal lecithotroph: a new age class in the ontogeny of reptiles. Herpetol. Monogr. 15:124–134.

Lance VA, AR Place, JS Grumbles, and DC Rostal. 2002. Variation in plasma lipids during the reproductive cycle of male and female desert tortoises, *Gopherus agassizzi*. J. Exp. Zool. 293:703–711.

Lance VA and DC Rostal. 2002. The annual reproductive cycle of the male and female desert tortoise: physiology and endocrinology. Chelonian Conserv. Biol. 4:302–312.

Landers JL. 1980. Recent research on the gopher tortoise and its implications. IN The Dilemma of the Gopher Tortoise—Is There a Solution? ed. R Franz and RJ Bryant, 8–14. Proceedings of the 1st Annual Meeting, Gopher Tortoise Council.

Landers JL and JL Buckner. 1981. The gopher tortoise: effects of forest management and critical aspects of its ecology. Technical Note No. 56, Southlands Experimental Forestry, Bainbridge, GA.

Landers JL, JA Garner, and WA McRae. 1980. Reproduction of the gopher tortoise (*Gopherus polyphemus*) in southwestern Georgia. Herpetologica 36:353–361.

Landers JL and DW Speake. 1980. Management needs of sandhill reptiles in southern Georgia. Proc. Annu. Conf. S.E. Assoc. Fish and Wildl. Agen. 34:515–529.

Landers JL, WA McRae, and JA Garner. 1982. Growth and maturity of the gopher tortoise in southwestern Georgia. Bull. FL. St. Mus. Biol. Sci. 27:81–110.

Lang JW and HV Andrews. 1994. Temperature-dependent sex determination in crocodilians. J. Exp. Zool. 270:28–44.

LaPré L. 2010. Timing is Everything for Renewable Energy. 35th Annual Desert Tortoise Council Symposium, Ontario, California. www.deserttortoise.org/abstract/210DTC SymposiumAbstracts .pdf.

Latch EK, WI Boarman, A Walde, and RC Fleischer. 2011. Fine-scale analysis reveals cryptic landscape genetic structure in desert tortoises. PLoS ONE 6:e27794.

Lau A. 2011. Spatial ecology of the gopher tortoise (*Gopherus polyphemus*) in coastal sand dune habitat: burrow site selection, home range and seasonal activity patterns. MS thesis, University of Florida, Gainesville.

Le M, CJ Raxworthy, WP McCord, and L Mertz. 2006. A molecular phylogeny of tortoises (Testudines: Testudinidae) based on mitochondrial and nuclear genes. Mol. Phylogenet. Evol. 40:517–31.

Lear CH, TR Bailey, PN Pearson, HK Coxall, and Y Rosenthal. 2008. Cooling and ice growth across the Eocene-Oligocene transition. Geology 38:251–254.

LeBlanc AM. (2009). Evolutionary and conservation implications of sex determination and hatchling depredation in Kemp's Ridley sea turtles. PhD diss., University of Alabama at Birmingham.

LeBlanc AM, KK Drake, KL Williams, MG Frick, TG Wibbels, and DC Rostal C. 2012. Nest temperatures and hatchling sex ratios from loggerhead turtle nests incubated under natural field conditions in Georgia, United States. Chelonian Conserv. Biol. 11(1):108–116.

Le Conte J. 1829. Description of the species of North American tortoises. Ann. Lycaeum Natur. Hist. 3:91–131.

Lee HH. 1963. Egg-laying in captivity by *Gopherus agassizii* Cooper. Herpetologica 19:62–65.

Lee PLM and G Hays 2004. Polyandry in a marine turtle: females make the best of a bad job. Proc. Nat. Acad. Sci. 101:6530–6535.

Legler JM. 1959. A new tortoise genus *Gopherus* from north central Mexico. U. Kansas Publ. Mus. Nat. Hist. 11:335–343.

Legler J and R Webb. 1961. Remarks on a collection of Bolson tortoises, *Gopherus flavomarginatus*. Herpetologica 17:26–37.

Leidy J. 1851. (On a new species of fossil tortoise). Proc. Acad. Nat. Sci. Phila. 5:172–173.

Lenth BE, RL Knight, and ME Brennan. 2008. The effects of dogs on wildlife communities. Nat. Areas J. 28:218–227.

Les HL, RT Paitz, T Ryan, and RM Bowden. 2007. Experimental test of the effects of fluctuating incubation temperatures on hatchling phenotype. J. Exp. Zool. Part A: Ecol. Genet. Physiol. 307(5):274–280.

Leu M, SE Hanser, and ST Knick. 2008. The human footprint in the West: a large-scale analysis of anthropogenic effects. Ecol. Appl. 18:1119–1139.

Lewis-Winokur V and RM Winokur. 1995. Incubation temperature affects sexual differentiation, incubation time and post hatching survival in desert tortoises (*Gopherus agassizii*). Can. J. Zool. 73:2091–2097.

Licht P, J Wood, DW Owens, and FE Wood. 1979. Serum gonadotropins and steroids associated with breeding activities in the green sea turtle *Chelonia mydas*. I. Captive animals. Gen. Comp. Endocrinol. 39:274–289.

Licht P, GL Breitenbach, and JD Congdon. 1985. Seasonal cycles in testicular activity, gonadotropin and thyroxin in the painted turtle *Chrysemys picta*, under natural conditions. Gen. Comp. Endocrinol. 59:130–139.

Lieberman SS and DJ Morafka. 1988. Ecological distribution of the Bolsón tortoise. IN The Ecogeography of the Mexican Bolsón Tortoise (*Gopherus flavomarginatus*): Derivation of its Endangered Status and Recommendations for Its Conservation, ed. D. J. Morafka and CJ McCoy, 31–46. Ann. Carnegie Mus. 57.

Lifson N and R McClintock. 1966. Theory of the use of the turnover rates of body water for measuring energy and material balance. J. Theor. Biol. 12:46–74.

Linley TA and HR Mushinsky. 1994. Organic composition and energy content of eggs and hatchlings of the gopher tortoise. IN Biology of North American Tortoises, ed. RB Bury and DJ Germano, 113–128. US Fish and Wildlife Research 13.

Linnaeus C. 1758. Systema naturæ per regna tria naturæ, secundum classes, ordines, genera, species, cum characteribus, differentiis, synonymis, locis. Editio decima, reformata, pp. [1–4], 825–1384. Holmiæ. (Salvius).

Linnaeus (as Linné) C. 1766. Systema naturæ per regna tria naturæ, secundum classes, ordines, genera, species, cum characteribus, differentiis, synonymis, locis. Editio duodecima, reformata. Holmiæ. (Salvius).

Liu C, PM Berry, TP Dawson, and RG Pearson. 2005. Selecting thresholds of occurrence in the prediction of species distributions. Ecography, 28:385–393.

Lohoefener R and L Lohmeier. 1981. Comparison of gopher tortoise (*Gopherus polyphemus*) habitats in young slash pine and old longleaf pine areas of southern Mississippi. J. Herpetol. 15:239–242.

Lohoefener R and L Lohmeier. 1984. The status of *Gopherus polyphemus* (Testudines, Testudinidae) west of the Tombigbee and Mobile Rivers. Report to US Fish and Wildlife Service in support of petition to list this population under the Endangered Species Act of 1973.

Lombard RE and TE Hetherington. 1993. Structural basis of hearing and sound transimission. IN The Skull: Vol. 3, Functional and Evolutionary Mechanisms, ed. J Hanken and BK Hall, 241–302. University of Chicago Press, Chicago.

Longleaf Alliance. 2011. The Longleaf Forest Served as the Walmart for Early Settlers, Lesson 15. The Longleaf Alliance, Andalusia, AL.

Longshore KM, JR Jaeger and JM Sappington. 2003. Desert tortoise (*Gopherus agassizii*) survival at two eastern Mojave Desert sites: death by short-term drought? J. Herpetol. 37:169–177.

Loomis FB 1909. Turtles from the upper Harrison beds. Am. J. Sci. ser. 4, 28:17–26.

Losos JB, DM Hillis, and HW Greene. 2012. Who speaks with a forked tongue? Science 338:1428–1429.

Loveridge A and EE Williams. 1957. Revision of the African tortoises and turtles of the suborder Cryptodira. Bull. Mus. Comp. Zool. 115:163–557.

Lovich JE. 1996. Possible demographic and ecologic consequences of sex ratio manipulation in turtles. Chelonian Conserv. Biol. 2:114–117.

Lovich J. 2000. *Pennisetum setaceum* Forsskal. IN Invasive Plants of California's Wildlands, ed. CC Bossard, JM Randall, and MC Hoshovsky, 258–262. University of California Press, Berkeley.

Lowe CH, PJ Lardner, and EA Halpern. 1971. Supercooling in reptiles and other vertebrates. Comp. Biochem. Physiol. 39A:125–135.

Lucas SG and GS Morgan. 1996. Pleistocene vertebrates from the Pecos River valley near Roswell, Chaves County, New Mexico. New Mexico Geol. 93–96.

Luckenbach RA. 1976. Population density, structure and feeding habits of the desert tortoise, *Gopherus agassizii*, in a low desert study area in southern Nevada. Proc. Desert Tortoise Council Symp., 22–37.

Luckenbach RA. 1982. Ecology and management of the desert tortoise (*Gopherus agassizii*) in California. IN North American Tortoises: Conservation and Ecology, ed. RB Bury and DJ Germano, 1–37. US Fish and Wildlife Service Research Report 12.

Lui Z, M Pagani, D Zinniker, R Deconto, M Huber, H Brinkhuis, SR Shah, RM Lecker, and A Pearson. 2009. Global cooling during the Eocene-Oligocene climate transition. Science 323:1187–1190.

Luke C, A Karl, and P Garcia. 1991. A status review of the desert tortoise. Report to City of Ridgecrest from Biosystems Analysis, Tiburon, CA.

Lynn WG. 1970. The Thyroid in Biology of the Reptilia, ed. C Gans and J Parsons, 201–234. Academic Press, London.

MacArthur RH and ER Pianka. 1966. On optimal use of a patchy environment. Am. Nat. 100:603–609.

MacDonald LA. 1996. Reintroduction of gopher tortoises (*Gopherus polyphemus*) to reclaimed phosphate land. Florida Institute of Phosphate Research. Bratow, FL + appendix. Publ No. 03-105-126.

MacDonald LA and HR Mushinsky. 1988. Foraging ecology of the gopher tortoise, *Gopherus polyphemus*, in a sandhill habitat. Herpetologica. 44:345–353.

MacKenzie DI, JD Nichols, JA Royle, KH Pollock, LL Bailey, and JE Hines. 2006. Occupancy Estimation and Modeling: Inferring Patterns and Dynamics of Species Occurrence. Academic Press, Amsterdam.

MacKenzie DI and JA Royle. 2005. Designing occupancy studies: general advice and allocating survey effort. J. Appl. Ecol. 42:1105–1114.

MacKenzie DS, CM VanPutte, and KA Leiner. 1998. Nutrient regulation of endocrine function in fish. Aquaculture. 161:3–25.

MacWilliams AC, RJ Hand, JR Roney, KR Adams, and WL Merrill. 2006. Una Investigación Arqueológica de los Sitios de Cultivo de Maíz Temprano en Chihuahua, México. Informe al Consejo de Arqueología, Instituto Nacional de Antropología e Historia.

Maher MJ. 1961. The effect of environmental temperature on metabolic response to thyroxine in the lizard, *Lacerta muralis*. Am. Zool. 1:461 (abst.).

Maher MJ. 1965. The role of the thyroid gland in the oxygen consumption of lizards. Gen. Comp. Endocrinol. 5:320–325.

Mallman MTO. 1994, Influência da temperatura de incubação na determinação sexual em *Geochelone carbonaria* (Spix, 1824) (Reptilia, Testudines, Testudinidae). Dissertação de Mestrado, Pontifícia Universidade Católica do Rio Grande do Sul.

Manel S, MK Schwartz, GP Luikart, and P Taberlet. 2003. Landscape genetics: combining landscape ecology and population genetics. Trends Ecol. Evol. 18(4):189–197.

Manno TG. 2008. Social networking in the Columbian ground squirrel, *Spermophilus columbianus*. Anim. Behav. 75:1221–1228.

Marlow RW and K Tollestrup. 1982. Mining and exploitation of natural mineral deposits by the desert tortoise, *Gopherus agassizii*. Anim. Behav. 30:475–478.

Marrinan RE. 2005. Early Mississippian faunal remains from Shields Mound (8DU12). Fla. Anthropol. 58:175–210.

Marschang RE, P Becher, H Posthaus, P Wild, H-J. Thiel, U. Muller-Diblies, EF Kaleta, and LN Bacciarini. 1999. Isolation and characterization of an iridovirus from Hermann's tortoises (*Testudo hermanni*). Archiv. of Virol. 144:1909–1922.

Marschang RE, JW Frist, M Gravendyck, and EF Kaleta. 2001. Comparison of 16 chelonid herpesviruses by virus neutralization tests and restriction endonuclease digestion of viral DNA. J. Vet. Med. B Infect. Dis. Vet. Public Health 48:393–399.

Marshall JE. 1987. The effects of nest predation on hatching success in gopher tortoises (*Gopherus polyphemus* Daudin 1802). MS thesis, University of South Alabama, Birmingham.

Martel A, S Blahak, H Vissenaekens, and F Pasmans. 2009. Reintroduction of clinically healthy tortises: the herpesvirus Trojan horse. J. Wildl. Dis. 45:218–220.

Martin BE. 1995. Ecology of the desert tortoise (*Gopherus agassizii*) in a desert-grassland community in southern Arizona. MS thesis, University of Arizona, Tucson.

Martinez-Cardenas, A. 2006. Evaluación del hábitat de la tortuga de Mapimí (*Gopherus flavomarginatus*) en la Reserva de la biosfera de Mapimí, Durango, México. Masters thesis, Instituto de Ecología, Xalapa, Veracruz.

Martınez-Silvestre A, N MajÓ, and A Ramis. 1999. Caso clínico: herpesvirosis en tortuga de desierto Americana (*Gopherus agassizii*). Clínica Verinaria de Pequños Animales 19:99–106.

Mason RT, and MR Parker. 2010. Social behavior and pheromonal communication in reptiles. J. Comp. Physiol. A 196:729–749.

Maxie MG. 2007. Urinary system. In Jubb, Kennedy, and Palmer's Pathology of Domestic Animals, Vol. 2, 5th ed., ed. MG Maxie, 425–522. Saunders/Elsevier, New York.

McArthur ED, SC Sanderson, and BL Webb. 1994. Nutritive quality and mineral content of potential desert tortoise food plants. Res. Pap. INT-473. USDA-Forest Service, Intermountain Research Station, Ogden, UT.

McAuliffe JR and EP Hamerlynck. 2010. Perennial plant mortality in the Sonoran and Mojave deserts in response to severe, multi year drought. J. Arid Environ. 74:885–896.

McCord RD. 1994. Fossil tortoises of Arizona. IN Fossils of Arizona. Proceedings 1994, Vol. 2, 83–89. Southwest Paleontol. Soc. & Mesa Southwest Mus. Mesa, AZ.

McCord, RD. 1997. Preliminary stratocladistic analysis of North American tortoises (genus *Gopherus*), ed., B Anderson, D Boaz, and RD McCord. Proc. Southwest Paleontol. Symp. Mesa Museum and Southwestern Paleontol. Soc., Vol. 1. Mesa, AZ.

McCord RD. 2002. The Sonoran Desert Tortoise. IN Natural History, Biology, and Conservation, ed, TR Van Devender, 52–66. University of Arizona Press. Tucson, AZ.

McCoy ED, RD Moore, HR Mushinsky, and SC Popa. 2011. Effects of rainfall and the potential influence of climate change on two congeneric tortoise species. Chelonian Conserv. Biol. 10:34–41.

McCoy ED and HR Mushinsky. 1991. A survey of gopher tortoise populations residing on twelve state parks in Florida. FL. Depart. Nat. Res. Tech. Rep. 1..

McCoy ED and HR Mushinsky. 1992a. Studying a species in decline: gopher tortoises and the dilemma of "correction factors." Herpetologica 48:402–407.

McCoy ED and HR Mushinsky. 1992b. Studying a species in decline: changes in populations of gopher tortoises on federal lands in Florida. FL Sci. 55:116–125.

McCoy ED and HR Mushinsky. 1995. The demography of *Gopherus polyphemus* in relation to size of available habitat. Florida Game and Freshwater Fish Commission Non-Game Wildlife Program Report, GFC 86–113. Tallahassee, FL.

McCoy ED and HR Mushinsky. 2007. Estimates of minimum patch size depend on the method of estimation and the condition of the habitat. Ecology 88:1401–1407.

McCoy ED, HR Mushinsky, JK Lindzey. 2005. Population consequences of upper respiratory tract disease on gopher tortoises. Final Report. Florida Fish and Wildlife Conservation Commission, Tallahassee.

McCoy ED, HR Mushinsky, and J Lindzey. 2006. Population declines of the gopher tortoise on protected lands. Biol. Conserv. 128:120–127.

McCoy ED, HR Mushinsky, and RD Moore. 2007. A future with small populations of the gopher tortoise. IN Urban Herpetology, ed. RE Jung and JC Mitchell. Herpetological Conservation 3. Society for the Study of Amphibians and Reptiles, Salt Lake City, UT.

McCoy ED, B Stys, and HR Mushinsky. 2002. A comparison of GIS and survey estimates of gopher tortoise habitat and numbers of individuals in Florida. Chelonian Conserv. Biol. 4:472–478.

McCraney WT, G Goldsmith, DK Jacobs, and AP Kinziger. 2010. Rampant drift in artificially fragmented populations of the endangered tidewater goby (*Eucyclogobius newberryi*). Mol. Ecol. 19:3315–3327.

McGinnis SM and WG Voigt. 1971. Thermoregulation in the desert tortoise, *Gopherus agassizii*. Comp. Biochem. Physiol. 40A:119–126.

McLachlan JS, JJ Hellman, and W Schwartz. 2007. A framework for debate of assisted migration in an era of climate change. Conserv. Biol. 21:297–302.

McLaughlin GS. 1990. Ecology of gopher tortoises (*Gopherus polyphemus*) on Sanibel Island, Florida. MS thesis, Iowa State University, Ames.

McLaughlin GS, ER Jacobson, DR Brown, CE McKenna, IM Schumacher, HP Adams, MB Brown, and PS Klein. 2000. Pathology of upper respiratory tract disease of gopher tortoises in Florida. J. Wildl. Dis. 36: 272–283.

McLuckie AM, MRM Bennion, and RA Fridell. 2007. Tortoise mortality within the Red Cliffs Desert Reserve following the 2005 wildfires. Rept. to Utah Divison of Wildlife Resources Publication no. 07–05.

McLuckie AM and RA Fridell. 2002. Reproduction in a desert tortoise (*Gopherus agassizii*) population on the Beaver Dam Slope, Washington County, Utah. Chelonian Conserv. Bi. 4:288–294.

McLuckie AM, DL Harstad, JW Marr, and RA Fridell. 2002. Regional desert tortoise monitoring in the Upper Virgin River Recovery Unit, Washington County, Utah. Chelonian Conserv. Bi. 4:380–386.

McLuckie AM, T Lamb, CR Schwable, and RD McCord. 1999. Genetic and morphological assessment of an unusual tortoise (*Gopherus agassizii*) population in the Black Mountains of Arizona. J. Herpetol. 33:36–44.

McLuckie AM, MA Ratchford, and RA Fridell. 2012. Regional desert tortoise monitoring in the Red Cliffs Desert Reserve, 2011. Utah Division of Wildlife Resources, Pub. no. 12–13. Salt Lake City.

McNabb FMA. 1992. Thyroid Hormones. Prentice-Hall, Englewood Cliffs, NJ, 283.

McPherson RJ, LR Boots, R MacGregor III, and KR Marion. 1982. Plasma steroids associated with seasonal reproductive changes in a multiclutched freshwater turtle, *Sternotherus odoratus*. Gen. Comp. Endocrin. 48: 440–451.

McRae WA, JL Landers, and JA Garner. 1981. Movement patterns and home range of the gopher tortoise. Am. Midl. Nat. 106:165–179.

Means B. 1982. Responses to winter burrow flooding of the gopher tortoise (*Gopherus polyphemus* Daudin). Herpetologica 38:521–525.

Means DB. 2006. Vertebrate faunal diversity of longleaf pine ecosystems. IN The Longleaf Pine Ecosystem, ed. S Jose, E Jokela, and D Miller, 157–213. Springer, New York, NY.

Medica PA, RB Bury, and F Turner. 1975. Growth of the Desert tortoise (*Gopherus agassizi*) in Nevada. Copeia 1975:639–643.

Medica PA, RB Bury, and R Luckenbach. 1980. Drinking and construction of water catchments by the desert tortoise, *Gopherus agassizii*, in the Mojave Desert. Herpetologica. 36(4):301–304.

Medica PA and SE Eckert. 2007. *Gopherus agassizii*, Desert Tortoise. Food/Mechanical Injury. Natural History Notes. Herpetol. Rev. 38:446–448.

Medica PA, CL Lyons, and FB Turner. 1981. A comparison of populations of the desert tortoise (*Gopherus agassizii*) in grazed and ungrazed areas in Ivanpah Valley, California. Bureau of Land Management Final Report.

Medica PA, KE Nussear, TC Esque, and MB Saethre. 2012. Long-term growth of desert tortoises (*Gopherus agassizii*) in a southern Nevada population. J. Herpetol. 46:213–220.

Mendonca M, R Beauman, and H Balbach. 2007. Burrow collapse as a potential stressor on the gopher tortoise (*Gopherus polyphemus*). Final report. US Army Corps of Engineers, Washington, DC.

Merchant-Lariosa H, S Ruiz-Ramirezb, N Moreno-Mendozaa, A Marmolejo-Valencia. 1997. Correlation among thermosensitive period, estradiol response, and gonad differentiation in the sea turtle, *Lepidochelys olivacea*. Gen. Comp. Endocrinol. 107:373–385.

Merriam JC. 1919. Tertiary mammalian fauna of the Mohave Desert. Bull. Dept. Geol. Univ. California, 450–533.

Mertens R. 1964. Über Reptilienbastarde III. Senck. Biol. 45:33–49.

Mertz DB. 1971. The mathematical demography of the California Condor population. Am. Nat.105:437–453.

Meylan PA and W Sterrer. 2000. *Hesperotestudo* (Testudines: Testudinidae) from the Pleistocene of Bermuda, with comments on the pylogenetic position of the genus. Zool. J. Linn. Soc.-Lond. 128:51–76.

Milanovich JR, WE Peterman, NP Nibbelink, and JC Maerz. 2010. Projected loss of a salamander diversity hotspot as a consequence of projected global climate change. PLOS One 5(8):e12189.

Miller L. 1779. Incones animalium et plantarum. Letterpress, London. 10 pp., 60 pls.

Miller L. 1932. Notes on the desert tortoise (*Testudo agassizii*). Trans. San Diego Soc. Nat. Hist. 7:187–208.

Miller PS. 2001. Preliminary population viability assessment for the gopher tortoise (*Gopherus polyphemus*) in Florida. Conservation Breeding Specialist Group, Apple Valley, Minnesota.

Minnich JE. 1977. Adaptive responses in water and electrolyte budgets of native and captive desert tortoises, *Gopherus agassizii*, to chronic drought. Proc. Desert Tortoise Council Symp. 1977:102–129.

Minnich RA and AC Sanders. 2000. Brassica tournefortii Gouan. In Invasive Plants of California's Wildlands, ed. C.C. Bossard, J.M. Randall, and MC. Hoshovsky, 68–72. University of California Press, Berkeley.

Minnich JE and MR Ziegler. 1977. Water turnover of free-living gopher tortoises, *Gopherus polyphemus*, in central Florida. Proc. Desert Tortoise Council 1977:130–171.

Mitchell MJ. 2005. Home range, reproduction, and habitat characteristics of the female gopher tortoise (*Gopherus polyphemus*) in southeast Georgia. MS thesis, Georgia Southern University, Statesboro.

Mitchell N and F Janzen. 2010. Temperature-dependent sex determination and contemporary climate change. Sex. Dev. 4:129–140.

Mollhausen B. 1858. Diary of a Journey from the Mississippi to the Coasts of the Pacific with a United States Government Expedition. Longman, Brown, Green, Longmans, and Roberts, London.

Mones A. 1989. Nomen dubium vs nomem vanum. Journ. Vert. Paleont. 9:232–234.

Moon JC, ED McCoy, HR Mushinsky, and SA Karl. 2006. Multiple paternity and breeding systems in the gopher tortoise, *Gopherus polyphemus*. J. Hered. 97:150–157.

Moore JA, M Stratton, and V Szabo. 2009. Evidence for year-round reproduction in the gopher tortoise (*Gopherus polyphemus*) in southeastern Florida. Bull. Peabody Mus. Nat. Hist. 50:387–392.

Mooser O. 1972. A new species of Pleistocene fossil tortoise, genus *Gopherus*, from Aguascalientes, Mexico. Southwest. Nat. 17:561–65.

Mooser O. 1980. Pleistocene fossil turtles from Aguascalientes, State of Aguascalientes. Revista 4:63–66.

Mooser O and WW Dalquest. 1975. Pleistocene mammals from Aguascalientes, central Mexico. J. Mammal. 56:78–82.

Morafka DJ. 1982. The status and distribution of the Bolsón tortoise (*Gopherus flavomarginatus*). IN North American Tortoises: Conservation and Ecology, ed. RB Bury and GJ Germano, 71–94. US Fish and Wildlife Service Research Report 12.

Morafka DJ. 1988. Historical biogeography of the Bolsón tortoise. IN The Ecogeography of the Mexican Bolsón Tortoise (*Gopherus flavomarginatus*): Derivation of its Endangered Status and Recommendations for its Conservation, ed. DJ Morafka and CJ McCoy, 47–72. Ann. Carnegie Mus. 57.

Morafka DJ. 1994. Neonates: missing links in the life histories of North American tortoises, IN Biology of North American Tortoises, ed. RB Bury and DJ Germano, 161–173. US Fish and Wildlife Research 13

Morafka DJ, GA Adest, G Aguirre, and M Recht. 1981. The ecology of the Bolson tortoise, *Gopherus flavomarginatus*. IN Ecology of the Chihuahuan Desert, ed. R Barbault and G. Halffter, 35–78. Instituto de Ecologia, Mexico.

Morafka DJ, G Aguirre, and GA Adest. 1989. *Gopherus Agassizii* desert tortoise. IN The Conservation Biology of Tortoises, ed. IR Swingland and MK Klemens, 10–13. Occasional papers of the IUCN Species Survival Commission (SSC) IUCN, Gland Switzerland.

Morafka DJ, GL Aguirre, and RW Murphy. 1994. Allozyme differentiation among gopher tortoises (*Gopherus*): conservation genetics and phylogenetic and taxonomic implications. Can. J. Zool. 72:1665–1671.

Morafka DJ and KH Berry. 2002. Is *Gopherus agassizii* a desert-adapted tortoise, or an exaptive opportunist? Implications for tortoise conservation. Chelonian Conserv. Biol. 4:263–287.

Morafka DJ, KH Berry, and EK Spangenberg. 1997. Predator-proof field enclosures for enhancing hatching success and survivorship of juvenile tortoises: a critical evaluation. IN Proceedings: Conservation, Restoration, and Management of Tortoises and Turtles—an International Conference, ed. J Van Abbema, 147–165. New York Turtle and Tortoise Society and WCS Turtle Recovery Program, New York.

Morafka DJ, EK Spangenberg, and VA Lance. 2000. Neontology of reptiles. Herpetol. Monogr. 14:353–370.

Morreale SJ, GJ Ruiz, JR Spotila, and EA Standora. 1982. Temperature dependent sex determination: current practices threaten conservation of sea turtles. Science 216:1245–1247.

Mrosovsky N. 1994. Sex ratios of sea turtles. J. Exp. Zool. 270:16–27.

Mrosovsky N and MH Godfrey. 1995. Manipulating sex ratios: turtle speed ahead! Chelonian Conserv. Biol. 1:238–240.

Mrosovsky N and J Pieau. 1991. Transitional range of temperature, pivotal temperatures and thermosensititive stages for sex determination in reptiles. Amphib. Reptilia. 12:169–179.

Mrosovsky N and J Provancha. 1992. Sex ratio of hatchling loggerhead sea turtles: data and estimates from a 5-year study. Can. J. Zool. 70:530–538.

Mrosovsky N and CL Yntema. 1980. Temperature dependence of sexual differentiation in sea turtles: implications for conservation practices. Biol. Cons. 18:271–280.

Mueller JM, KR Sharp, KK Zander, DL Rakestraw, KR Rautenstrauch, and PE Lederle. 1998. Size-specific fecundity of the desert tortoise (*Gopherus agassizii*). J. Herpetol. 32:313–319.

Murphy RW. 1993 The phylogenetic analysis of allozyme data: invalidity of coding alleles by presence/absence and recommended procedures. Biochem. Syst. Ecol. 21:25–38.

Murphy R, K Berry, T Edwards, A Leviton, A Lathrop, and JD Riedle. 2011. The dazed and confused identity of Agassiz's land tortoise, *Gopherus agassizii* (Testudines: Testudinidae) with the description of a new species and its consequences for conservation. ZooKeys 113:39–71.

Murphy RW, KH Berry, T Edwards, and AM McLuckie. 2007. A genetic assessment of the recovery units for the Mojave population of the desert tortoise, *Gopherus agassizii*. Chelonian Conserv. Biol. 6:229–251.

Murphy RW and KD Doyle. 1998. Phylophenetics: frequencies and polymorphic characters in genealogical estimation. Syst. Biol. 47:737–761.

Murphy RW and NR Lovejoy. 1998. Punctuated equilibrium or gradualism in the lizard genus *Sceloporus*? Lost in plesiograms and a forest of trees. Cladistics 14:950103.

Murphy RW and FR Méndez de la Cruz. 2010. The herpetofauna of Baja California and its associated islands. IN A Conservation Assessment and Priorities in Conservation of Mesoamerican Amphibians and Reptiles, ed. LD Wilson, JH Townsend, and JD Johnson, 238–273. Eagle Mountain Publishing, LC, Eagle Mountain, UT.

Murray RC. 1997. *Gopherus agassizii,* desert tortoise diet. Herpetol. Rev. 28:87.

Murray RC, CR Schwalbe, SJ Bailey, SP Cuneo, and SD Hart. 1996. Reproduction in a population of the desert tortoise, *Gopherus agassizii,* in the Sonoran Desert. Herpetol. Nat. Hist. 4:83–88.

Mushinsky HR, and ED McCoy. 1994. Comparison of gopher tortoise populations on islands and on the mainland in Florida. IN Biology of North American Tortoises, ed. RB Bury and DJ Germano, 39–47. US Fish and Wildlife Research 13.

Mushinsky HR, ED McCoy, JE Berish, RE Ashton Jr., and DS Wilson. 2006. *Gopherus polyphemus*—gopher tortoise. IN Biology and Conservation of Florida Turtles, ed. PA Meylan, 350–375. Chelonian Res. Monogr. 3.

Mushinsky HR, TA Stilson, and ED McCoy. 2003. Diet and dietary preferences of the juvenile gopher tortoise. Herpetologica 59:477–485.

Mushinsky HR, DS Wilson, and ED McCoy. 1994. Growth and sexual dimorphism of *Gopherus polyphemus* in central Florida. Herpetologica 50:119–128.

Naegle SR. 1976. Physiological responses of the desert tortoise, *Gopherus agassizii*. MS thesis, University of Nevada, Las Vegas.

Nagy KA. 1977. Cellulose digestion and nutrient assimilation in *Sauromalus obesus,* a plant-eating lizard. Copeia. 1977:355–362.

Nagy KA. 2000. Energy Costs of Growth in Neonatal Reptiles. Herpetol. Monogr. 14:378–387.

Nagy KA, BT Henen, DB Vyas, and IR Wallis. 2002. A condition index for the desert tortoise (*Gopherus agassizii*). Chelonian Conserv. Biol. 4:425–429.

Nagy KA and PA Medica 1986. Physiological ecology of desert tortoises in Southern Nevada. Herpetologica 42:73–92.

Nagy KA, DJ Morafka, and RA Yates. 1997. Young desert tortoise survival: energy, water, and food requirements in the field. Chelonian Conserv. Biol. 2:396–404.

Nagy KA, MW Tuma, and LS Hillard. 2011. Shell hardness measurement in juvenile desert tortoises, *Gopherus agassizii*. Herpetological Review 42:191–195.

NCDC. 2012a. National Climatic Data Center, National Oceanic and Atmospheric Administration. Annual climatological summaries. Climate Data Online. www.ncdc.noaa.gov/cdo-web/.

NCDC. 2012b. National Climatic Data Center, National Oceanic and Atmospheric Association. GHCND (Global Historical Climatology Network), monthly summaries. Climate Data Online. www.ncdc.noaa.gov/cdo-web/.

Niblick HA, DC Rostal, and T Classen. 1994. Role of male-male interactions and female choice in the mating system of the desert tortoise, *Gopherus agassizii*. Herpetol. Monogr. 8:124–132.

Nichols JD, JE Hines, JR Sauer, FW Fallon, JE Fallon, and PJ Heglund. 2000. A double-observer approach for estimating detection probability and abundance from point counts. Auk 117:393–408.

Nichols RA. 1989. Mitochondrial DNA clues to gopher tortoise dispersal. Trends Ecol. Evol. 4:192–193.

Nichols UG. 1953. Habits of the desert tortoise, *Gopherus agassizii*. Herpetologica 9:65–69.

Nichols UG. 1957. The desert tortoise in captivity. Herpetologica 13:141–144.

Niño-Ramírez A, RY Benavides-Ruiz, A Guerra-Pérez, and D Lazcano-Villareal. 1999. Distribución y estructura poblacional de la tortuga de Berlandier (*Gopherus [=Xerobates] berlandieri*) en México. Universidad Autónoma de Nuevo León, Facultad de Ciencias Biológicas. Informe final SNIB-CONABIO proyecto No. H093. México DF.

Nomani SZ, RR Carthy, and MK Oli. 2008. Comparison of methods for estimating abundance of gopher tortoises. Appl. Herpetol. 5:13–31.

Nussear KE. 2004. Mechanistic investigation of the distributional limits of the desert tortoise *Gopherus agassizii*. PhD diss., EECB, University of Nevada, Reno.

Nussear KE, TC Esque, DF Haines, and CR Tracy. 2007. Desert tortoise hibernation: temperatures, timing, and environment. Copeia 2007:378–386.

Nussear KE, TC Esque, RD Inman, L Gass, KA Thomas, CSA Wallace, JB Blainey, DM Miller, and RH Webb. 2009. Modeling habitat of the desert tortoise (*Gopherus agassizii*) in the Mojave and parts of the Sonoran Deserts of California, Nevada, Utah, and Arizona. US Geological Survey open-file report 1102:18.

Nussear KE and CR Tracy. 2007. Can modeling improve estimation of desert tortoise population densities? Ecol. Appl. 17:579–86.

Nussear KE, CR Tracy, PA Medica, DS Wilson, RW Marlow, and PS Corn. 2012. Translocation as a conservation tool for desert tortoises: survivorship, reproduction, and movements. J. Wildl. Manage. 76:1341–1353.

O'Connor MP. 1999. Physiological and ecological implications of a simple model of heating and cooling in reptiles. J. Therm. Biol. 24:113–136.

O'Connor MP. 2000. Extracting operative temperatures from temperatures of physical models with thermal inertia. J. Therm. Biol. 25:329–343.

O'Connor MP, JS Grumbles, RH George, LC Zimmerman, and JR Spotila. 1994. Potential hematological and biochemical indicators of stress in free-ranging desert tortoises and captive tortoises exposed to a hydric stress gradient. Herpetol. Monogr. 8:5–26.

O'Connor MP, LC Zimmerman, EM Dzialowski, and JR Spotila. 2000. Thick-walled physical models improve estimates of operative temperatures for moderate to large-sized reptiles. J. Therm. Biol. 25:293–304.

Oftedal OT. 2002. Nutritional ecology of the desert tortoise in the Mojave and Sonoran deserts. IN The Sonoran Desert Tortoise: Natural History, Biology, and Conservation, ed. TR Van Devender, 194–241. University of Arizona Press. Tucson, AZ.

Oftedal OT, S Hillard, and DJ Morafka. 2002. Selective spring foraging by juvenile desert tortoises (*Gopherus agassizii*) in the Mojave Desert: Evidence of an adaptive nutritional strategy. Chelonian Conserv. Biol. 4:341–352.

Okamoto CL. 2002. An experimental assessment of color, calcium, and insect dietary preferences of captive juvenile desert tortoises *Gopherus agassizii*. Chelonian Conserv. Biol. 4:359–365.

Origgi FC, CH Romero, D Bloom, PA Klein, JM Gaskin, SJ Tucker, and ER Jacobson. 2004. Experimental transmission of a herpesvirus in Greek tortoises (*Testudo graeca*). Vet. Path. 41:50–61.

Osentoski MF and T Lamb. 1995. Intraspecific phylogeography of the gopher tortoise, *Gopherus polyphemus:* RFLP analysis of amplified mtDNA segments. Mol. Ecol. 4:709–718.

Osorio SR and RB Bury. 1982. Ecology and status of the desert tortoise (*Gopherus agassizii*) on Tuburon Island, Sonora. IN North American Tortoises: Conservation and Ecology, ed. RB Bury and DJ Germano, 39–49. US Fish and Wildlife Service Research Report 12.

Ott JA, MT Mendonca, C Guyer, and WK Michener. 2000. Seasonal changes in sex and adrenal steroid hormones of gopher tortoises (*Gopherus polyphemus*). Gen. Comp. Endocr. 117:299–312.

Ottley JR, and VM Velázques Solis. 1989. An extant, indigenous tortoise population in Baja California Sur, Mexico, with the description of a new species of *Xerobates* (Testudines: Testudinidae). Great Basin Natur. 49:496–502.

Overton WS. 1971. Estimating the numbers of animals in wild populations. IN Wildlife Management Techniques. 3rd ed., ed. RH Giles, 403–436. The Wildl. Soc., Washington, DC.

Packard GC, MJ Packard, and L Benigan. 1991. Sexual differentiation, growth, and hatching success by embryonic painted turtles incubated in wet and dry environments at fluctuating temperatures. Herpetologica 47, 125–132.

Packard GC, MJ Packard, K Miller, and TJ Boardman. 1987. Influence of moisture, temperature and substrate on snapping turtle eggs and embryos. Ecology 68:983–993.

Palmer KS, DC Rostal, JS Grumbles, and M Mulvey. 1998. Long-term sperm storage in the desert tortoise (*Gopherus agassizii*). Copeia 1998:702–705.

Patino-Martinez J, A Marco, L Quinones, and L Hawkes. 2012. A potential tool to mitigate the impacts of climate change to the Caribbean leatherback sea turtle. Global Change Biol. 18:401–411.

Patterson R. 1982. The distribution of the desert tortoise (*Gopherus agassizii*). IN North American Tortoises: Conservation and Ecology, ed. RB. Bury and DJ Germano, 51–55. US Fish and Wildlife Research Report 12.

Patterson R and B Brattstrom. 1972. Growth in captive *Gopherus agassizzi*. Herpetologica 28:169–171.

Pavao-Zuckerman B and VM LaMotta. 2007. Missionization and Economic Change in the Pimería Alta: The Zooarchaeology of San Agustín de Tucson. Int. J. Hist. Archaeol. 11:241–268.

Peet RK and DJ Allard. 1993. Longleaf pine vegetation of the southern Atlantic and eastern Gulf Coast regions: a preliminary classification. IN Proceedings of the Tall Timbers Fire Ecology Conference 18, ed. SM Hermann, 45–81. Tall Timbers Research Station, Tallahassee, FL.

Penick DN, J Congdon, JR Spotila, and JB Williams. 2002. Microclimates and energetics of free-living box turtles, *Terrapene carolina*, in South Carolina. Physiol. Biochem. Zool. 75:57–65.

Penninck DG, JS Stewart, J Paul-Murphy, P Pion. 1991. Ultrasonography of the California desert tortoise (*Xerobates agassizi*): anatomy and application. Vet. Radiol. 32:112–116.

Pepper C. 1963. The truth about the tortoise. Desert Magazine 26:10–11.

Perez-Heydrich C, MK Oli, and MB Brown. 2012. Population-level influence of a recurring disease on a long-lived wildlife host. Oikos 121:377–388.

Peterson CC. 1994. Different rates and causes of high mortality in two populations of the threatened desert tortoise *Gopherus agassizii*. Biol. Conserv. 70:101–108.

Peterson CC. 1996a. Ecological energetics of the desert tortoise (*Gopherus agassizii*): effects of rainfall and drought. Ecology 77:1831–1844.

Peterson CC. 1996b. Anhomeostasis: seasonal water and solute relations in two populations of the desert tortoise (*Gopherus agassizii*) during chronic drought. Physiol. Zool. 69:1324–1358.

Pettan-Brewer KCB, ML Drew, E Ramsay, FC Mohr, and LJ Lowenstine. 1996. Herpesvirus particles associated with oral and respiratory lesions in a California desert tortoise (*Gopherus agassizii*). J. Wildl. Dis. 32:521–526.

Pianka ER. 1988. Evolutionary Ecology. 4th ed. Harper and Row, New York. 468 pp.

Picco AM, AP Karam, and JP Collins. 2010. Pathogen host switching in commercial trade with management recommendations. EcoHealth 7:252–256.

Pieau C. 1972. Effects de la temperature sur le developpement des glandes genitales chez les embryons de dux chelonians, *Emys orbicularis* L. et *Testudo graeca*. L.C.R. Acad. Sci. Paris D274:719–722.

Pieau C and M Dorizzi. 1981. Determination of the temperature sensitive stages of sexual differentiation of the gonads in embryos of the turtle, *Emys orbicularis*. J. Morphol. 170:373–382.

Pike DA. 2006. Movement patterns, habitat use, and growth of hatchling tortoises, *Gopherus polyphemus*. Copeia 2006:68–76.

Pike DA and RA Seigel. 2006. Variation in hatchling tortoise survivorship at three geographic localities. Herpetologica 62:125–131.

Place AR and VA Lance. 2004. The temperature-dependent sex determination drama: same cast, different stars. IN Temperature dependent sex determination in vertebrates, ed. N Valenzuela and V Lance, 99–110. Smithsonian Books, Washington, DC.

Porter WP and DM Gates. 1969. Thermodynamic equilibria of animals with environment. Ecol. Monogr. 39:227–244.

Pough FH. 1973. Lizard energetics and diet. Ecology. 54:837–844.

Pough FH. 1983. Amphibians and Reptiles as Low Energy Systems. IN Behavioral Energetics, ed. WP Aspey and SI Lustick, 141–188. Ohio State University Press, Columbus.

Powell LA. 2007. Approximating variance of demographic parameters using the delta method: a reference for avian biologists. Condor 109:949–954.

Preston RE. 1979. Late Pleistocene cold-blooded vertebrate faunas from the mid-continental United States. I. Reptilia, Testudines, Crocodilia. Claude W Hibbard Memorial 6. Museum of Paleontology, University of Michigan.

Pritchard PCH. 1994. Cladism: the great delusion. Herpetol. Rev. 25:103–110.

Puckett C and R Franz. 1991. Gopher tortoise: a species in decline. Gopher Tortoise Council Special Publication through the Institute of Food and Agricultural Sciences. University of Florida, Gainesville.

Quinn RL, BD Tucker, and J Krigbaum. 2008. Diet and mobility in Middle Archaic Florida: stable isotopic and faunal evidence from the Harris Creek archaeological site (8Vo24), Tick Island. J. Archaeol. Sci. 35:2346–2356.

Quitmyer IR. 2001. Zooarchaeological analysis. Phase III. IN Mitigative Excavation at the Lake Monroe Outlet Midden (8VO53), Florida, ed. E. Horvath, 6.1–6.25. US Department of Transportation Federal Highway Administration, Washington DC.

Rabinow P. 1996. Making PCR: A Story of Biotechnology. University of Chicago Press, Chicago.

Rafinesque CS. 1832. Descriptions of two new genera of soft shell turtles of North America. Atlantic J. Friend Knowledge 1:64–65.

Rainboth WJ, DG Buth, and FB Turner. 1989. Allozyme variation in Mojave populations of the desert tortoise, *Gopherus agassizi* [*sic*]. Copeia 1989:115–123.

Ramírez RG, GFW Haenlein, CG García-Castillo, MA Núñez-Gonzalez. 2004. Protein, lignin and mineral contents and in situ dry matter digestibility of native Mexican grasses consumed by range goats. Small Ruminant Res. 52:261–269.

Rautenstrauch KR, ALH Rager, and DL Rakestraw. 1998. Winter behavior of desert tortoises in southcentral Nevada. J. Wildl. Manag. 62:98–104.

Reed JM, N Fefferman, RC Averill-Murray. 2009. Vital rate sensitivity analysis as a tool for assessing management actions for the desert tortoise. Biol. Conserv. 142:2710–2717.

Reitz EJ. 1991. Animal use and culture change in Spanish Florida. IN Animal Use and Culture Change, ed. P Crabtree and K Ryan, 62–78. MASCA Research Papers in Science and Archaeology, Suppl. to Vol. 8. University of Pennsylvania Museum, Philadelphia.

Retallech GJ. 1997. Neogene expansion of the North American prairie. Palaios 12:380–390.

Reynoso VH and M Montellano-Ballesteros. 2004. A new giant turtle of the genus *Gopherus* (Chelonia: Testudinidae) from the Pleistocene of Tamaulipas, Mexico, and a review of the phylogeny and biogeography of gopher tortoises. J. Vertebr. Paleontol. 24:822–837.

Rhen T and A Schroeder. 2010 Molecular Mechanisms of Sex Determination in Reptiles. Sex. Dev. 4:16–28.

Rhen T, K Metzger, A Schroeder, and R Woodward. 2007. Expression of putative sex-determining genes during the thermosensitive period of gonad development in the snapping turtle, *Chelydra serpentina*. Sex. Dev. 1:255–270.

Richter SC, JA Jackson, M Hinderliter, D Epperson, CW Theodorakis, and SM Adams. 2011. Conservation genetics of the largest cluster of federally threatened gopher tortoise colonies with implications for species management. Herpetologica 67:406–419.

Rick CM and RI Bowman. 1961. Galapagos tomatoes and tortoises. Evolution. 15:407–417.

Riedle JD, RC Averill-Murray, CL Lutz, and DK Bolen. 2008. Habitat use by desert tortoises (*Gopherus agassizii*) on alluvial fans in the Sonoran Desert, south-central Arizona. Copeia 2008:414–420.

Robbins CT. 1983. Wildlife Feeding and Nutrition. Academic Press, Orlando, FL.

Roberson JB, BL Burge, and P Hayden. 1985. Nesting observations of free-living desert tortoises (*Gopherus agassizii*) and hatching success of eggs protected from predators. Proc. Desert Tortoise Counc. Symp. 1985:91–99.

Rose FL and FW Judd. 1975. Activity and home range size of the Texas tortoise, *Gopherus berlandieri,* in South Texas. Herpetologica 31:448–456.

Rose FL and FW Judd. 1982. Biology and status of Berlandier's tortoise (*Gopherus berlandieri*). IN North American Tortoises: Conservation and Ecology, ed. RB Bury and DJ Germano 57–70. US Fish and Wildlife Research Report 12.

Rose FL and FW Judd. 1989. *Gopherus berlandieri*. Berlandier's tortoise, Texas tortoise. IN The Conservation Biology of Tortoises, ed. IR Swingland and MW Klemens. Occ. Pap. IUCN/SSC 5, Tortoise and Freshwater Specialists Group. IUCN, Gland, Switzerland, 8–9.

Rose FL and FW Judd. 1991. Egg size versus carapace-xiphiplastron aperture size in *Gopherus berlandieri*. J. Herpetol. 25:248–250.

Rose FL, FW Judd, and MF Small. 2011. Survivorship in two coastal populations of *Gopherus berlandieri*. J. Herpetol. 45:75–78.

Rose FL, J Koke, R Koehn, and D Smith. 2001. Identification of the etiological agent for necrotizing scute disease in the Texas tortoise. J. Wildl. Dis. 37:223–228.

Rostal DC, JS Grumbles, VA Lance, and JR Spotila. 1994a. Non-lethal sexing techniques for hatchling and immature desert tortoises (*Gopherus agassizii*). Herpetol. Monogr. 8:83–87.

Rostal DC and DN Jones. 2002. Population biology of the gopher tortoise (*Gopherus polyphemus*) in southeast Georgia. Chelonian Conserv. Biol. 4:479–487.

Rostal DC, VA Lance, JS Grumbles, and AC Alberts. 1994b. Seasonal reproductive cycle of the desert tortoise (*Gopherus agassizii*) in the Mojave Desert. Herpetol. Monogr. 8:72–82.

Rostal DC, TR Robeck, DW Owens, and DC Kraemer. 1990. Ultrasound imaging of ovaries and eggs in Kemp's ridley sea turtles (*Lepidochelys kempi*). J. Zoo. Wildl. Med. 21:27–35.

Rostal DC, T Wibbels, JS Grumbles, VA Lance, and JR Spotila. 2002. Chronology of sex determination in the desert tortoise (*Gopherus agassizii*). Chelonian Conserv. Biol. 4:313–318.

Rowe CL. 2008. "The calamity of so long life": life histories, contaminants, and potential emerging threats to long-lived vertebrates. Bioscience 58:623–631.

Royle JA and JD Nichols. 2003. Estimating abundance from repeated presence-absence data or point counts. Ecology 84:777–790.

Royle JA and KV Young. 2008. A hiererarchical model for spatial capture-recapture data. Ecology 89:2281–2289.

Ruby DE and HA Niblick. 1994. A behavioral inventory of the desert tortoise: development of an ethogram. Herpetol. Monogr. 8:88–102.

Ruby DE, LC Zimmerman, SJ Bulova, CJ Salice, MP O'Connor, and JR Spotila. 1994. Behavioral responses and time allocation differences in desert tortoises exposed to environmental stress in semi-natural enclosures. Herpetol. Monogr. 8:27–44.

Sayre NF and RL Knight. 2010. Potential effects of United States-Mexico border hardening on ecological and human communities in the Malpai Borderlands. Conserv. Biol. 24:345–348.

Scalise JL. 2011. Food habits and selective foraging by the Texas tortoise (*Gopherus berlandieri*). MS thesis, Texas State University, San Marcos.

Schamberger ML and FB Turner. 1986. The application of habitat modeling to the desert tortoise (*Gopherus agassizii*). Herpetologica 42:134–138.

Schmidt-Nielsen K. 1964. Desert Animals: Physiological Problems of Heat and Water. Oxford University Press, New York.

Schmidt-Nielsen K and PJ Bentley. 1966. Desert tortoise *Gopherus agassizii:* cutaneous water loss. Science 154:911.

Schneider PB. 1980. A comparison of three methods of population analysis of the desert tortoises, *Gopherus agassizi*. IN Proc. Desert Tortoise Council Symp., 156–162.

Schneider JS and GD Everson. 1989. The desert tortoise (*Xerobates agassizii*) in the prehistory of the southwestern Great Basin and adjacent areas. J. Calif. Great Basin Anthropol. 11:175–202.

Schoener TW 1971. Theory of feeding strategies. Ann. Rev. Ecol. Syst. 2:369–404.

Schumacher J. 2012. Chelonians (turtles, tortoises, and chelonians). IN Zoo Animal and Wildlife Immobilization and Anesthesia, ed. G West, D Heard, and N Caulkett, 259–266. Blackwell Publishing, Ames, IA.

Schumacher IM, MB Brown, ER Jacobson, BR Collins, and PA Klein. 1993. Detection of antibodies to a pathogenic mycoplasma in desert tortoises (*Gopherus agassizii*) with upper respiratory tract disease. J. Clin. Micro. 31:1454–1460.

Schumacher IM, DC Rostal, R Yates, DR Brown, ER Jacobson, and PA Klein. 1999. Persistence of maternal antibodies against *Mycoplasma agassizii* in desert tortoise hatchlings. Am. J. Vet. Res. 60:826–831.

Schwanz LE and FJ Janzen. 2008. Climate change and temperature-dependent sex determination: can individual plasticity in nesting phenology prevent extreme sex ratios? Physiol. Biochem. Zool. 81:826–834.

Schwartz DL, ed. 2008. Life in the Foothills: Archaeological Investigations in the Tortolita Mountains of Southern Arizona. Anthropological Papers 46. Center for Desert Archaeology, Tucson.

Schwartz TS and SA Karl. 2005. Population and conservation genetics of the gopher tortoise (*Gopherus polyphemus*). Conserv. Gen. 6:917–928.

Schwartz TS and SA Karl. 2008. Population genetic assignment of confiscated gopher tortoises. J. Wildl. Manage. 72: 254–259.

Schwarzkopf L. 1993. Costs of reproduction in water skinks. Ecology 74:1970–1981.

Schweigger AF. 1812. Prodromus Monographia Cheloniorum auctore

Schweigger. Königsberg. Arch. Naturwiss. Mathem. 1:271–368, 406–458.

Seager R, T Mingfang, I Held, Y Kushnir, J Lu, G Vecchi, H Huang, N Harnik, A Leetmaa, N Lau, C Li, J Velez, and N Naik. 2007. Model projections of an imminent transition to a more arid climate in southwestern North America. Science 316:1181–1184.

Seba A. 1734–1765. Locupletissimi rerum naturalium thesauri accurate description et iconibus artificiosissimus expressio, per universam physics historiam. Janssonio-Waesbergios, & Wetstenium, & Gul. Smith, Amsterdam.

Seber, GAF. 1965. A note on the multiple-recapture census. Biometrika 52:249–259.

Secretaría de Medio Ambiente y Recursos Naturales (SEMARNAT). 1994. Poder ejecutivo diario oficial de la Federación, Vol. 488 no. 10, Mexico, DF, Monday 16 May 1994.

Secretaría de Medio Ambiente y Recursos Naturales (SEMARNAT). 2000. Norma Oficial Mexicana, NOM-059-ECOL-2000. Protección ambiental. Especies natives de México de flora y fauna silvestres. Categorías de riesgo y específicamente para su inclusión, exclusión o cambio. Lista de especies en riesgo, México, DF, Mexico.

Secretaría de Medio Ambiente y Recursos Naturales (SEMARNAT). 2002. Norma Oficial Mexicana NOM-059-ECOL-2001, Protección ambiental—Especies nativas de México de flora y fauna silvestres—Categorías de riesgo y especificaciones para su inclusíon, exclusíon o cambio-Lista de especies en reisgo. Diarío Oficial de la Federación 582(4)2a sección:1–80.

Seigel RA and CK Dodd. 2000. Manipulation of turtle populations for conservation: halfway technologies or viable options? IN Turtle Conservation, ed. MW Klemens, 218–238. Smithsonian Institution Press, Washington, DC.

Seigel RA, RB Smith, and NA Seigel. 2003. Swine flu or 1918 pandemic? Upper respiratory tract disease and the sudden mortality of gopher tortoises (*Gopherus polyphemus*) on a protected habitat in Florida. J. Herpetol. 37:137–144.

Seltzer MD and KH Berry. 2005. Laser ablation ICP-MS profiling and semiquantitative determination of trace element concentrations in desert tortoise shells: documenting the uptake of elemental toxicants. Sci. Total Environ. 339:253–265.

Shaffer HB, NN FitzSimmons, A Georges, and AGH Rhodin, eds. 2007. Proceedings of a Workshop on Genetics, Ethics, and Taxonomy of Freshwater Turtles and Tortoises. Chelonian Research Monogr. 4.

Shaffer HB, P Meylan, and ML Mcknight. 1997. Tests of turtle phylogeny: molecular, morphological, and paleontological approaches. Syst. Biol. 46:235–268.

Sheffer RJ, PW Hedrick, and AL Velasco. 1999. Testing for inbreeding and outbreeding depression in the endangered Gila topminnow. Anim. Conserv. 2:121–129.

Sheppard DH and SM Yoshida. 1971. Social behavior in captive Richardson's ground squirrels. J. Mammal. 52:793–799.

Sherer LM. 1967. The name Mojave, Mohave: a history of its origin and meaning. Southern Calif. J. 49:1–36.

Shilling DG, TA Bewick, JF Gaffney, SK McDonald, CA Chase, and ERL Johnson. 1997. Ecology, physiology, and management of cogongrass (*Imperata cylindrica*). Florida Institute of Phosphate Research, Pub. no.03–107–140. 128 pp.

Shine R. 1999. Why is sex determined by nest temperature in many reptiles? Trends Ecol. Environ. 14:186–189.

Shmida A. 1985. Biogeography of the desert flora. IN Ecosystems of the World 12A: Hot Deserts and Shrublands, ed. Evanari, I. Noy-Meir, and D. W. Goodall, 23–77. Elsevier, Amsterdam.

Simpson GG. 1961. Principles of Animal Taxonomy. Columbia University Press, New York.

Sinclair CS, PJ Dawes, and RA Seigel. 2010. Genetic structuring of gopher tortoise (*Gopherus polyphemus*) populations on the Kennedy Space Center, FL, USA. Herpetol. Conserv. Biol. 5:189–195.

Sinervo B, F Mendez-de-la-Cruz, DB Miles. 2010. Erosion of lizard diversity by climate change and altered thermal niches. Science 328:894–899.

Sleeman JM and J Gaynor. 2000. Sedative and cardiopulmonary effects of medetomidine and reversal with atipamezole in desert tortoises (*Gopherus agassizii*). J. Zoo. Wildl. Med. 31:28–35.

Small CR and LA Macdonald. 2001. Reproduction and growth in relocated and resident gopher tortoises (*Gopherus polyphemus*) on reclaimed phosphate mined land. Florida Institute for Phosphate Research, Publ No. 93-03-105R.

Smith HM and RB Smith. 1979. Synopsis of the Herpetofauna of Mexico. Guide to Mexican turtles, Bibliographic add. III, Vol. 6. John Johnson, North Bennington, VT.

Smith KR, JA Hurley, and RA Seigel. 1997. Reproductive biology and demography of gopher tortoises (*Gopherus polyphemus*) from the western portion of their range. Chelonian Conserv. Biol. 2:596–600.

Smith LL. 1995. Nesting ecology, female home range and activity, and population size-class structure of the gopher tortoise, *Gopherus polyphemus*, on the Katherine Ordway Preserve, Putnam County, Florida. Bull. FL. Mus. Nat. Hist. 37:97–126.

Smith LL, TD Tuberville, and RA Seigel. 2006. Workshop on the ecology, status and management of the gopher tortoise (*Gopherus polyphemus*). Joseph W. Jones Ecological Research Center, 16–17 January 2003: final results and recommendations. Chelonian Conserv. Biol. 5:326–330.

Smith RB, DR Breininger, and VL Larson. 1997. Home range characteristics of radiotagged gopher tortoises on Kennedy Space Center, Florida. Chelonian Conserv. Biol. 2:358–362.

Smith RB, RA Seigel, and KR Smith. 1998. Occurrence of upper respiratory tract disease in gopher tortoise populations in Florida and Mississippi. J. Herpet. 32:426–430.

Smith RB, TD Tuberville, AL Chambers, KM Herpich, and JE Berish. 2005. Gopher tortoise burrow surveys: external characteristics, burrow cameras, and truth. Appl. Herpetol. 2:161–170.

Sokal O. 1971. Lithophagy and geophagy in reptiles. J. Herpetol. 5:69–71.

Sorrie, BA, JB Gray, and PJ Crutchfield. 2006. The vascular flora of the longleaf pine ecosystem of Fort Bragg and Weymouth Woods, North Carolina. Castanea 71:129–161.

Spalinger DE and NT Hobbs. 1992. Mechanisms of foraging in mammalian herbivores: new models of functional response. Am. Nat. 140:325–348.

Speake DW and RH Mount. 1973. Some possible ecological effects of rattlesnake roundups in the southeastern coastal plain. Proc. Annu. Conf. S.E. Assoc. Game and Fish Comm. 27:267–277.

Speakman JR. 1997. Doubly Labeled Water: Theory and Practice. Chapman and Hall, New York.

Spinks PQ, HB Shaffer, JB Iverson, and WP McCord. 2004. Phylogenetic hypotheses for the turtle family Geoemydidae. Mol. Phylogenet. Evol. 32:164–182.

Spotila JR, OH Soule, and DM Gates. 1972. The biophysical ecology of the alligator, energy budgets and climate spaces. Ecology 43:1094–1102.

Spotila JR, LC Zimmerman, CA Binckley, JS Grumbles, DC Rostral, A List Jr., EC Beyer, KM Phillips, and SJ Kemp. 1994. Effects of incubation conditions on sex determination, hatchling success, and growth of hatchling desert tortoises, *Gopherus agassizii*. Herpetol. Monogr. 8:103–116.

Stain WS. 1996. Balcan mammalian fauna and Pleistocene formations Hudspeth County, Texas. Bull. Tex. Mem. Mus. 10:1–31.

Steinberg MB, AL Finelli, RW Gerwien, and HB John-Alder. 1993. Behavior effects of thyroxine in a lizard *Ameiva undulate teiidae,* Physiol. Zool. 66:148–165.

Stejneger L. 1893. Annotated list of the reptiles and batrachians collected by the Death Valley Expedition in 1891, with descriptions of new species. N. Amer. Fauna 7:159–228.

Stephens DW and JR Krebs. 1986. Foraging Theory. Princeton University Press, New Jersey.

Stevens JM and JS Fehmi. 2009. Competitive effect of two nonnative grasses on a native grass in southern Arizona. Invas. Plant Sci. Manag. 2:379–385.

Stevens PA, WJ Sutherland, and RP Freckleton. 1999. What is the Allee effect? Oikos 87:185–190.

Stitt EW. 2004. Demography, reproduction, and movements of desert tortoises (*Gopherus agassizii*) in the Rincon Mountains, Arizona. MS thesis, University of Arizona.

Stitt EW and C Davis. 2003. *Gopherus agassizii,* desert tortoise: caliche mining. Herpetol. Rev. 34:57.

Stober JM and LL Smith. 2010. Total counts versus line transects for estimating abundance of small gopher tortoise populations. J. Wildl. Manage. 74:1595–1600.

Stoleson SH, RS Felger, G Ceballos, C Raish, MF Wilson, and A Burquez. 2005. Recent history of natural resource use and population growth in northern Mexico. IN Biodiversity, Ecosystems, and Conservation in Northern Mexico, ed. JE Cartron, G Ceballos, and RS Felger, 52–85. Oxford University Press.

Styrsky JN, C Guyer, H Balbach, and A Turkmen. 2010. The relationship between burrow abundance and area as a predictor of gopher tortoise population size. Herpetologica 66:403–410.

Suzan H, G Malda, DT Patten, and GP Nabhan. 1999. Effects of exploitation and park boundaries on legume trees in the Sonoran Desert. Conserv. Biol. 13:1497–1501.

Switak KH. 1973. California's desert tortoise. Pac. Disc. 26:9–15.

Taylor RW. 1982a. Seasonal aspects of the reproductive biology of the gopher tortoise, *Gopherus polyphemus.* PhD diss., University of Florida, Gainesville. 102 pp.

Taylor RW. 1982b. Human predation on the gopher tortoise (*Gopherus polyphemus*) in north-central Florida. Bull. FL St. Mus. Biol. Sci. 28:79–102.

Taylor RW and ER Jacobson. 1981. Hematology and serum chemistry of the gopher tortoise, *Gopherus polyphemus.* Comp. Biochem. Physiol. 2A:425–428.

Tedford RH, LB Albright III, AD Barnosky, I Ferrusquia-Villafranca, RM Hunt, JE Storer, CC Swisher III, MR Voorhies, SD Webb, DP Whistler. 2004. Mammalian biochronology of the Arikareean through Hemphillian interval (Late Oligocene through early Pliocene epochs). IN Late Cretaceous and Cenozoic Mammals of North America, ed. MO Woodburne, 169–231. Columbia University Press, New York.

Templeton AR, K Shaw, E Routman, and SK Davis. 1990. The genetic consequences of habitat fragmentation. Ann. Mo. Bot. Gard. 77:13–27.

Tenneson M. 1985. Crawling out of limbo. Int. Wildl. 15(4):30–34.

Tessier N, S Paquette, and FJ Lapointe. 2005. Conservation genetics of the wood turtle (*Glyptemys insculpta*) in Quebec, Canada. Can. J. Zool. 83:765–772.

Thomas KA and P Guertin. 2007. Southwest Exotic Mapping Program 2007: Occurrence Summary and Maps of Select Invasive, Non-native Plants in Arizona. US Geol. Sur., Southwest Biological Science Center, Sonoran Desert Research Station and University of Arizona, School of Natural Resources. 85 pp.

Thomas L, ST Buckland, EA Rexstad, JL Laake, S Strindberg, SL Hedley, JRB Bishop, TA Marques, and KP Burnham. 2010. Distance software: design and analysis of distance sampling surveys for estimating population size. J. Appl. Ecol. 47:5–14.

Thomson RC and HB Shaffer. 2010. Sparse supermatrices for phylogenetic inference: taxonomy, alignment, rogue taxa, and the phylogeny of living turtles. Syst. Biol. 59:42–58.

Timmerman WW and RE Roberts. 1994. *Gopherus polyphemus* (gopher tortoise) maximum size. Herpetol. Review 25:64.

Tom J. 1988. The daily activity pattern, microhabitat, and home range of hatchling Bolson tortoises, *Gopherus flavomarginatus.* MS thesis, California State University, Los Angeles.

Tom J. 1994. Microhabitats and use of burrows of Bolson tortoise hatchlings. IN Biology of North American Tortoises, ed. RB Bury and DJ Germano, 139–146. Dep. US Fish and Wildlife Research 13.

Tracy CR. 1982. Biophysical modeling in reptilian physiology and ecology. IN Biology of the Reptilia, ed. C Gans and FH Pough, 275–321. Academic Press, New York.

Tracy CR and CR Tracy. 1995. Estimating age of desert tortoises (*Gopherus agassizii*) from scute rings. Copeia 1995:964–966.

Tracy CR, RC Averill-Murray, WI Boarman, DJ Delehanty, JS Heaton, ED McCoy, DJ Morafka, KE Nussear, BE Hagerty, and PA Medica. 2004. Desert Tortoise Recovery Plan Assessment. US Fish and Wildlife Service, Reno, NV.

Tracy CR, KE Nussear, TC Esque, K. Dean-Bradley, CR Tracy, LA DeFalco, KT Castle, LC Zimmerman, RE Espinoza, and AM Barber. 2006. The importance of physiological ecology in conservation biology. Integ. Comp. Biol. 46:1191–1205.

Traphagen M, S Hillard, R Kiester, J Juvik, T Leuteritz, L Hernandez, J Laundre, E Rodriguez, E Goode, M Rodrigues, M McCullough, M Phillips, and J Truett. August 2007. Recommendation for IUCN Red Listing Endangered Status for the Bolson tortoise (*Gopherus flavomarginatus*). Unpublished.

Trevino MA, ME Haro, SL Barrett, and CR Schwalbe. 1994. Preliminary desert tortoise surveys in central Sonora, Mexico. Proc. Desert Tortoise Council Symp., 379–388.

Trevino E, DJ Morafka, and G Aguirre. 1997. A second reserve for the Bolson tortoise, *Gopherus flavomarginatus,* at Rancho Sombreretillo, Chihuahua, Mexico. IN Proceedings: Conservation, Restoration, and Management of Tortoises and Turtles—An International Conference, ed. J Van Abbema, 417–420. New York Turtle and Tortoise Society.

Trowbridge CC. 1952. Untitled letter. FL Wildl. 5(9):4.

Troyer K. 1982. Transfer of fermentative microbes between generations in a herbivorous lizard. Science 216:540–542.

Troyer K. 1991. Role of microbial cellulose degradation in reptile nutrition. IN Biosynthesis and Biodegradation of Cellulose, ed. CH Haigler and PJ Weimer, 311–325. Marcel Dekker, New York.

True FW. 1881. On the North American land tortoises of the genus *Xerobates.* Proc. US Natl. Mus. 4:434–449.

Truett J and M Phillips. 2009. Beyond historic baselines: restoring Bolson tortoises to Pleistocene range. Ecol. Restor. 27(2):144–151.

Tuberville TD, KA Buhlmann, HE Balbach, SH Bennett, JP Nestor, JW Gibbons, and RR Sharitz. 2007. Habitat selection by the gopher tortoise (*Gopherus polyphemus*). Report to US Army Corps of Engineers, Construction Engineering Research Laboratory. ERDC/CERL TR-07-01.

Tuberville TD, EE Clark, KA Buhlmann, and JW Gibbons. 2005. Translocation as a conservation tool: site fidelity and movement of repatriated gopher tortoises (*Gopherus polyphemus*). Anim. Conserv. 8:349–358.

Tuberville TD, JW Gibbons, and HE Balbach. 2009. Estimating viability of gopher tortoise populations. US Army Corps of Engineers, Washington, DC.

Tuberville TD, TM Norton, BD Todd, and JS Spratt. 2008. Long-term apparent survival of translocated gopher tortoises: a comparison of newly released and previously established animals. Biol. Conserv. 141:2690–2697.

Tuberville TD, TM Norton, BJ Waffa, C Hagen, and TC Glenn. 2011. Mating system in a gopher tortoise population established through multiple translocations: apparent advantage of prior residence. Biol. Conserv. 144:175–183.

Turner FB and KH Berry. 1984. Appendix 3: Methods used in analyzing desert tortoise populations. IN The Status of the Desert Tortoise (*Gopherus agassizii*) in the United States, ed. KH Berry, 118–153. Report to the US Fish and Wildlife Service from the Desert Tortoise Council. Order no. 11310–0083–81.

Turner FB, KH Berry, DC Randall, and GC White. 1987a. Population ecology of the desert tortoise at Goffs, California, 1983–1986. Report to Southern California Edison Company, Rosemead, CA.

Turner FB, PA Medica, and CL Lyons. 1980. A comparison of populations of the desert tortoise (*Gopherus agassizii*) in grazed and ungrazed areas in Ivanpah Valley, California. Bureau of Land Management Report.

Turner FB, PA Medica, and CL Lyons. 1984. Reproduction and survival of the desert tortoise (*Scaptochelys agassizii*) in Ivanpah Valley, California. Copeia 1984:811–820.

Turner FB, P Hayden, BL Burge, and JB Roberson. 1986. Egg production by the desert tortoise (*Gopherus agassizii*) in California. Herpetol. 42:93–104.

Turner FB, PA Medica, and RB Bury. 1987b. Age-size relationships of desert tortoises (*Gopherus agassizzi*) in southern Nevada. Copeia 1987:974–979.

Turner RM. 1994. Mojave desertscrub. IN Biotic Communities of the American Southwest—United States and Mexico, ed. DE Brown, 157–168. University of Utah Press, Salt Lake City.

Turner RM and DE Brown. 1994. Sonoran desertscrub. IN Biotic Communities of the American Southwest—United States and Mexico, ed. DE Brown, 181–221. University of Utah Press, Salt Lake City.

Turtle Conservation Coalition. 2011. Turtles in Trouble: The World's 25+ Most Endangered Tortoises and Freshwater Turtles—2011, ed. AG Rhodin, AD Walde, BD Horne, PP van Dijk, T Blanck, and R Hudson. IUCN/SSC Tortoise and Freshwater Turtle Specialist Group, Turtle Conservation Fund, Turtle Survival Alliance, Turtle Conservancy, Chelonian Research Foundation, Conservation International, Wildlife Conservation Society, and San Diego Zoo Global, Lunenburg, MA.

Ultsch GR. 1989. Ecology and physiology of hibernation and overwintering among freshwater fishes, turtles, and snakes. Biol. Rev. 64:435–515.

Ultsch GR and JF Anderson. 1986. The respiratory microenvironment within the burrows of Gopher tortoises (*Gopherus polyphemus*). Copeia 1986:787–795.

Ultsch GR and JF Anderson. 1988. Gas exchange during hypoxia and hypercarbia of terrestrial turtles: a comparison of a fossorial species (*Gopherus polyphemus*) with a sympatric nonfossorial species (*Terrepene carolina*). Physiol. Zool. 61:142–152.

Une Y, K Uemura, Y Nakano, J Kamiie, T Ishibashi, and Y Nomura. 1999. Herpesvirus infection in tortosies (*Malacochersus tornieri* and *Testudo horsfieldii*). Vet. Pathol. 36:624–627.

Ureña-Aranda CA and AE de los Monteros. 2012. The genetic crisis of the Mexican Bolson Tortoise (*Gopherus flavomarginatus:* Testudinidae). Amphibia-Reptilia 3:45–53.

US Census Bureau. 2010. US census 2010. http://2010.census.gov/2010census.

US District Court (USDC). 2009. Order summary judgment motions. Case 3:06-cv-04884-SI. Center for Biological Diversity, et al., Plaintiffs v. BLM. US District Court for the Northern District of California. San Francisco, CA. www.biologicaldiversity.org/programs/public_lands/deserts/california_desert_conservation_area/pdfs/WEMO_NECO_case_order_9–28–09.pdf.

US District Court (USDC). 2011. Order re: remedy. Case 3:06-cv-04884-SI. Center for Biological Diversity, et al., Plaintiffs v. BLM. US District Court for the Northern District of California. San Francisco, CA.

US Fish and Wildlife Service (USFWS). 1980. Endangered and threatened wildlife and plants: listing as threatened with critical habitat for the Beaver Dam Slope population of the desert tortoise in Utah. Federal Register 52:25376–25380.

US Fish and Wildlife Service (USFWS). 1985. Endangered and threatened wildlife and plants: finding on desert tortoise petition. Federal Register 50:49868–49870.

US Fish and Wildlife Service. 1987. Endangered and threatened wildlife and plants: determination of threatened status for the gopher tortoise (*Gopherus polyphemus*) finding on gopher tortoise petition. Federal Register 50:49868–49870.

US Fish and Wildlife Service (USFWS). 1990a. Gopher tortoise recovery plan. US Fish and Wildlife Service, Jackson, MS.

US Fish and Wildlife Service (USFWS). 1990b. Endangered and threatened wildlife and plants: determination of threatened status for the Mojave population of the desert tortoise. Federal Register 55:12178–12191.

US Fish and Wildlife Service (USFWS). 1994a. Desert tortoise (Mojave population) recovery plan. Portland, OR, US Fish and Wildlife Service.

US Fish and Wildlife Service (USFWS). 1994b. Endangered and threatened wildlife and plants: determination of critical habitat for the Mojave population of the desert tortoise. Federal Register 59:5820–5866.

US Fish and Wildlife Service (USFWS). 2006. Range-wide Monitoring of the Mojave Population of the Desert Tortoise: 2001–2005. USFWS, Desert Tortoise Recovery Office, Reno, NV. www.fws.gov/nevada/desert_tortoise/documents/reports/rangewide_monitoring_report_20061024.pdf.

US Fish and Wildlife Service (USFWS). 2009a. Endangered and threatened wildlife and plants: 90-day finding on a petition to list the eastern population of gopher tortoise (*Gopherus polyphemus*) as threatened. Federal Register 74:46401–46406.

US Fish and Wildlife Service (USFWS). 2009b. Guidelines for the establishment, management, and operation of gopher tortoise conservation banks. US Fish and Wildlife Service, Washington, DC.

US Fish and Wildlife Service (USFWS). 2010a. Endangered and threatened wildlife and plants: 12-month finding on a petition to list the Sonoran population of the desert tortoise as endangered or threatened; proposed rule. Federal Register 75:78094–78146.

US Fish and Wildlife Service (USFWS). 2010b. Mojave population of the desert tortoise (*Gopherus agassizii*). 5-year review: summary and evaluation. Reno, NV, Desert Tortoise Recovery Office, US Fish and Wildlife Service, September 30, 2010.

US Fish and Wildlife Service (USFWS). 2011a. Endangered and threatened wildlife and plants: 12-month finding on a petition to list the gopher tortoise as threatened in the eastern portion of its range. Federal Register 76:45130–45162.

US Fish and Wildlife Service (USFWS). 2011b. Revised recovery plan for the Mojave population of the desert tortoise (*Gopherus agassizii*). US Fish and Wildlife Service, Sacramento, CA. 227 pp.

US Fish and Wildlife Service (USFWS). 2011c. Health Assessment Procedures for the Desert Tortoise (*Gopherus agassizii*): A Handbook Pertinent to Translocation. US Fish and Wildlife Service, Desert Tortoise Recovery Office, Reno, NV.

Van Devender, TR 2002a. Natural history of the Sonoran tortoise in Arizona: life in a rock pile. IN The Biology of the Sonoran Desert

Tortoise: Natural History, Biology, and Conservation, ed. TR Van Devender, 3–28. University of Arizona Press. Tucson, AZ.

Van Devender, TR. 2002b. Cenozoic environments and the evolution of the gopher tortoises (Genus *Gopherus*). IN The Sonoran Desert Tortoise: Natural History, Biology, and Conservation, ed. TR Van Devender, 29–51. University of Arizona Press, Tucson, AZ.

Van Devender TR, RC Averill-Murray, TC Esque, PA Holm, VM Dickinson, CR Schwalbe, EB Wirt, and SL Barrett. 2002. Grasses, mallows, desert vine, and more: diet of the desert tortoise in Arizona and Sonora. IN The Sonoran Desert Tortoise: Natural History, Biology, and Conservation, ed. TR Van Devender, 159–193. University of Arizona Press, Tucson, AZ.

van Dijk PP and O Flores-Villela. 2007. *Gopherus flavomarginatus.* IN IUCN 2010. IUCN Red List of Threatened Species. Version 2010.4. www.iucnredlist.org.

Van Lear DH, WD Carroll, PR Kapeluck, and R Johnson. 2005. History and restoration of the longleaf pine-grassland ecosystem: implications for species at risk. Forest Ecol. Manag. 211:150–165.

Vogt RC. 1994. Temperature controlled sex determination as a tool for turtle conservation. Chelonian Conserv. Biol. 1:159–162.

Voigt WG. 1975. Heating and cooling rates and their effects upon heart rate and subcutaneous temperatures in the desert tortoise, *Gopherus agassizii.* Comp. Biochem. Physiol. 52A:527–531.

Voigt WG and CR Johnson. 1976. Aestivation and thermoregulation in the Texas tortoise, *Gopherus berlandieri.* Comp. Biochem. Physiol. 53A:41–44.

Voigt WG and CR Johnson. 1977. Physiological control of heat exchange rates in the Texas tortoise, *Gopherus berlandieri.* Comp. Biochem. Physiol. 56A:495–498.

von Seckendorff Hoff K, and RW Marlow. 2002. Impacts of vehicle road traffic on desert tortoise populations with consideration of conservation of tortoise habitat in southern Nevada. Chelonian Conserv. Biol. 4:449–456.

Wake DB, MH Wake and CD Specht. 2011. Homoplasy: From detecting pattern to determining process and mechanisms in evolution. Science 331: 1032–1035.

Wake MH. 1993. The skull as locomotor organ. IN The Skull, Vol. 3, Functional and Evolutionary Mechanisms, ed. J Hanken and BK Hall, 197–240. University of Chicago Press, Chicago.

Waldbauer, GP and S Friedman. 1991. Self-selection of optimal diets by insects. Annu. Rev. Entomol. 36:43–63.

Walde AD, ML Harless, DK Delaney, and LL Pater. 2006. *Gopherus agassizii,* desert tortoise. Diet. Herpetol. Rev. 37:77–78.

Walde AD, DK Delaney, ML Harless, and LL Pater. 2007. Osteophagy by the desert tortoise, *Gopherus agassizii.* Southwest. Nat. 52:147–149.

Walker J and RK Peet. 1984. Composition and species diversity of pine-wiregrass savannas of the Green Swamp, North Carolina. Vegetatio 55:163–179.

Walker K. 1985. Kingsley Plantation and subsistence patterns of the southeastern coastal slave. IN Indians, Colonists, and Slaves: Essays in Memory of Charles H. Fairbanks, ed. KW Johnson, JM Leader, and RC Wilson, 11–34. University of Florida Press, Gainesville.

Wallis IR, BT Henen, and KA Nagy. 1999. Egg size and annual egg production by female desert tortoises (*Gopherus agassizii*): the importance of food abundance, body size, and date of egg shelling. J. Herpetol. 33:394–408.

Ware S, C Frost, and PD Doerr. 1993. Southern mixed hardwood forest: the former longleaf pine forest. IN Biodiversity of the southeastern United States: Lowland Terrestrial Communities, ed. WH Martin, SG Boyce, and AC Echternacht, 447–493. John Wiley & Sons, New York.

Weaver WG. 1970. Courtship behavior in *Gopherus berlandieri.* Bull. FL State Mus.15:1–43.

Wendland LD, LA Zacher, PA Klein, DR Brown, D Decomvitz, R Littell, and MB Brown. 2007. Improved enzyme-linked immunosorbent assay to reveal *Mycoplasma agassizii* exposure: a valuable tool in the management of environmentally sensitive tortoise populations. Clinical and Vac. Immunol. 14:1190–1195.

Wendland L, H Balbach, M Brown, JD Berish, R Littell, and M Clark. 2009. Handbook on Gopher Tortoise (*Gopherus polyphemus*). US Army Corps of Engineers, Engineer Research and Development Center, CERL TR-09-01.

Westgate JW. 1989. Mass occurrence of the giant gopher tortoise (*Gopherus hexagonatus*) in the late Pleistocene Beaumont Formation, Willacy County, Texas. J. Vert. Paleont. 9(3 Supplement): 44A.

Westhouse RA, ER Jacobson, RK Harris, KR Winter, and BL Homer. 1996. Respiratory and pharyngo-esophageal iridovirus infection in a gopher tortoise (*Gopherus polyphemus*). J. Wildl. Dis. 32:682–686.

Westoby M. 1974. An analysis of diet selection by large generalist herbivores. Am. Nat. 108:290–304.

Wever EG. 1979. The Reptile Ear. Princeton University Press, Princeton, NJ.

White GC, DR Anderson, KP Burnham, and DL Otis. 1982. Capture-recapture and removal methods for sampling closed populations. LA-8787-NERP. Los Alamos National Laboratory, Los Alamos, NM.

Wibbels T. 2003. Critical approaches to sex determination in sea turtles. The Biology of Sea Turtles 2:103–134.

Wibbels T. 2007. Sex determination and sex ratios in ridley turtles. In Biology and Conservation of Ridley Sea Turtles, ed. P. Plotkin, 167–189. Johns Hopkins University Press, Baltimore, MD.

Wibbels T, JJ Bull, and D Crews. 1991a. Chronology and morphology of temperature-dependent sex determination. J. Exp. Zool. 260:371–381.

Wibbels T, RE Martin, DW Owens, and MS Amoss Jr. 1991b. Female biased sex ratio of immature loggerhead sea turtles inhabiting the Atlantic coastal waters of Florida. Can. J. Zool. 69:2973–2977.

Wibbels T, DW Owens, P Licht, CJ Limpus, PC Reed, and MS Amoss Jr. 1992. Serum gonadotropins and gonadal steroids associated with ovulation and egg production in sea turtles. Gen. Comp. Endocrinol. 87:71–78.

Wibbels T, JJ Bull, and D Crews. 1994. Temperature-dependent sex determination: a mechanistic approach. J. Exp. Zool. 270:71–78.

Wibbels T, D Rostal, and R Byles. 1998. High pivotal temperature in the sex determination of the olive ridley sea turtle from Playa Nancite, Costa Rica. Copeia. 1998:1086–1088.

Wiley EO. 1978. The evolutionary species concept reconsidered. Syst. Biol. 27:17–26.

Williams DA and MF Osentoski. 2007. Genetic considerations for the captive breeding of tortoises and freshwater turtles. Chelonian Conserv. Biol. 6(2):302–313.

Williams EE. 1950a. *Testudo cubensis* and the evolution of western hemisphere tortoises. B. Am. Mus. Nat. Hist. 95:1–36.

Williams EE. 1950b. Variation and selection in the cervical central articulations in living turtles. B. Am. Mus. Nat. Hist. 94:505–561.

Williams EE. 1952. A new fossil tortoise from Mona Island, West Indies, and a tentative arrangement of the tortoises of the world. B. Am. Mus. Nat. Hist. 99:541–560.

Wilson DS. 1991. Estimates of survival for juvenile gopher tortoises, *Gopherus polyphemus.* J. Herpetol. 25:376–379.

Wilson DS, DJ Morafka, CR Tracy, and KA Nagy. 1999a. Winter activity of juvenile desert tortoises (*Gopherus agassizii*) in the Mojave Desert. J. Herpetol. 33:496–501.

Wilson DS, HR Mushinsky, and RA Fisher. 1997. Species profile: gopher tortoise (*Gopherus polyphemus*) on military installations in the southeastern United States. Technical report SERDP-97-10, United States Army Corps of Engineers, Vicksburg, MS.

Wilson DS, HR Mushinsky, and ED McCoy. 1994. Home range, activity and use of burrows of juvenile gopher tortoises in central Florida. IN Biology of North American Tortoises, ed. RB Bury and DJ Germano.147–160. US Fish and Wildlife Research 13.

Wilson DS, KA Nagy, CR Tracy, DJ Morafka, and RA Yates. 2001. Water balance in neonate and juvenile desert tortoises, *Gopherus agassizii*. Herpetol. Monogr. 15:158–170.

Wilson DS, CR Tracy, KA Nagy, and DJ Morafka. 1999b. Physical and microhabitat characteristics of burrows used by juvenile desert tortoises (*Gopherus agassizii*). Chelonian Conserv. Biol. 3:448–453.

Wilson DS, CR Tracy, and CR Tracy. 2003. Estimating age of turtles from growth rings: a critical evaluation of the technique. Herpetologica 59:178–194.

Wilson EO. 1975. Sociobiology: The New Synthesis. Harvard University Press, Cambridge, MA.

Wilson KR and DR Anderson. 1985. Evaluation of two density estimators for small mammal population size. J. Mammal. 66:13–21.

Wirt EB and PA Holm. 1997. Climatic effects on survival and reproduction of the desert tortoise (*Gopherus agassizii*) in the Maricopa Mountains, Arizona. IIPAM Report I92035 to Arizona Game and Fish Department, Phoenix.

Witz BW, DS Wilson, and MD Palmer. 1991. Distribution of *Gopherus polyphemus* and its vertebrate symbionts in three burrow categories. Am. Midl. Nat. 126:152–158.

Witz BW, DS Wilson, and MD Palmer. 1992. Estimating population size and hatchling mortality of *Gopherus polyphemus*. FL Sci. 55:14–19.

Wolf KN, CA Harms, and JF Beasley. 2008. Evaluation of five clinical chemistry analyzers for use in health assessment in sea turtles. J. Amer. Vet. Med. Assoc. 233:470–475.

Woodburne MO. 2004. Late Cretaceous and Cenozoic Mammals of North America: Stratigraphy and Geochronology. Columbia University Press, New York.

Woodbury AM. 1952. Hybrids of *Gopherus berlandieri* and *G. agassizii*. Herpetologica 8:33–36.

Woodbury AM and R Hardy. 1948. Studies of the desert tortoise, *Gopherus agassizii*. Ecol. Monogr. 18:145–200.

Wright JS. 1982. Distribution and population biology of the gopher tortoise, *Gopherus polyphemus*, in South Carolina. MS thesis, Clemson University.

Yager LY, MG Hinderliter, CD Heise, and DM Epperson. 2007. Gopher tortoise response to habitat management by prescribed burning. J. Wildl. Manage. 71:428–434.

Yntema CL. 1968. A series of stages in the embryonic development of *Chelydra serpentina*. J. Morphol. 125:219–252.

Yntema CL. 1979. Temperature levels and periods of sex determination during incubation of eggs of *Chelydra serpentina*. J. Morphol. 159:17–27.

Yntema CL and N Mrosovsky. 1982. Critical periods and pivotal temperatures for sexual differentiation in loggerhead sea turtles. Can. J. Zool. 60:1012–1016.

Young JK, KA Olson, RP Reading, S Amgalanbaatar, and J Berger. 2011. Is wildlife going to the dogs? Impacts of feral and free-roaming dogs on wildlife populations. BioScience 61:125–132.

Yu Z, R Dickstein, WE Magee, and JR Spotila. 1998. Heat shock response in the salamanders *Plethodon jordani* and *Plethodon cinereus*. J. Therm. Biol. 23:259–265.

Yurk H, L Barrett-Lennard, JKB Ford, and CO Matkin. 2002. Cultural transmission within maternal lineages: vocal clans in resident killer whales in southern Alaska. Anim. Behav. 63:1103–1119.

Zachos, JM and LR Kump. 2005. Carbon cycle feedback and the initiation of Antarctic glaciations in the earliest Oligocene. Glob. Planet. Change 47:51–66.

Zimmerman LC, MP O'Connor, SJ Bulova, JR Spotila, SJ Kemp, and. CJ Salice. 1994. Thermal ecology of desert tortoises in the eastern Mojave Desert: seasonal patterns of operative and body temperatures, and microhabitat utilization. Herpetol. Monogr. 8:45–59.

Zimmerman LC and CR Tracy. 1989. Interactions between the environment and ectothermy and herbivory in reptiles. Physiol. Zool. 62:374–409.

Zylstra E, J Capps, and B Weise. 2005. Field observations of interactions between the desert tortoise and the Gila monster. Sonoran Herpetol. 18:30–31.

Zylstra ER and RJ Steidl. 2009. Habitat use by Sonoran desert tortoises. J. Wildl. Manage. 73:747–754.

Zylstra ER, RJ Steidl, and D Swann. 2010. Evaluating survey methods for monitoring a rare vertebrate, the Sonoran desert tortoise. J. Wildl. Manage. 74:1311–1318.

INDEX